George Minchin Minchin

Uniplanar kinematics of solids and fluids

With applications to the distribution and flow of electricity

George Minchin Minchin

Uniplanar kinematics of solids and fluids
With applications to the distribution and flow of electricity

ISBN/EAN: 9783337108243

Printed in Europe, USA, Canada, Australia, Japan

Cover: Foto ©berggeist007 / pixelio.de

More available books at **www.hansebooks.com**

Clarendon Press Series

UNIPLANAR KINEMATICS

OF

SOLIDS AND FLUIDS

G. M. MINCHIN

London

HENRY FROWDE

OXFORD UNIVERSITY PRESS WAREHOUSE

7 PATERNOSTER ROW

Clarendon Press Series

UNIPLANAR KINEMATICS

OF

SOLIDS AND FLUIDS

WITH

APPLICATIONS TO THE DISTRIBUTION AND
FLOW OF ELECTRICITY

BY

GEORGE M. MINCHIN, M.A.

PROFESSOR OF APPLIED MATHEMATICS
IN THE ROYAL INDIAN ENGINEERING COLLEGE, COOPER'S HILL

Oxford
AT THE CLARENDON PRESS
1882

[*All rights reserved*]

PREFACE.

THE present work aims at supplying a deficiency in the course of Mathematical Physics usually pursued by the higher class students in our Colleges and Universities. In the majority of cases, I think, little attention is paid to the special study of Kinematics, and, as a consequence, the student enters into the study of Kinetics with somewhat misty notions of *acceleration* and other leading conceptions which belong to the province of Kinematics.

In view of the great progress which has been made in this country in Experimental Physics, under the leadership of Faraday, Thomson, and Clerk Maxwell, it appears to me very desirable to include in the ordinary course of higher mathematics as much as possible of the application of mathematics to physical problems—at the expense, no doubt, of its applications to manifest unrealities. Hence a large portion of the work deals with the theory of fluid motion and electric distribution and flow.

Anything like a complete treatise on general (or three-dimensional) Kinematics would be a much more difficult task, and would be addressed to a very limited class of students. When, however, the leading results for motion in one plane are properly mastered, the student will have little difficulty in understanding what they become when a third co-ordinate is introduced. Thus much as to the general aim of the work. The subjects treated of call for a few special remarks.

In Chapter I considerable attention is devoted to the Instantaneous Centre (of velocities), and in Chapter II much prominence is given to the Instantaneous Acceleration Centre, with the accompanying Centrodes, some results with regard to which will probably be new to most readers. Applications of these results are made to several problems, the solutions of which are thus simplified. Professor Wolstenholme kindly placed his *Book of Mathematical Problems* at my disposal, and I have freely availed myself of his permission to select illustrative questions.

Chapter III deals largely with Roulettes, and aims at bringing under one general principle the discussion of many isolated results.

Some of the theorems in this chapter are, I believe, new; and for several of them I am indebted to the kindness of Mr. McCay, Fellow of Trinity College, Dublin. My obligations are very great to such an elegant and skilful geometer for the assistance which he has rendered.

In Chapter IV I have advocated a reversion to the use of the term *force of inertia* which was employed and clearly explained by Newton, and which is employed by many French physicists, including Delaunay. The term was shown by Newton to convey a definite physical idea, to which, it seems to me, greater prominence ought to attach, now that the notion of actions propagated through a medium has been brought into such importance by the experiments of Faraday and the great work of Clerk Maxwell. The expression 'effective force,' used by D'Alembert, seems but a poor substitute for Newton's *vis inertiæ*.

Chapter V follows closely the order and method of the chapter on three-dimensional strain in my *Statics*.

Chapter VI contains the subjects which, undoubtedly, are the most difficult of treatment in the book; and I could not venture, with any degree of confidence, to publish it without the criticism of some able physicist. Happily, Professor O. J. Lodge and his brother, Mr. Alfred Lodge, Fereday Fellow of St. John's College, Oxford, rendered their invaluable aid, and many alterations and improvements in the original MS. of the chapter are the result of their criticism.

I hope that the student will find, in particular, the section on Conjugate Functions an assistance in his reading of Clerk Maxwell.

Lamb's admirably simple *Treatise on the Motion of Fluids* has been of much assistance to me in this chapter.

I must thank my friend Professor G. Carey Foster for the free use of his papers on Electrical Flow, and his revision of the pages dealing with this subject.

Some Notes treating more generally of kinematical questions are added, in the belief that they will be of assistance to the more advanced student. I may particularly mention Note D, on *Vectors and their Derivatives*.

Captain Eagles has rendered me efficient service in revising the proofs, and I hereby acknowledge my obligations to him.

G. M. MINCHIN.

COOPER'S HILL, *November*, 1882.

TABLE OF CONTENTS.

		PAGE
CHAPTER I.	Displacement and Velocity . .	1
CHAPTER II.	Acceleration of Velocity . . .	52
CHAPTER III.	General Theorem of Epicycloidal Motion	79
CHAPTER IV.	Mass-Kinematics of Solid Bodies . .	106
CHAPTER V.	Analysis of Small Strains	121
CHAPTER VI.	Kinematics of Fluids	137
Section I.	General Properties	137
Section II.	Multiply-Connected Spaces	193
Section III.	Motion due to Sources and Vortices. Electrical Flow	200
Section IV.	Conjugate Functions	226

NOTES:—

A. 'Centrifugal Force' 255
B. Strain Invariants 256
C. Conjugate Functions 257
D. Vectors and their Derivatives 259
E. Flow of Electricity in one plane . . . 263
F. Current Power 263

INDEX 265

ERRATA.

Page 5, line 5 for PO read Po, inside the brackets.

For O read P from line 20, p. 17, to line 6, p. 18.

In Art. 112, and accompanying figure, read $\psi, \psi-a, \psi-2a, \ldots$ instead of $\psi, \psi+a, \psi+2a, \ldots$; since the sum of the separate stream functions must be constant along the resultant stream lines. The figure, as it stands, suits the discussion of equipotential lines (p. 180), the additive property being attended to.

ADDITIONAL ERRATA.

In Ex. 4, p. 34, after 'straight towards it,' read 'with constant velocity, β.'

,, Ex. 5, p. 50, for $\omega^2 y^2$ read $\dfrac{\omega^2}{\mu^2} y^2$; and for a in last line of answer read μa.

,, p. 107, observe that the force of inertia of a moving particle is defined with regard to its *magnitude* only. Considered as a vector, its component along the axis of x would be $-m \dfrac{d^2 x}{dt^2}$.

,, line 2 from end, p. 134, for 'along' read 'at right angles to.'

,, ,, 3, p. 141, for $2\pi r^2$ read $2\pi^2 r$.

,, ,, 8, p. 189, for f_0 read $\dfrac{1}{f_0}$.

,, ,, 3, p. 204, for 'magnetic pole,' read 'magnetised edge.'

,, ,, 21, p. 261, for the sign $-$ outside bracket, read $+$.

CHAPTER I.

Displacement and Velocity.

1. Uniplanar Motion. By *uniplanar motion*, or one-plane motion, is understood in the following pages motion which takes place in one plane or parallel to one plane. Thus, a cylinder rolling down an inclined plane, its axis always remaining horizontal, has uniplanar motion, because the motions of all the particles of the cylinder take place in parallel planes.

2. Linear Velocity. When a point moves in any manner, its velocity has a component in every direction. The velocity in any direction may be defined as *the rate of describing space, per unit of time, in that direction.*

Linear velocities follow the same laws of composition and resolution as Forces in Statics; and with these (such as the parallelogram and polygon of velocities) the student is assumed to be already familiar.

If a point, P, moves in any manner in the plane of the paper, its velocity may be completely defined (both in magnitude and in direction) by its components along two fixed axes Ox and Oy, just as the position of P at any instant may be completely defined by its two coordinates in these directions.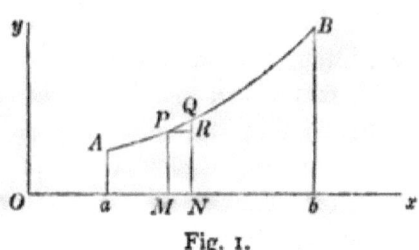

Fig. 1.

The resultant velocity of P takes place along the tangent at P to the path AB, along which P is moving.

The magnitude of this resultant velocity is $\dfrac{ds}{dt}$, where $s=$ the length of the arc of the path measured from some fixed point, A, on it up to P; for if, in the time Δt, P moves from P to a very close point Q, the velocity is the limiting value of $\frac{\Delta s}{\Delta t}$ when the length PQ (or Δs) is indefinitely diminished. The velocity in the direction Ox is the limit of $\frac{PR}{\Delta t}$, or of $\frac{\Delta x}{\Delta t}$; and the velocity in the direction Oy is the limit of $\frac{QR}{\Delta t}$ or of $\frac{\Delta y}{\Delta t}$.

The velocities parallel to the axes of x and y are therefore

$$\frac{dx}{dt} \text{ and } \frac{dy}{dt}.$$

3. Diagram of Space described. If a point moves with constant velocity, u, whether in a right line or in any curve whatever, the length of path described in time t is $u \cdot t$. If the velocity of the point is perpetually changing, and if v is its velocity at any instant, we may assume that the velocity remains constant and equal to v during an indefinitely small time, Δt. Now the length described (whatever be the path) in this time is $v \cdot \Delta t$; so that the whole length described in any interval of time is

$$\int v \, dt$$

taken throughout this interval.

This may be graphically represented as follows. Draw two rectangular axes, Ox and Oy (fig. 1), and take the first as axis of times and the second as axis of velocities; i.e. measure off OM to represent the time t, and MP, perpendicular to it, to represent v, the magnitude of the velocity at the time t. Let MN represent Δt; then $v \cdot \Delta t$ is represented by the area of the little rectangle $MPQN$. By laying off the different times along Ox, and drawing at the extremity of each abscissa a perpendicular (as above) to represent the corresponding velocity, we trace out *a curve, AB, of velocities*; and it is clear from the above that the area included between this curve, the axis Ox, and two extreme ordinates, Aa and Bb, will represent, on the scale adopted, the length of path (whether straight or curved) travelled over by the moving point between the time represented by Oa and the time represented by Ob.

4. Angular Velocity. When P moves in any manner, the line joining P to a point O revolves about O, or, in other words, changes its direction in fixed space. The direction of OP may be measured by the angle, POx, which OP makes with any fixed line Ox. This angle is to be expressed in absolute or *circular measure*. Denote, then, its circular measure by θ. If after the interval of time Δt the point P comes to Q, and if the circular measure of the angle QOx is $\theta + \Delta\theta$, the circular measure of the small angle QOP through which OP has turned is $\Delta\theta$; and the limit of the ratio $\dfrac{\Delta\theta}{\Delta t}$, viz.

$$\frac{d\theta}{dt},$$

is called the *angular velocity* of P about O.

Angular velocity means *rate of describing angle* (in circular measure) *per unit of time*, just as *linear velocity* means rate of describing space per unit of time.

If P moves in a circle whose centre is O, the linear velocity of P is at every instant equal to its angular velocity about the centre multiplied by the length of the radius,—or

$$v = r\omega,$$

where v = linear velocity, ω = angular velocity about the centre, r = radius of the circle.

Even when P moves in a circle, this equation does not hold between the linear velocity and the angular velocity about a point which is not the centre.

For if the direction of the resultant velocity is PQ (or rather the tangent to the path of P at P), and if PR is drawn perpendicular to OP and OQ, we have

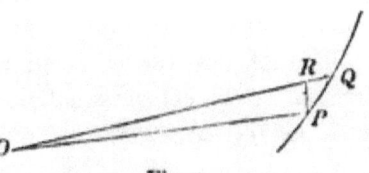

Fig. 2.

$$PR = OP \times \Delta\theta,$$

where $\Delta\theta$ is the circular measure of the small angle QOP.

Now $PR = PQ \sin OQP = v \sin OQP \times \Delta t$, so that if $OP = r$,

$$r \Delta \theta = v \sin OQP \times \Delta t,$$

$$\therefore \frac{d\theta}{dt} = \frac{v}{r} \sin \phi,$$

where ϕ is the angle between the radius vector OP, and the tangent at P.

Hence it is only when the direction of velocity is at right angles to the radius vector OP drawn through a fixed point O, that—

linear velocity = angular velocity × distance.

The velocity of the moving point P perpendicular to the radius vector, OP, is the limiting value of $\frac{PR}{\Delta t}$, or of $\frac{r\Delta\theta}{\Delta t}$; i.e. this component of velocity is $r\frac{d\theta}{dt}$.

The velocity component along the radius vector is the limit of $\frac{RQ}{\Delta t}$, i.e. $\frac{dr}{dt}$.

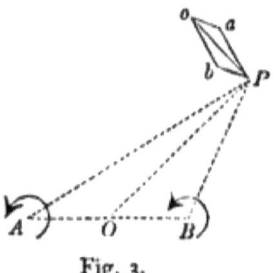

Fig. 3.

5. Composition of Angular Velocities. An angular velocity ω round a fixed point A combined with an angular velocity ω' round a fixed point B is equivalent to an angular velocity $\omega + \omega'$ round a fixed point, O, dividing the line AB so that $\frac{AO}{OB} = \frac{\omega'}{\omega}$.

For, let P (fig. 3) have the angular velocities ω and ω' about A and B; then it has linear velocities equal to $\omega \cdot PA$ and $\omega' \cdot PB$ along perpendiculars to PA and PB respectively. Now the resultant of any two directed magnitudes (whether they are forces or velocities) measured along PA and PB and represented by $\omega \cdot PA$ and $\omega' \cdot PB$ is a magnitude of the same kind represented by $(\omega + \omega') \cdot PO$, where O divides AB so that $\frac{AO}{OB} = \frac{\omega'}{\omega}$ (see *Statics*, Art. 20). If instead of being measured *along* PA and PB the directed magnitudes are measured perpendicularly to them, or making any constant angle with them, the resultant will be perpendicular to PO or inclined at the supposed angle to PO.

Hence the resultant of the two velocities (represented by Pa and Pb, perpendiculars to PA and PB, and equal to $\omega \cdot PA$ and $\omega' \cdot PB$) is a velocity (represented by PO) equal to $(\omega+\omega') \cdot PO$ and perpendicular to PO. Hence the two angular velocities are equivalent to an angular velocity $\omega+\omega'$ about O.

These angular velocities compound, therefore, exactly like two parallel forces proportional to ω and ω' acting at A and B.

If the angular velocities are in opposite senses, the point O must be taken on AB produced.

If $\omega' = -\omega$, the point O is at infinity and the resulting angular velocity is zero. The motion of P in this case is a velocity equal to $\omega \cdot AB$ in a direction perpendicular to AB. For, the resultant of a velocity $\omega \cdot PA$ along PA and a velocity $\omega \cdot PB$ along BP produced is a velocity parallel to AB and equal to $\omega \cdot AB$, as is easily seen. Turning through a right angle the lines representing these motions, we get the resultant of the rotatory motions of P, with equal and opposite angular velocities, round A and B to be a motion perpendicular to AB with velocity $\omega \cdot AB$.

Hence all points which have these equal and opposite angular velocities round A and B have exactly the same velocity, $\omega \cdot AB$, in magnitude and direction; so that if a lamina had these angular velocities, the resulting motion would be a simple *motion of translation* of the lamina in a direction perpendicular to AB.

Angular velocities and motions of translation in kinematics are therefore analogous to forces and couples, respectively, in Statics.

Again (in exact analogy with *Statics*, p. 93), an angular velocity of a lamina round any point B is equivalent to an equal angular velocity, in the same sense, round any point A, together with a motion of translation.

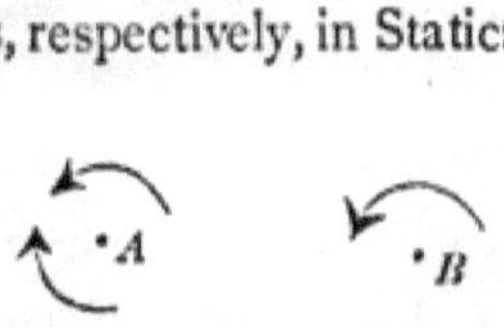

Fig. 4.

For, let two equal and opposite angular velocities round A (fig. 4) be introduced, each equal to that about B. This will not alter the motion. Then we may take the angular velocity about B with the equal and opposite angular velocity about A as equivalent to a motion of translation with velocity $\omega \cdot AB$; and in addition we have the angular velocity about A of same sign as that about B. Therefore, etc.

Examples.

1. A point P moves in a circle with constant linear velocity; find its angular velocity at any instant about any given point.

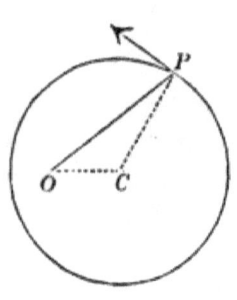

Fig. 5.

Let C (fig. 5) be the centre of the circle, O the point about which the angular velocity at any instant is required, $CP = a$, $OC = \dfrac{a}{n}$, $v = P$'s linear velocity; and let $OP = r$. Then the component linear velocity of P perpendicular to OP is $v \cdot \cos OPC$; therefore the angular velocity about O is
$$\frac{v \cdot \cos OPC}{r}.$$
But $\cos OPC = \dfrac{r^2 + \left(1 - \dfrac{1}{n^2}\right)a^2}{2ar}$; therefore the angular velocity about O is $\dfrac{v}{2a}\left[1 + \left(1 - \dfrac{1}{n^2}\right)\dfrac{a^2}{r^2}\right]$.

The angular velocity about O is therefore variable with the distance OP; but if O is on the circumference, the angular velocity becomes $\dfrac{v}{2a}$, i.e. constantly half the angular velocity about the centre—which is obvious without calculation by the proposition that the angle at the centre is double the angle at the circumference.

2. A point moves along a right line with constant velocity, prove that its angular velocity about any fixed point varies inversely as the square of the distance from this point.

3. Prove that in general the angular velocity of a point, P, moving with velocity v in any curve about any fixed point, O, is $\dfrac{pv}{r^2}$, where $r = OP$, and $p = $ perpendicular from O on tangent at P to the curve.

4. A point moves in an ellipse so that its angular velocity about one focus varies inversely as the square of the distance from this focus; find its angular velocity, in any position, about the other focus.

Ans. If the angular velocity about the first focus is $\dfrac{h}{r^2}$, the angular velocity about the second is $\dfrac{h}{rr'}$, at the same instant, the focal distances being r and r'. Hence if the ellipse (like a planet's orbit) be one the square of whose eccentricity can be neglected, the angular velocity about the second focus would be approximately constant.

5. Find the curve such that if a point describes it with constant velocity, the angular velocity of the point about a given fixed point shall vary inversely as the distance.

Ans. An equiangular spiral having the fixed point for pole.

6. A point moves in a parabola; compare, in any position, its angular velocities about the vertex and focus.

Ans. If r is the distance of the point from the focus, ω and ω' the angular velocities about the focus and vertex, and m one-fourth of the latus rectum, $\dfrac{\omega'}{\omega} = \dfrac{r}{r+3m}$.

7. A point describes an ellipse in such a manner that its angular velocity about the centre varies inversely as the square of the distance; show that the sum of the reciprocals of its angular velocities about the foci is constant.

[Use the expression in example 3.]

8. In the same case show that the sum of the reciprocals of the angular velocities about the extremities, whether of the axis major or of the axis minor, is constant.

9. A point moves in an ellipse so that its velocity varies inversely as the perpendicular from one focus on the tangent; show that the velocity can be resolved into two (oblique) components, each of constant magnitude — one perpendicular to the focal radius vector and the other perpendicular to the major axis. (Prof. Adams.)

[The velocity is directly proportional to the perpendicular from the other focus on the tangent; therefore, etc.]

10. If at every instant the velocity of a point along its path is proportional to the time, find the length described in any interval.

[Use a velocity diagram, Art. 3. It will be a right line.]

6. Harmonic Motion. If a point, R, (fig. 6) moves continuously round in a circle with constant velocity, the foot, P, of the perpendicular from the point on any diameter of the circle moves backwards and forwards along the diameter with a motion which is called a simple harmonic motion.

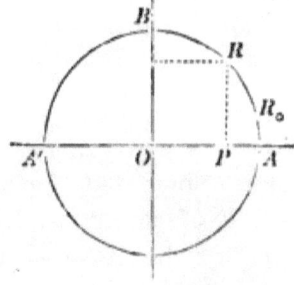

Fig. 6.

Let O be the centre of the circle, and let the time be

reckoned from the instant at which R occupied any position, R_0; then if $a = \angle R_0OA$, a = radius of circle, ω = angular velocity of R about O, and $OP = x$, we have

$$x = a \cos(\omega t + a),$$

as the type of harmonic motion. If T is the whole time taken by R to go round the circle, $\omega = \dfrac{2\pi}{T}$, and

$$x = a \cos\left(\dfrac{2\pi t}{T} + a\right). \qquad (1)$$

In the same way, if OB is the diameter perpendicular to OA, and if RQ is the perpendicular on OB; OQ being denoted by y, we have

$$y = a \sin\left(\dfrac{2\pi t}{T} + a\right). \qquad (2)$$

The time T is obviously also the time taken by P to complete its excursion, i.e. the time taken to move from A to A' and back again to A; and T is called the *period* of the motion.

The velocity of P at any instant is, of course,

$$-\dfrac{2\pi a}{T} \sin\left(\dfrac{2\pi t}{T} + a\right);$$

and this, which vanishes at A and A', attains a maximum at O.

Similarly for the motion of Q.

If we imagine another point, R', to start from R_0', so that $\angle R_0'OA = a'$, when R starts from R_0, and to go round the circle in the same time as R, its projection, P', on OA will have the harmonic motion

$$x' = a \cos\left(\dfrac{2\pi t}{T} + a'\right),$$

and the extreme limits of the excursion of P' will be A and A', just as before; and P' will occupy all the positions occupied by P—but, of course, *at different times*. What essentially distinguishes the displacement (from O) of P' from that of P is the angle $\dfrac{2\pi t}{T} + a'$, as compared with the angle $\dfrac{2\pi t}{T} + a$; and the displacement of the projection of a point which started from

A at the origin from which time is reckoned would be $a \cos \frac{2\pi t}{T}$. The angles $\frac{2\pi t}{T} + a$ and $\frac{2\pi t}{T} + a'$ which at each instant thus essentially distinguish the two previous displacements from each other and from this last are therefore called the *phases* of those motions.

The expression, $a \sin \left(\frac{2\pi t}{T} + a \right)$, for the displacement along the perpendicular axis, can be written

$$y = -a \cos \left(\frac{2\pi t}{T} + a + \frac{\pi}{2} \right),$$

whose phase is $\frac{2\pi t}{T} + \frac{\pi}{2} + a$. Thus we see that if the position of a moving point, R, be compounded, according to the parallelogram law, from two displacements along two rectangular lines, the displacements differing in phase by $\frac{\pi}{2}$, the motion of the point will take place in a circle; or *two simultaneous rectangular harmonic motions of the same period and amplitude* (presently defined) *and differing in phase by* $\frac{\pi}{2}$ *are equivalent to circular motion*.

Harmonic motions play an important part in Physical Optics, the motions being performed by particles of the luminiferous ether. The excursions of the particles from their positions of rest (in the ether when it is unstrained, i.e. not transmitting light) are, of course, extremely small; and the motions are called *vibrations*. We shall speak of them as such in what follows.

The maximum excursion of the harmonic vibration

$$x = a \cos \left(\frac{2\pi t}{T} + a \right)$$

is a, and this magnitude is called the *amplitude* of the vibration. The angle a is sometimes called the epoch angle, or simply the *epoch*.

The three essential characteristics of any vibration are, therefore, its *amplitude*, its *phase*, and its *period*, T. Another magnitude which plays an important part in the vibration is the *mean square of the velocity in the time of a complete vibration*. This is, of course,

$$\frac{\int_0^T (\frac{dx}{dt})^2 dt}{T} \quad \text{or} \quad \frac{2\pi^2 a^2}{T^2},$$

which is half the square of the maximum velocity in the vibration.

7. Superposed Equiperiodic Rectilinear Vibrations. Suppose several disturbing causes to impress simultaneously on a particle at O a corresponding number of harmonic motions of the same period, all superposed in the direction OA. Then if O' is the position of the particle at any time t, and if $OO' = \xi$, we have

$$\xi = a_1 \cos\left(\frac{2\pi t}{T} + a_1\right) + a_2 \cos\left(\frac{2\pi t}{T} + a_2\right)$$
$$+ a_3 \cos\left(\frac{2\pi t}{T} + a_3\right) + \ldots \quad (1)$$

where $a_1, a_2, a_3, \ldots$ are the amplitudes, and $a_1, a_2, a_3, \ldots$ the epochs of the several independent vibrations.

This equation for ξ is of the form

$$\xi = A \cos\left(\frac{2\pi t}{T} + \psi\right), \quad (2)$$

for it gives

$$\xi = (a_1 \cos a_1 + a_2 \cos a_2 + \ldots) \cos \frac{2\pi t}{T}$$
$$- (a_1 \sin a_1 + a_2 \sin a_2 + \ldots) \sin \frac{2\pi t}{T},$$

and comparing this with (2), we have

$$A \cos \psi = a_1 \cos a_1 + a_2 \cos a_2 + \ldots = \Sigma a \cos a,$$

for shortness;

$$A \sin \psi = a_1 \sin a_1 + a_2 \sin a_2 + \ldots = \Sigma a \sin a.$$

Hence
$$A = \sqrt{(\Sigma a \cos a)^2 + (\Sigma a \sin a)^2}, \quad (a)$$
$$\tan \psi = \frac{\Sigma a \sin a}{\Sigma a \cos a}, \quad (b)$$

where obviously A is the amplitude of the resultant vibration, and ψ its epoch, or $\frac{2\pi t}{T} + \psi$ its phase at any instant.

These values of the resultant amplitude and phase, when compared with the values of the resultant of any number of coplanar forces and the tangent of its angle of direction, show that *amplitudes and epochs of any number of equiperiodic vibrations, superposed in the same right line, answer exactly in their composition to magnitudes and directions of coplanar concurrent forces.*

Just as in the composition of forces it is only the angle *between* two forces that is important, and not the separate angles which their directions make with an arbitrary line, so here it is only *difference* of epoch, or of phase, that is important; and any one of the vibrations may be arbitrarily selected as that of zero epoch.

We have already seen how the phase of a rectilinear vibration $x = a \cos\left(\frac{2\pi t}{T} + a\right)$ can be exhibited as an angle subtended at the centre of a circle having the whole excursion $2a$ as diameter.

To compound amplitudes and phases of such vibrations *graphically*, we draw lines equal to the different amplitudes each in the direction given by the corresponding phase or corresponding epoch; and thus the graphic methods of Leibnitz and the polygon of forces can be applied (see *Statics*, pp. 15, 17).

Hence the resulting agitation will be nil in certain cases; and, in particular, *when two vibrations of equal amplitude and period, and differing in phase by* π, *are superposed.*

8. Graphic Circular Superposition. Taking the case of two equiperiodic vibrations in the same right line, the resultant position of the moving point at every instant may be graphically represented by the following method. Let the circle (fig. 6) in Art. 6 completely represent in amplitude, phase, etc. the vibration (a, T, a), and draw another circle having the same centre and completely representing the second vibration (a', T, a'). Let R' be the position at time t of the point which is supposed to travel over this second circle, R being, as before, the position of the point travelling over the first. Then the diagonal, Op, of the *rigid* parallelogram whose two adjacent sides are OR and OR' will be the resultant amplitude of the two vibrations, and its extremity p will determine by its motion in a circle (in the manner explained in Art. 6) a simple harmonic motion which is at every instant the resultant of the two superposed motions.

This is perfectly obvious from the graphic construction given at the end of last Article.

9. Harmonic Motions of Different Periods. If two harmonic motions of *different* periods, T, T', are superposed in the same right line, the resultant motion will be expressed by the equation

$$\xi = a \cos\left(\frac{2\pi t}{T} + a\right) + a' \cos\left(\frac{2\pi t}{T'} + a'\right), \qquad (1)$$

instead of equation (1), Art. 7.

For simplicity put n and n' for $\frac{2\pi}{T}$ and $\frac{2\pi}{T'}$, respectively. Then, in general, the resultant vibration may be put into the form

$$\xi = A \cos(mt + \psi) \qquad (2)$$

in an infinite number of ways; but A and ψ will be no longer constant magnitudes, but functions of t.

In particular, let us choose m equal to $\frac{1}{2}(n+n')$. Then putting $\phi = \frac{1}{2}(n-n')t$, we have

$$\xi = a \cos(mt + a + \phi) + a' \cos(mt + a' - \phi),$$

which gives

$$A^2 = a^2 - 2aa' \cos(a - a' + 2\phi) + a'^2,$$

$$\tan \psi = \frac{a \sin a + a' \sin a' + (a \cos a - a' \cos a') \tan \phi}{a \cos a + a' \cos a' - (a \sin a - a' \sin a') \tan \phi},$$

so that the resultant motion is presented in the form of simple harmonic motion with periodically varying amplitude and phase.

Two superposed simple harmonic motions of different periods may be compounded by the graphic circular method of Art. 8. Let the uniform circular motions of R and R' represent the two harmonic motions (a, T, a) and (a', T', a'); then this case differs from the case of two equiperiodic motions in that the parallelogram $OR\rho R'$ is not rigid, the angle ROR' varying with the time. The parallelogram must now be *jointed* at its four vertices; the point ρ will not trace out a circle or move uniformly, but its distance at any instant from a fixed diameter properly chosen will represent the resultant displacement.

Fourier's Theorem.

This is the principle of Sir W. Thomson's *Tidal Clock*; the two radii OR, OR' represent the amplitudes of the Lunar and Solar tides, and their periods of revolution are the periods of the tides.

10. Fourier's Theorem. A function of a variable, x, is said to be a *periodic function* of the variable if its values repeat themselves for values of the variable differing by a constant or by any integer multiple of this constant. Thus, the simple harmonic function $a\cos(nx+a)$ is such, its period being 2π. Generally, if
$$\phi(x+n\lambda) = \phi(x), \qquad (1)$$
whatever x may be, n being any integer and λ a constant, $\phi(x)$ is a periodic function of x, its period being λ. Now the object of Fourier's Theorem is to show that—*Any arbitrary periodic function can be expressed as the sum of a number of simple harmonic functions.*

Suppose that $\phi(x)$ is any periodic function of period λ, and let it not become infinite within its period.

Assume

$$\phi(x) = A_0 + A_1\cos 2\pi\frac{x}{\lambda} + A_2\cos 2\pi\frac{2x}{\lambda} + \ldots + A_n\cos 2\pi\frac{nx}{\lambda} + \ldots$$
$$+ B_1\sin 2\pi\frac{x}{\lambda} + B_2\sin 2\pi\frac{2x}{\lambda} + \ldots + B_n\sin 2\pi\frac{nx}{\lambda} + \ldots \qquad (2)$$

where $A_0, A_1, \ldots B_1, \ldots$ are all constant coefficients which are to be determined.

Multiply both sides of this equation by $2\cos 2\pi\frac{mx}{\lambda}$, and then integrate both sides from $x = 0$ to $x = \lambda$.

The coefficient of A_m alone will remain, as is at once obvious; and we have
$$A_m = \frac{2}{\lambda}\int_0^\lambda \phi(x)\cos 2\pi\frac{mx}{\lambda}dx. \qquad (3)$$

Similarly, if we multiply both sides of (2) by $2\sin 2\pi\frac{mx}{\lambda}$, and integrate, we have
$$B_m = \frac{2}{\lambda}\int_0^\lambda \phi(x)\sin 2\pi\frac{mx}{\lambda}dx; \qquad (4)$$

while
$$A_0 = \frac{1}{\lambda}\int_0^\lambda \phi(x)dx. \qquad (5)$$

The coefficients, $A_0, A_1, \ldots$ are therefore constant quantities depending simply on the form of the function ϕ; and we have finally

$$\phi(z) = \frac{1}{\lambda}\int_0^\lambda \phi(x)dx + \sum_1^\infty \left(A_n \cos 2\pi\frac{nz}{\lambda} + B_n \sin 2\pi\frac{nz}{\lambda}\right), \ldots \text{(a)}$$

(the variable being, for clearness, denoted by z), the suffixes signifying that n is to receive all integer values from 1 to ∞, and the coefficients having the values determined in (3), (4), (5).

The typical term $A_n \cos 2\pi\frac{nz}{\lambda} + B_n \sin 2\pi\frac{nz}{\lambda}$ may, of course, be put into the usual simple harmonic form $A \cos(2\pi\frac{nz}{\lambda} + a)$; and thus the Theorem enunciated is proved. Observe that the periods of the successive component simple harmonic vibrations —*wave-lengths* as we shall call them, for a reason to be presently explained—are λ, $\tfrac{1}{2}\lambda$, $\tfrac{1}{3}\lambda \ldots$

[For a complete discussion of the proof of Fourier's Theorem the student is referred to Thomson and Tait's *Nat. Phil.* vol. i. pp. 55, etc.; Price's *Infin. Cal.*, vol. ii.; or Donkin's *Acoustics*, pp. 51, etc.]

Fourier's Theorem has a very wide application in Physics—and notably in the cases of periodic motion which present themselves in the theory of Sound. A body—a tuning fork, for example—when thrown into vibration throws the air into some kind of periodic motion; and Fourier's Theorem shows that this motion can be exhibited as the superposition of a (theoretically infinite) number of simple harmonic vibrations whose wave-lengths are represented by λ, $\tfrac{1}{2}\lambda$, $\tfrac{1}{3}\lambda$, ... and whose amplitudes are represented by the values of $\sqrt{A_n^2 + B_n^2}$ just given. It is pointed out by Donkin (*Acoustics*, p. 60) that Fourier's Theorem and the Ear of an animal perform exactly the same function, the first analytically and the second physically—each resolves a vibration into a number of *simple harmonic* vibrations; but whereas Fourier's Theorem takes account of an infinite number of these simple harmonic vibrations, with diminishing wave-lengths and gradually dying amplitudes, the Ear is capable of appreciating comparatively few of them.

We may put Fourier's Theorem to the purely analytical use

Fourier's Theorem.

of expressing the ordinate of a given curve within given limits as the sum of a number of simple harmonic values.

Thus, take the arc, OA, of a parabola cut off by a line OA perpendicular to the diameter and at a distance h from the vertex.

Taking O as origin and OA as axis of x, the equation of the curve, if $OA = 2c$, is $y = h\left(\dfrac{2x}{c} - \dfrac{x^2}{c^2}\right)$.

Assuming $y = A_0 + \Sigma\left(A_n \cos 2\pi \dfrac{nx}{2c} + B_n \sin 2\pi \dfrac{nx}{2c}\right)$,

we have $A_0 = \dfrac{h}{2c}\int_0^{2c}\left(\dfrac{2x}{c} - \dfrac{x^2}{c^2}\right)dx = \dfrac{2}{3}h$.

$$A_n = \dfrac{h}{c}\int_0^{2c}\left(\dfrac{2x}{c} - \dfrac{x^2}{c^2}\right)\cos\pi\dfrac{nx}{c}dx = -\dfrac{4h}{n^2\pi^2}.$$

$$B_n = \dfrac{h}{c}\int_0^{2c}\left(\dfrac{2x}{c} - \dfrac{x^2}{c^2}\right)\sin\pi\dfrac{nx}{c}dx = 0.$$

Hence $y = \dfrac{2}{3}h - \dfrac{4h}{\pi^2}\sum_1^\infty \dfrac{1}{n^2}\cos\dfrac{n\pi x}{c}$.

As another example (see Donkin's *Acoustics*, p. 54) take a broken line, consisting of a right line OA and a right line AC. Take the right line OC as axis of x; let $OC = \lambda$; from A let fall AN perpendicular to OC, and let $ON = a$, $AN = b$. It is required to express the ordinate of the broken line OAC in a Fourier series.

From $x = 0$ to $x = a$, we have $y = \dfrac{b}{a}x$, and from $x = a$ to $x = \lambda$, we have $y = \dfrac{b}{\lambda - a}(\lambda - x)$.

Assume $y = A_0 + \sum_1^\infty\left(A_n\cos 2\pi\dfrac{nx}{\lambda} + B_n\sin 2\pi\dfrac{nx}{\lambda}\right)$ (ϕ)

Then

$$A_n = \dfrac{2}{\lambda}\cdot\dfrac{b}{a}\int_0^a x\cos 2\pi\dfrac{nx}{\lambda}dx + \dfrac{2}{\lambda}\cdot\dfrac{b}{\lambda - a}\int_a^\lambda(\lambda - x)\cos 2\pi\dfrac{nx}{\lambda}dx$$

$$= \dfrac{b\lambda^2}{4n^2\pi^2 a(\lambda - a)}\left(\cos\dfrac{2n\pi a}{\lambda} - 1\right).$$

Similarly $B_n = \dfrac{b\lambda^2}{4n^2\pi^2 a(\lambda - a)}\sin\dfrac{2n\pi a}{\lambda}$, and $A_0 = \tfrac{1}{2}b$. Substituting these values in (ϕ), and forming a single harmonic term with A_n and B_n, we have

$$y = \tfrac{1}{2}b + \frac{b\lambda^2}{\pi^2 a(\lambda-a)} \sum_1^\infty \frac{1}{n^2} \sin\frac{n\pi a}{\lambda} \sin\frac{2n\pi(x-\tfrac{1}{2}a)}{\lambda}.$$

It is obvious that in all cases the first term, A_0, in Fourier's series is the *mean* value of y within the given limits.

EXAMPLES.

1. Exhibit the ordinate of the right line $y = m\left(x - \dfrac{\lambda}{2}\right)$ between the limits $x = 0$ and $x = \lambda$ in a Fourier series.

Ans. $y = -\dfrac{m\lambda}{\pi}\left\{\dfrac{1}{1}\sin\dfrac{2\pi x}{\lambda} + \dfrac{1}{2}\sin\dfrac{4\pi x}{\lambda} + \dfrac{1}{3}\sin\dfrac{6\pi x}{\lambda} + \ldots\right\}.$

2. Exhibit the ordinate of the curve $y = 2a \sinh\dfrac{x}{c}\cos\dfrac{x}{c}$ between $x = 0$ and $x = \dfrac{1}{2}\pi c$ in a Fourier series.

Ans. $A_n = \dfrac{16a}{\pi(1+16n^2)}\sinh^2\dfrac{\pi}{4}$; $B_n = \dfrac{-32an}{\pi c(1+16n^2)}\sinh\dfrac{\pi}{2}.$

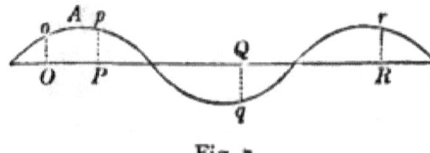

Fig. 7.

11. Wave Disturbance. Suppose a disturbance of every particle on the right line Oz (fig. 7) to take place perpendicularly to Oz; let O be chosen as origin, and let $OP = z =$ the distance from O of a particle at P. Let ξ be at any time the displacement of P from its position of rest, and suppose the value of ξ, as dependent upon both z and t, to be given by the equation

$$\xi = a\sin\frac{2\pi}{\lambda}(vt-z), \qquad (a)$$

where λ is a constant linear magnitude and v a constant (expressing necessarily a velocity).

We might with equal propriety have chosen the form $a\cos\dfrac{2\pi}{\lambda}(vt-z)$ for ξ; but the above is chosen because we wish to reckon the time, t, from the instant at which our origin particle, O, was in its undisturbed position; a change to the *cosine* form would amount simply to a change in the origin of time.

Now consider what the values of ξ are at the *same time* and at *different distances*. They are exactly the same at the distance z as at the distances $z+\lambda$, $z+2\lambda$, $z+3\lambda$,......; and they are the same in magnitude but opposite in sign at the distances z and $z+\tfrac{1}{2}\lambda$.

Now let us draw a curve representing all the displacements of particles at the same time. The displacement of O is $a\sin\dfrac{2\pi vt}{\lambda}$, and is represented by Oo (fig. 7); that of P by Pp; and we have just seen that the particles at distances λ, 2λ, 3λ, .. from P are in exactly the same state of disturbance as P.

Let $PR=\lambda$; then, in other words, all the disturbances of particles from R onwards are exact reproductions of the disturbances between R and P. Let $PQ=\tfrac{1}{2}\lambda$; then at Q we have a particle whose disturbance, Qq, is the same in magnitude as that of P but different in sign.

The history of the disturbances at different points at *the same time* is therefore represented by some such curve as that figured.

Next let us confine our attention to one particle, O, and represent its disturbances at *different times*.

It is quite obvious that if z is constant and t variable, we shall get from the expression $a\sin\dfrac{2\pi}{\lambda}(vt-z)$ exactly the same series of values as we got from it when t was taken constant and z variable; in other words—*the curve which represents the history of the displacements of all particles at the same time represents also the history of the displacement of any one particle at different times.*

And the values of the displacement of O will reproduce themselves after an interval of time equal to $\dfrac{\lambda}{v}$.

The distance λ which intervenes between two particles which are always in the same phase is called the *length of the wave*.

We may now put the matter thus:—ξ is at the time t the same at O and R, and after this instant its value changes at both points, being, however, always of equal amounts at both; ξ will again be the same at R as it was at time t after the interval $\frac{\lambda}{v}$; we may suppose this second disturbance ξ at R to have travelled up to R from O; and since it has gone over OR, or λ, in time $\frac{\lambda}{v}$, v is the velocity with which the disturbance travels.

When this disturbance has got up to R, O is again in exactly the same state of disturbance as at time t; i.e., O has just gone through all its cycle of disturbances. Hence—*every particle takes exactly the same time to go through all its phases of disturbance as that taken by any given disturbance to travel over a wave length.*

When any one disturbance (value of ξ) has advanced a wave length, the wave is said to have travelled over a wave length.

A point (such as A) at which a disturbed particle is at its maximum distance from its position of rest is called a *crest* of the wave.

Two particles, such as Q and P, which are *always* in opposite phases are separated by $\frac{\lambda}{2}$, or half a wave length; for if $OQ = z'$, the difference of phase of Q and P is $\frac{2\pi}{\lambda}(z'-z)$, and if this $= \pi$, or $z'-z = \frac{1}{2}\lambda$, the value of ξ for $Q = -$ that for P.

In the expression $a \sin \frac{2\pi}{\lambda}(vt-z)$, z is called the *retardation* of the vibration.

A retardation of a whole wave length (or difference of phase equal to 2π) is tantamount to no retardation at all.

12. Composition of two Rectangular Vibrations. Suppose a particle at O (fig. 1) to receive a displacement ξ along Ox and a displacement η perpendicular to Ox, the periods of both being the same, the axes Ox and Oy being in the plane of the paper, while an axis Oz may be imagined as drawn upwards from, and perpendicular to, the plane of the paper; and suppose that

$$\xi = a \sin\left(2\pi \frac{t}{T} + a\right),$$

$$\eta = b \sin\left(2\pi \frac{t}{T} + \beta\right).$$

Composition of Rectangular Vibrations.

Eliminating t, we get

$$\frac{\xi^2}{a^2} - \frac{2\xi\eta}{ab}\cos\delta + \frac{\eta^2}{b^2} = \sin^2\delta,$$

where $\delta = a - \beta =$ difference of phases of the two constituent vibrations. Thus the resultant vibration is one in an ellipse having the undisturbed position of O for centre.

To determine *the direction of vibration* in this ellipse, find the value of $\frac{d\eta}{dt}$ when $\xi = a$. Now $\frac{d\eta}{dt} = \frac{2\pi b}{T}\cos\left(2\pi\frac{t}{T} + \beta\right)$; and if $\xi = a$, $2\pi\frac{t}{T} + a = \frac{\pi}{2}$;

$$\therefore \frac{d\eta}{dt} = \frac{2\pi b}{T}\cos\left(\frac{\pi}{2} - \delta\right),$$

and $\frac{d\eta}{dt}$ will be positive if $\delta > 0$ and $< \frac{\pi}{2}$, i.e., the motion is from the positive axis of ξ to that of η if $\delta > 0$, $< \frac{\pi}{2}$. The retardation of the η component behind the ξ component is $\frac{\lambda\delta}{2\pi}$, so that if this retardation is between 0 and a quarter of a wave length the motion will be from the positive axis of ξ to that of η.

Cor. 1. Two rectangular vibrations of same wave length and period, one being a quarter of a wave length in advance of the other compound an elliptic vibration, the axes of the ellipse being in the directions of the constituent vibrations. For then $\delta = \frac{\pi}{2}$, and the equation of the ellipse is $\frac{\xi^2}{a^2} + \frac{\eta^2}{b^2} = 1$.

Cor. 2. If the amplitudes, a, b, of the constituent rectangular vibrations are equal, the resultant vibration is, in general, elliptic; but if the phases differ by $\frac{\pi}{2}$, i.e., if the retardation be $\frac{1}{4}\lambda$, the resultant vibration is circular.

If the *periods* of the two rectangular components are not the same, the equations will be

$$\xi = a\sin\left(2\pi\frac{t}{T} + a\right), \qquad \eta = b\sin\left(2\pi\frac{t}{T'} + \beta\right),$$

or, for shortness, let us write

$$\xi = a\sin nt, \qquad \eta = b\sin(mt + \epsilon). \qquad (a)$$

The Cartesian equation of the curve described by the disturbed particle is obtained by eliminating t from these equations.

Now it is clear that if n and m have a common multiple, this curve will be closed, the particle performing complete revolutions in the curve in a time equal to the least common multiple of the two constituent vibrations; but if the periods are incommensurable, the particle will never return to its original position—it describes a curve which never closes.

Figures of the curves which will be described for different values of ϵ and $\dfrac{m}{n}$ will be found in Thomson and Tait (p. 51) and in several works on Physics.

Clifford proposes to study all cases of motion expressed by equations (a) by converting the motion (which, of course, is uniplanar) into motion on a cylinder. This is done by introducing a third component harmonic motion along a line perpendicular to the plane of the two components ξ, η, and choosing this third motion, ζ, so that it produces, when taken in conjunction with one of the components—η, suppose— uniform circular motion. In this case we should have

$$\zeta = b \cos (mt + \epsilon),$$

and the total resulting motion is motion in a circle perpendicular to the axis of ξ combined with motion along this axis; which comes to the same thing as imagining a cylinder described on the circle, and a generating line to be carried round it with uniform angular velocity m, while a particle performs the harmonic motion ξ along this generating line. The curve of positions on the cylinder is obviously obtained by wrapping round it the wave curve which is the curve of positions of the motion $\xi = a \sin nt$; and the given motion, expressed by equations (a), is obtained by projecting the cylindrical motion on a plane perpendicular to ζ. Keeping the same cylinder, with the same curve of positions wrapped round it, if we project the cylindrical motion on *different* planes passing through the axis of the cylinder, we obtain the same effect as is produced by varying ϵ in the equations (a).

[See Clifford's *Kinematic*, p. 33.]

13. Resolution of a Rectilinear Vibration. If in the preceding Article the two rectangular vibrations have the same phase (i.e., if there is no retardation), or $\alpha = \beta$, the resultant vibration is rectilinear; for then

$$\frac{\eta}{\xi} = \frac{b}{a},$$

which is the equation of a right line; and the amplitude of the resultant is $\sqrt{a^2 + b^2}$.

If the difference of phase is π, instead of 0 (i.e., if the retardation is λ), the resultant vibration is again rectilinear; for then $\frac{\eta}{\xi} = -\frac{b}{a}$.

Conversely, a rectilinear vibration of amplitude a along any line OA (fig. 8) can be resolved into two rectilinear vibrations of amplitudes

$$a \cos \alpha \text{ and } a \sin \alpha,$$

along any two rectangular lines, Ox and Oy, the phases of the components and that of the resultant being all the same, and α being the angle between Ox and OA. Reckoning from Ox as an initial line, the angle α is called the *azimuth* of the vibration in OA.

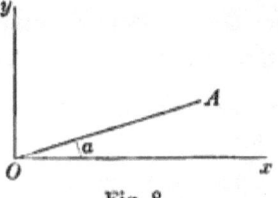

Fig. 8.

Again, *any rectilinear vibration is equivalent to two circular vibrations, the rotations in these being in opposite senses.*

Suppose a vibration $\xi = a \sin \frac{2\pi t}{T}$ along the axis of x, and a vibration $\eta = 0$—in other words, no vibration—along the axis of y. We may write these

$$\xi = \tfrac{1}{2} a \sin \frac{2\pi t}{T} + \tfrac{1}{2} a \sin \frac{2\pi t}{T} \qquad = \xi_1 + \xi_2, \text{ suppose};$$

$$\eta = \tfrac{1}{2} a \sin \left(\frac{2\pi t}{T} + \frac{\pi}{2}\right) + \tfrac{1}{2} a \sin \left(\frac{2\pi t}{T} + \frac{3\pi}{2}\right) = \eta_1 + \eta_2.$$

Take the vibration (ξ_1, η_1) separately; it is obviously circular, since

$$\xi_1 = \tfrac{1}{2} a \sin \frac{2\pi t}{T}, \qquad \eta_1 = \tfrac{1}{2} a \cos \frac{2\pi t}{T},$$

$$\therefore \xi_1^2 + \eta_1^2 = \tfrac{1}{4} a^2.$$

And by finding the value of $\frac{d\eta_1}{dt}$ when $\xi_1 = \tfrac{1}{2}a$, we see that the sense of rotation in the circle is from the positive part of the axis of η to that of the axis of ξ.

In the same way, (ξ_2, η_2) is an equal circular vibration, the sense of rotation being the reverse of the preceding. Consequently the (ether) particle which describes the vibration along the axis of x (i.e., along any line) may be supposed to be agitated by—or to be compounding—two equal and opposite circular vibrations.

This mode of resolving a rectilinear vibration is important in the theory of the *rotatory polarisation of quartz*. Thus, suppose that the quartz retards the circular vibration (ξ_1, η_1) more than it retards the circular vibration (ξ_2, η_2) of opposite rotatory sense; and let the amount of this retardation be $\frac{1}{n}\lambda$, where λ is a wave length and n any number.

Then the vibrations (ξ_1, η_1) and (ξ_2, η_2) on emergence from the quartz may be written

$$\xi_1 = \tfrac{1}{2}a \sin\left(\frac{2\pi t}{T} - \frac{2\pi}{n}\right); \qquad \xi_2 = \tfrac{1}{2}a \sin\frac{2\pi t}{T},$$

$$\eta_1 = \tfrac{1}{2}a \cos\left(\frac{2\pi t}{T} - \frac{2\pi}{n}\right); \qquad \eta_2 = -\tfrac{1}{2}a \cos\frac{2\pi t}{T}.$$

The resultant disturbance of the ether particle at the place of exit of the light from the quartz has then for components

$$\xi = a \cos\frac{\pi}{n} \sin\left(\frac{2\pi t}{T} - \frac{\pi}{n}\right),$$

$$\eta = a \sin\frac{\pi}{n} \sin\left(\frac{2\pi t}{T} - \frac{\pi}{n}\right);$$

and since $\frac{\eta}{\xi} = \tan\frac{\pi}{n}$, the vibration at emergence is rectilinear, of the same amplitude, a, as at incidence, but along a line making an angle $\frac{\pi}{n}$ with its original direction.

The same method of resolution of a rectilinear vibration into two oppositely rotating circular vibrations will serve to explain the phenomenon observed first by Faraday—viz., that *the plane of polarisation of a plane polarised ray of light is rotated as the ray travels through a strong magnetic field*—if we adopt the view that magnetism is due to (or consists of) *whirls* in the ether— the very same ether whose oscillatory rectilinear motion at each point along the ray of light constitutes this light. For the sense of the motion in the magnetic whirl must coincide with that of one of the constituent luminous circular vibrations and oppose the motion of the other circular vibration. One of these will, therefore, be retarded and the other accelerated, with exactly the rotatory result above obtained.

14. Definition of a Wave in general. Consider at any one instant the state of an agitated medium—air, water, ether, &c. Every disturbed particle is at this instant deviated by a certain amount from its undisturbed position, and each particle is in a *particular phase* of its vibration. Now all those which are in exactly the *same* phase lie on some surface, closed or unclosed. We must here be very particular about what constitutes *same phase*. Previously we remarked (Art. 11) that a difference of phase of 2π is tantamount to no difference; but now when we wish to make a distinction between particles in different parts of the medium, we must take account of even this difference. Reckoning the time from the moment at which the disturbance under consideration originated, take all those agitated particles which are in *exactly the same phase* of vibration; then the surface-locus of these is the *Surface of a Wave*.

15. Transverse and Longitudinal Vibrations. Consider any one particle on the surface of any one wave. The displacement of this particle may take place—

(*a*) along the normal to the wave surface, or

(*b*) somewhere in the tangent plane to this surface.

These are the only cases that present themselves in physical investigations, though, doubtless, other directions of displacement are possible.

Case (*a*) is that of the vibration, for example, of *sound*, the disturbed medium being air; Case (*b*) is pre-eminently that of *light*, the disturbed medium being the luminiferous ether. The vibrations in the first case are called *longitudinal* and in the second *transverse*. A stone thrown into still water will cause vibrations of the second sort, the ripples being curves containing particles in the same phase—i.e., each ripple is a curve on a wave surface, the wave surface extending below into the water and being broken at the contact of the water and the superincumbent air.

At a great distance from the origin of disturbance all waves become approximately *plane* waves.

If in fig. 7, Oz is a line of air particles, disturbed so that Oz is normal to a wave surface, the displacements, Oo, Pp,... of the particles will take place *along* Oz and not perpendicular to it, as in the case of light.

This (longitudinal) vibration will be attended necessarily by successive *condensations* and *rarefactions* along the line Oz—particles in one place being crowded together, while particles in other places are more thinly distributed than in the undisturbed state. Every point along Oz will sooner or later be a position of maximum condensation and also of maximum rarefaction.

The expression for the distance ζ between any particle P at the time t and its undisturbed position will be of the form

$$\zeta = a \sin \frac{2\pi}{\lambda}(vt - z),$$

and the values of ζ may be represented by measuring them (as in Art. 11) *perpendicular* to their direction Oz; so that the wavy curve in fig. 7 still suits the case of sound—remembering that it is now only a graphic mode of representing displacements which the curve measures in a direction at right angles to their proper direction.

Exactly the same definitions, properties depending on phase difference, etc., as before hold.

In the case of the transverse vibrations of the ether, condensations and rarefactions need not necessarily take place.

EXAMPLES.

1. The surface of still water is agitated by wave disturbances proceeding from two fixed points A_1 and A_2; find the locus of points of no disturbance.

Let the amplitudes of the waves coming from A_1 and A_2 be a_1 and a_2

respectively, and let r_1 and r_2 be the distances of any point, P, on the surface from A_1 and A_2. Now the resultant amplitude and phase at any disturbed point are got by compounding two forces whose magnitudes are proportional to a_1 and a_2 whose angles of direction are the phases (or epoch angles) of the two disturbances at P (Art. 7). The angle between the forces is $\frac{2\pi(r_2-r_1)}{\lambda}$. But if there is no disturbance at P, the forces above mentioned must be equal and opposite in the same right line—i.e., we must have

$$a_1 = a_2, \text{ and} \tag{1}$$

$$\frac{2\pi r_2}{\lambda} = \frac{2\pi r_1}{\lambda} + (2n+1)\pi, \text{ or}$$

$$r_2 - r_1 = (2n+1)\frac{\lambda}{2} \tag{2}$$

where n is any integer. [Observe that two forces will be opposed in the same right line if the direction-angle of one equals that of the other increased by any odd multiple of π.] Hence in order that there may be points of no disturbance, the amplitudes of the two sets of waves must be equal. If this is so, all points on the curve whose equation is (2) are points of no disturbance. This curve is a hyperbola; and by varying n, we get a series of hyperbolas having A_1 and A_2 for foci.

2. The surface of still water is agitated by wave disturbances proceeding from three fixed points, A_1, A_2, A_3; find the points of zero disturbance.

Let the amplitudes of the waves coming from A_1, A_2, A_3 be a_1, a_2, a_3, respectively. Then the resultant amplitude and phase at any point, P, whose distances from the centres of disturbance are r_1, r_2, r_3, are got by drawing lines, Oa_1, Oa_2, Oa_3 (fig. 9), proportional to a_1, a_2, a_3 and having direction angles (measured from any initial line) $\frac{2\pi r_1}{\lambda}$, $\frac{2\pi r_2}{\lambda}$, $\frac{2\pi r_3}{\lambda}$. If P is a point of zero disturbance these lines will answer to an equilibrating system of forces. Obviously, therefore, in order that there may be zero-disturbance at any point, the amplitudes of the interfering vibrations must be such that the sum of any two exceeds the third.

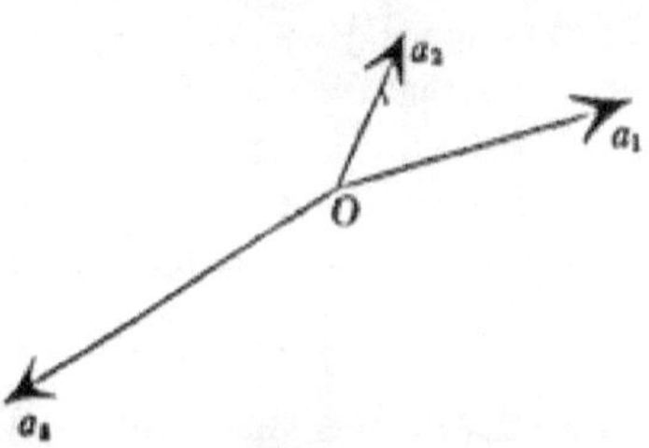

Fig. 9.

Denote the angles $a_2 O a_3$, $a_3 O a_1$, $a_1 O a_2$ by a_1, a_2, a_3. Now

$$\frac{2\pi}{\lambda}(r_3 - r_2) = a_1; \quad \frac{2\pi}{\lambda}(r_2 - r_1) = 2\pi - a_3; \quad \frac{2\pi}{\lambda}(r_3 - r_1) = a_2,$$

any one of which equations follows from the other two.

Hence the points of intersection of the hyperbola'

$$r_3 - r_2 = \frac{a_1}{2\pi}\lambda \tag{1}$$

with the hyperbola

$$r_2 - r_1 = \frac{a_1}{2\pi}\lambda \qquad (2)$$

are points of zero disturbance. But now observe that the direction angle of (say) Oa_1 may be just as well $\frac{2\pi r_2}{\lambda} \pm 2n\pi$, where n is any integer, as $\frac{2\pi r_2}{\lambda}$, so that we can have

$$r_2 - r_1 = \frac{a_1}{2\pi}\lambda + n\lambda, \qquad (3)$$

and therefore all the points on the fixed hyperbola (2) which lie on the series of hyperbolas obtained by varying n in (3) are points of rest on the surface of the water. But in the same way as r_2 may be altered to $r_2 \pm n\lambda$, any other distance, as r_3, may be similarly altered; and thus we shall obtain a series of hyperbolas

$$r_3 - r_1 = \frac{a_2}{2\pi}\lambda + m\lambda \qquad (4)$$

round A_1 and A_3 as foci.

Hence finally we obtain a *net-work of points of rest* by taking all the points of intersection of the series of hyperbolas (3) obtained by varying n with the series of hyperbolas (4) obtained by varying m. Of course these points are continuously at rest, and not merely so at a particular time.

[We have in the solution assumed that the amplitude of each wave remains constant—a supposition which is allowable if we consider only points not very far removed from the origins of disturbance.]

3. If the motion of a point consists of a harmonic vibration of period T, and x denotes its distance at any time from its mean position, show that

$$\frac{d^2x}{dt^2} = -\frac{4\pi^2}{T^2}\cdot x.$$

[Conversely an equation of motion of this form indicates harmonic vibration of period T.]

4. Prove that the resultant of any number of simple harmonic vibrations, in different directions, with different phases, but of the same period, is elliptic harmonic motion. [Arts. 13 and 12.]

5. Prove that the motion expressed by the components (α), Art. 12, when $m = 2n$ and $\epsilon = \frac{\pi}{2}$, is oscillatory motion in a parabolic arc.

16. Relative Motion. When any two points, O and P, move in any manner, we must distinguish three different motions which take place, viz., the *absolute* motion of P (i. e., its motion in fixed space), the *absolute* motion of O, and their *relative* motion.

The *relative motion* of two points, *whether both moving or not* is the motion which each *appears* to the other to possess.

If a point, P, moves round a circle with a velocity either constant or variable, the centre, O, of the circle, although fixed, appears to P to move round P in a circle of the same radius, and with a velocity either constant or varying exactly as the

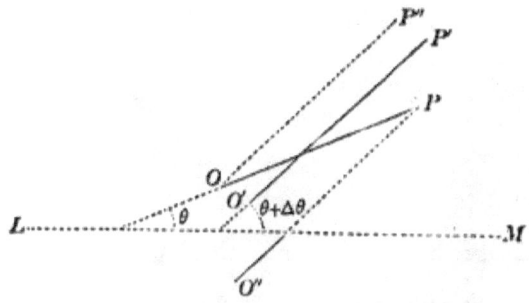

Fig. 10.

velocity of P varies. [Two observers may be supposed to be stationed at P and O.]

Let there be two points, whether both moving or not; then the angular velocity of the first with respect to the second is exactly the same as the angular velocity of the second with regard to the first, and it is measured at any instant by the rate per unit of time at which the line joining them revolves,—i.e., by the rate at which this line describes angle (in circular measure) with a fixed line in space.

Let O and P be the points at any instant, and after a very short time let them come to O' and P', respectively, the distance between them being altered or unaltered.

Let LM be a fixed line in the plane of motion. An observer at O might measure the direction of one at P by the angle θ which OP makes with LM; and an observer at P might measure the direction of one at O by the same angle. If OP'' is drawn equal and parallel to $O'P'$, the point P will appear to the observer at O, at the end of the motions considered, to occupy the position P'', and the change of P's direction is measured by the angle $P''OP$, which is $\Delta\theta$.

If Δt is the small time occupied by the motions OO' and PP', the limiting value of the ratio $\dfrac{\Delta\theta}{\Delta t}$, or $\dfrac{d\theta}{dt}$, is the angular velocity of P about O in the positions O and P.

Exactly in the same way, if PO'' is drawn equal and parallel to $P'O'$, the point O will, at the end of the time Δt, appear to P to occupy the position O'', i.e., it will have revolved about P through the angle OPO'', which is also $\Delta\theta$; so that the angular velocity of O about P is the same thing exactly as the angular velocity of P about O.

To find their *relative linear velocities*, we observe that P appears to O to have described the little straight path PP'' in the time Δt, so that the limiting value of the ratio $\dfrac{PP''}{\Delta t}$ represents the relative linear velocity of P with respect to O.

In the same way the limiting value of the ratio $\dfrac{OO''}{\Delta t}$ represents the relative linear velocity of O with respect to P; and it is obvious that PP'' is equal and parallel to OO'', and that they are in *opposite senses*.

Two points may be moving in such a way that at some instant each *appears stationary* to the other. This will happen, if at the end of the small interval Δt, the line joining them remains parallel to its direction at the beginning of the interval; or, in other words, when for a short time the line joining them does not revolve. When this is the case *the components of their absolute velocities perpendicular to the line joining them are equal and in the same sense.*

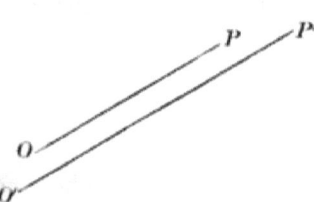

Fig. 11.

For if O', P' are the positions occupied by O and P at the end of a very short time Δt, and if $O'P'$ is parallel to OP, the component of OO' perpendicular to OP is equal to that of PP' in the same direction; and $\dfrac{OO'}{\Delta t}$ and $\dfrac{PP'}{\Delta t}$ are in the limit, the absolute velocities of O and P.

Thus, a figure will easily show that one planet must at a

certain time appear stationary to another, and their positions at this time can be roughly indicated without calculation.

The *relative path*, or *relative orbit*, as it is called, of one moving point with respect to another *is a curve obtained by drawing from a fixed origin lines parallel and equal to the simultaneous distances of the two points*.

Thus, let the moving point O (fig. 12) occupy positions O_1, O_2, O_3, ... along any path in fixed space at the same times that the moving point P occupies positions P_1, P_2, P_3, ... along any other path in fixed space, the time intervals between these simultaneous positions being very small. Then, starting from any fixed point A, draw AP_1', AP_2', AP_3', ... equal and parallel to O_1P_1, O_2P_2, O_3P_3, ... and we shall obtain a number of points P_1', P_2', P_3', ... sufficiently close and numerous, if the time intervals are small enough, to enable us to draw a new curve, perhaps widely different in shape and equation from either of the absolute paths, and this is the curve which P *appears* to O to describe.

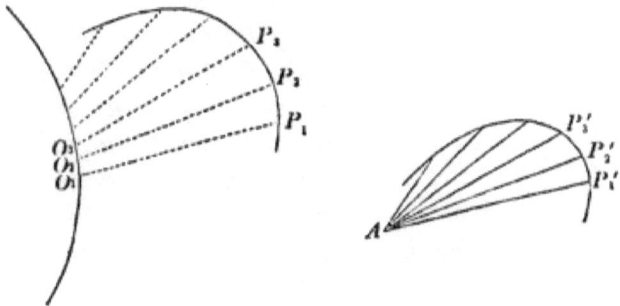

Fig. 12.

Obviously, also the rate at which P_1' moves along this relative path is exactly the relative velocity of P with respect to O. [Consider fig. 10, and the definition of relative velocity before given.]

In particular, consider O to be the centre of a circular wheel, which rolls along a horizontal plane without slipping, and P to be any point on its rim. The *absolute* path of O is a horizontal right line, the *absolute* path of P is a cycloid, but the relative path (of O with respect to P, or of P with respect to O) is a circle.

17. Problem. *Given the absolute velocity of a moving point, O, and the absolute velocity of another moving point, P, to find the relative velocity of P with respect to O.*

This problem has been already solved, but for clearness we reproduce the solution. Let OO' (fig. 13) represent the magnitude and direction of the absolute velocity of O. [OO' may either be drawn *equal* to the velocity of O, or equal to this velocity × Δt. If the latter, O' is *actually* the position occupied by O at the end of Δt; if the former, O will, of course, never be at O' unless O moves in the right line OO', and continues to do so.] Let PP' represent in magnitude and direction the absolute velocity of P. Then drawing PQ equal and parallel to OO', but in the opposite sense, the relative velocity of P with respect to O will consist of the two components PP' and PQ, which give as a resultant PP'', the diagonal of the parallelogram determined by PP' and PQ.

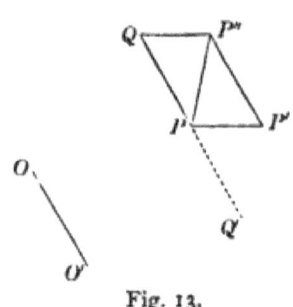

Fig. 13.

In other words, *reverse the velocity of O on P, and compound it with the absolute velocity of P; and the resultant of these is the relative velocity of P with respect to O.*

18. Problem. *Given the absolute velocity of a moving point, O, and the relative velocity of another moving point, P, with respect to O, to find the absolute velocity of P.*

In other words, given OO' and PP'' in the above figure, to find PP'.

At P draw PQ' equal and parallel to OO' and in the same sense; then compound PQ' and PP'' and we obtain PP'; so that we *give P the absolute velocity of O* (not reversed), *and compound it with the relative velocity of P with respect to O, and the resultant is the absolute velocity of P.*

Such construction is evident from common sense.

19. Problem. *Given the absolute motion of one moving point, O, and the relative path, together with the law of its description, of another moving point, P, with respect to O; find the absolute path of P.*

From each point, O_1, O_2,... (fig. 12) of the path of O draw a line, O_1P_1, O_2P_2, ... equal and parallel to the corresponding radius vector, AP_1', AP_2',.. of the relative path, and the locus of the extremity of this radius vector is obviously the absolute path of P.

As a particular case, let the relative path be a right line passing through the moving point O and described with constant velocity, β. The point P always appears to O to be coming straight towards it, so that if a is the angle made with the axis of x by the line of relative motion (which is that joining the initial positions of the points) and if $c =$ initial distance between the points, the co-ordinates of P at any time will be

$$x+(c-\beta t)\cos a, \quad y+(c-\beta t)\sin a,$$

if x and y are the co-ordinates of O.

20. Cartesian equation of Relative Path of two Points. Let the co-ordinates of the point O with reference to rectangular axes fixed in space be (a, β), those of P with reference to the same axes (x, y), and those of P with reference to axes drawn through O parallel to the fixed axes (x', y'). Then

$$\left.\begin{array}{l} x = a + x', \\ y = \beta + y', \end{array}\right\} \qquad (1)$$

and in (fig. 12) the co-ordinates of P' with reference to axes drawn through A parallel to the fixed-space axes will be obviously (x', y'), so that these are the co-ordinates of the point in the relative orbit.

The relation between x' and y'—i.e., the equation of the relative path—is to be found by eliminating x, y, a, β between the above equations and those which will be given as defining the absolute motions of O and P—such as, for example,

$$x = f_1(t), \quad y = f_2(t), \quad a = \phi_1(t), \quad \beta = \phi_2(t),$$

where t is the *time*, which must be also eliminated.

21. Components of Relative Velocity of two Points. The components, parallel to the fixed axes, of the relative velocity are $\dfrac{dx'}{dt}$ and $\dfrac{dy'}{dt}$, where

$$\frac{dx'}{dt} = \frac{dx}{dt} - \frac{da}{dt},$$

$$\frac{dy'}{dt} = \frac{dy}{dt} - \frac{d\beta}{dt}.$$

22. Path of a moving point relatively to a moving plane. Quite distinct from the path of the moving point P relatively to a moving point Q is the path of P relative to a moving plane space containing Q and P. This latter would be got by drawing *any* two axes in the moving plane—i.e., fixed in the moving plane—and laying down at each instant the co-ordinates which P has at that instant with reference to these axes. The curve thus obtained would be the path of P relatively to the moving plane; and, of course, the axes may be drawn at any point of the plane. Suppose that A (fig. 14) is the point chosen in the moving plane through which to draw axes, $A\xi$, $A\eta$, fixed in the plane, and let Ox and Oy be axes fixed in space. Then the motion of the moving plane will be defined by the values (a, β) of the co-ordinates of A and the angle, ϕ, which $A\xi$ makes with Ox.

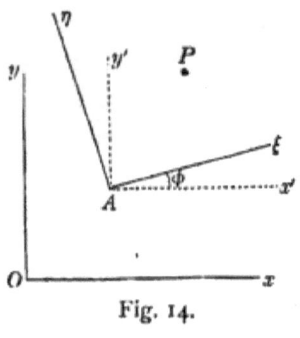

Fig. 14.

Obviously if (ξ, η) are the co-ordinates of the moving point P with reference to $A\xi$ and $A\eta$, and (x, y) the co-ordinates of P with reference to Ox and Oy, we have

$$\left.\begin{array}{l}\xi = (x-a)\cos\phi + (y-\beta)\sin\phi, \\ \eta = -(x-a)\sin\phi + (y-\beta)\cos\phi,\end{array}\right\} \qquad (a)$$

and in addition we shall be given some such equations as

$$x = f_1(t),\ y = f_2(t),\ a = \phi_1(t),\ \beta = \phi_2(t),\ \phi = F(t),$$

where t is the time. Hence the relation between ξ and η may be found.

23. Velocity of a moving point relatively to a moving plane. In the last case $\dfrac{d\xi}{dt}$ and $\dfrac{d\eta}{dt}$ are the components of the velocity of P relatively to the moving plane, or, what is the same thing, relatively to the moving system of axes $A\xi$ and $A\eta$.

Examples.

They are known by differentiating the values of ξ and η just given when the other quantities are assigned as functions of t. Obviously for the purpose of finding the resultant relative velocity we may take the axes $A\xi$ and $A\eta$ parallel to Ox and Oy, i.e., we may put $\phi = 0$ *after differentiating* the general expressions (a).

EXAMPLES.

1. Two points, P and Q (fig. 15), starting at the same instant from given positions, move each in a right line with constant velocity; find their relative angular velocity at any instant, and their relative path.

Let their lines of motion meet in O; suppose them to have been originally (i.e., when $t = 0$) at p and q; let $Op = a$, $Oq = b$, $u =$ velocity of P, $v =$ velocity of Q, $a =$ angle POQ, and $\theta =$ the angle OPQ. Then $\frac{d\theta}{dt}$ is their relative angular velocity. If t is the time at which they are at P and Q

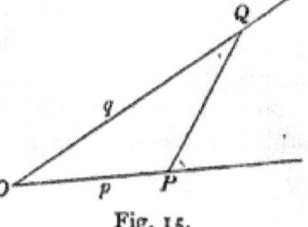

Fig. 15.

$$OP = a + ut, \quad OQ = b + vt, \text{ and}$$
$$(a + ut) \sin \theta = (b + vt) \sin (\theta + a),$$
$$\therefore \frac{a + ut}{b + vt} = \cos a + \sin a \cot \theta.$$

Differentiating with respect to t,
$$\frac{bu - av}{(b + vt)^2} = -\frac{\sin a}{\sin^2 \theta} \frac{d\theta}{dt}.$$

But if $r = PQ$, we have $r \sin \theta = (b + vt) \sin a$,
$$\therefore \frac{d\theta}{dt} = \frac{av - bu}{r^2} \sin a.$$

At any instant the direction of the relative velocity of Q with respect to P is got by reversing P's velocity on itself and on Q, and compounding this reversed velocity with Q's absolute velocity. If at Q we draw a parallelogram with adjacent sides equal to v along OQ and u parallel to PO, its diagonal passing through Q is the direction of relative velocity, i.e., the tangent to the relative path at Q. But this diagonal, in all positions of P and Q, makes a constant angle with OQ, viz.,

$$\tan^{-1} \frac{v - u \cos a}{u \sin a}.$$

Hence the relative path is such that at all its points its tangents are in the same direction. It is therefore a right line; and since the magnitude of the diagonal of the above parallelogram is always the same, the relative velocity (or velocity in the relative path) is constant.

2. A man, M, is about to start along a straight road AMB with uniform velocity v in the sense MB; a dog, D, is at the same instant to start along a straight path with uniform velocity u, for the purpose of meeting the man; what must be the direction of this path and the velocity of the dog?

Ans. At D draw a right line DR parallel to MB, the length of DR representing v on any scale and its sense being opposite to that of v; draw DM; from R draw a right line RP meeting DM in any point P. Then if the dog starts along a path parallel to RP with a velocity represented by the length RP, he will meet the man.

3. If in the last problem the dog is to overtake the man in a given time, how must he start?

Ans. The relative path being in all cases DM, this distance will be described in the given time, t, with the relative velocity; therefore if $DM = c$, the relative velocity will be $\frac{c}{t}$. Hence in DM take the point P so that the length DP represents $\frac{c}{t}$ while, as before, DR represents v, and the direction, RP, of the required path of the dog is determined. If DM makes an angle a with the man's path, the velocity of the dog must be

$$\frac{\sqrt{c^2 + 2cvt \cos a + v^2 t^2}}{t}.$$

4. One point, M, moves along a right line with constant acceleration (see next Chapter), while another point, D, appears to M to be always coming straight towards it; what is the absolute path of D?

Ans. A parabola. [Take the line of motion of M as axis of x, and the initial position of M as origin. Then for M we have $x = at + bt^2$, where a and b are constants. Hence the co-ordinates of D are

$$at + bt^2 + (c - \beta t) \cos a, \quad (c - \beta t) \sin a.]$$

24. Displacement of a Rigid Body. We now proceed to consider the displacements of a rigid body, *the motions being all supposed to take place parallel to one plane.*

We shall suppose the plane of motion to be that of the paper, and our figures will represent merely *sections* of the moving bodies, and not figures in three dimensions. The exact position of a rigid body will be known if we know the positions of any *three* points in it, provided that these three are not in one right line; but if the motion of the body is uniplanar, the exact position of the body will be known if we know the positions of any *two* points in it. Thus, however complicated may be the curve of section of the body, we may consider its displacements as depending simply on those of the *right line* joining any two points, O and P, selected in the plane of section. Let fig. 16 represent the body in one position, and fig. 17 the same body displaced in any manner, the displacement being either great or small.

Then the body may be brought from the first to the second position in several ways, because the right line, OP, may be brought into its second position in several ways. For example, we may move every point parallel to the direction PP' through a distance equal to PP', thus bringing P into position, and then rotate the line about P' through an angle equal to that between the directions OP and O'P'. These two motions will bring the line OP, and every point, A, in the body into position.

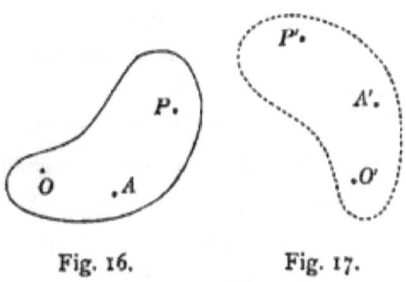

Fig. 16.   Fig. 17.

Again, we may move every point parallel to OO' and through a distance equal to OO', thus bringing O into position, and then rotate the line about O' so as to bring P into position at P'. This angular motion is exactly the same in magnitude and sense as before.

Definition.—A motion of a body whereby all points in the body move in the *same direction* and through the *same distance* is called *a motion of translation*.

Thus, in the first case the body receives a motion of translation represented in direction and amount by PP'; and in the second it receives a motion of translation represented by OO'. These motions of translation are not the same either in magnitude or in direction; but it must be carefully observed that in each case *the motion of rotation is the same*.

In the same way, the body might receive the motion of

translation, AA', which brings any point, A, into position, and a subsequent rotation about A' of the same amount as before.

Of course if the second position differs widely from the first the points, P, O, ... may have moved to their second positions, P', O', ... by very complicated paths. We merely say here that the second position *may be* attained by the aforesaid motions of translation and rotation.

A body, therefore, may be brought from any one position to any other by two things: *firstly, by a motion of translation whereby any one point is brought directly into position: and, secondly, by a motion of rotation about this point*[1].

The first motion depends on the point selected; the second is the same whatever point be selected.

25. Rotation in General. A body is said to *rotate* whenever the line joining any two points, or particles, in the body changes the angle which it makes with a fixed line in space. Thus the line $O'P'$ in fig. 10 has rotated because the line has altered the angle which it made with a fixed line in space. The amount of rotation is the difference between the angles which $O'P'$ and OP make with LM.

The angular velocity of the body at any instant is the rate, per unit of time, at which a line, OP, fixed in the body, increases the angle which it makes with any line fixed in space.

We say here that the body rotates,—about what does it rotate? The answer is, about *every point of the body*. The result of the displacement in the figure is that an observer at O sees P, A, and every point in the body rotated about him in a sense opposite to that of watch-hand rotation through an angle equal to that between the directions OP and $O'P'$. Similarly an observer at A sees every point rotated about him through this

---

[1] Strictly speaking, we ought to say 'a motion of rotation about *an axis* through the point perpendicular to the plane of displacement,' since the angular velocity of a body takes place about *an axis* and not about *a point*. Since, however, we are considering only uniplanar motion, we shall (as in the analogous case of the moments of uniplanar forces in Statics) speak of angular velocity about a point to avoid circumlocution. (See Statics, Art. 72.)

same angle and in the same sense; and with regard to all other points the body has similarly revolved.

When a rigid body moves in any manner parallel to one plane its angular velocity at any instant is the same about every point in it.

Thus, for example, when a wheel rolls along a plane, its angular velocity about the centre is exactly the same as its angular velocity about any point on the rim.

The fact that the Moon always presents the same face to the Earth shows that a lunar day (i.e., the time taken by the Moon to perform a complete revolution about its axis) is equal in length to a lunar month.

26. Pure Rotation. A rigid body is said to have a *pure rotation* about an axis at any instant, if at that instant every point in the body is moving in a plane perpendicular to the axis, and in a direction at right angles to the perpendicular let fall from the point on the axis; in other words, if every point in the body is at the instant describing an arc of a circle whose centre is on the axis and plane perpendicular thereto. The motion of which we here speak is, moreover, the *absolute* motion of each point, or its motion in fixed space. Such a motion about the axis may take place only at one instant, or it may continue to take place about the axis. In the former case, the axis is called an *instantaneous axis*, and, in the latter, a *permanent axis*. If a hole is bored through a body, and an axis be driven through it, and then fixed in space, any displacement of the body is necessarily a pure rotation about the axis. A pendulum, simple or compound, is a familiar example. As we are considering only motion which is parallel to one plane, we may speak of the *point* instead of the *axis* about which the body has a motion of pure rotation. A wheel rolling, without sliding, along a plane has at any instant a pure rotation about its point of contact with the plane, i.e., this point is for the instant at rest, and every point in the wheel (whether on its rim or its spokes) is moving, with reference to fixed space, at right angles to the line joining that point to the point of contact of the wheel with the plane. But the motion of pure rotation about this point of contact is not permanent, since the point of the body in contact with the plane is perpetually changing. The point of contact is, therefore, only an *instantaneous centre* of pure rotation.

It must not be assumed that, because every point in the wheel is rotating about the centre of the wheel, this latter is the instantaneous centre; for, in the first place, as has been pointed out, there is the same rotation about *every* point in the body as about the centre, and, in the next place, the direction of *absolute* motion of any point is not at right angles to the line joining it to the centre of the wheel.

The rule for finding the position of the instantaneous centre is, evidently,—*find the directions of the* ABSOLUTE *motions of any two points, O and P, in the body; erect perpendiculars to these directions at O and P, and the point of intersection of these perpendiculars is the instantaneous centre.*

27. Body Locus and Space Locus. An important distinction must be made between the curve or locus traced out *in the body* by the successive positions of a point, and the curve or locus which the same point traces out in fixed space. This distinction has already been virtually pointed out; for it is the same as that between the fixed-space locus of P (given by an equation between x and y) and the locus of P relatively to the moving plane (given by an equation between ξ and η) in Art. 22.

Consider the very simple case of a wheel rolling along a fixed plane, and imagine a piece of chalk to be placed at each instant at the point of contact of the wheel with the plane; and, moreover, imagine this piece of chalk to make a mark on the wheel and also a mark on the plane. After any amount of motion has taken place, examine the body and also the plane. What curve do the chalk marks make in the rolling body? Evidently the curve formed by the rim of the wheel—a circle, if the wheel is a circular one. What curve do the chalk marks make on the fixed plane? Evidently a right line. The locus traced out in the body by the successive positions of the point of contact is, therefore, quite different from the locus traced out in space by them.

Instead of using the point of contact as a tracing point, we may use a point defined in any other way. For example, a circular lamina rolls along a plane; what locus is traced out in the body, and what locus is traced out in fixed space by a point which is always taken half-way between the point of contact and the centre? Obviously, the locus traced out in the body—the *body locus*, we shall call it—is a circle concentric with the lamina, and of half its radius, while the *space locus* is a right line parallel to the plane. What locus is traced out in the body in this same case by a point taken on the radius to the point of contact at a distance from the latter, equal to three times the radius? In such a case as this, in which the point is outside the *actual* physical limits of the body, we imagine the body as indefinitely prolonged on all sides by means of a thin rigid membrane which can be marked by the piece of chalk imagined to be placed at the successive positions of the tracing point.

A line, or curve, is therefore *fixed in the body* when it always passes through the same particles of the body.

A locus traced out by the successive positions of an instantaneous centre of pure rotation has received the special name of a *centrode*, a word derived from the Greek, and signifying 'the path of the centre.' We shall have therefore, in all cases, both a *body centrode* and a *space centrode*.

28. Fundamental Theorem. We now proceed to prove that in all cases of uniplanar motion, i.e., motion parallel to one plane, *there is at every moment an instantaneous axis of pure rotation.*

Let G be a point in the body whose absolute motion takes place at any instant in the direction GT with velocity v, and let ω be the angular velocity of the body about G at that instant. We must now find the direction of the absolute motion of another point, P.

Draw Pp perpendicular to PG and proportional to $\omega \, PG$; then (last example) Pp represents in magnitude and direction the velocity of P relative to G. Again, draw Pq parallel to GT and

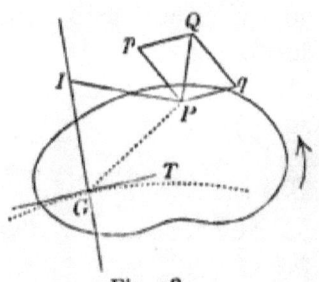

Fig. 18.

proportional on the same scale to v; then, completing the parallelogram $PqQp$, the diagonal PQ represents in magnitude and direction the absolute velocity of P (§ 18). Hence, drawing GI perpendicular to GT and PI perpendicular to PQ, their point of intersection, I, is the instantaneous centre (§ 26).

To find the distance GI, we have

$$\frac{GI}{GP} = \frac{\sin GPI}{\sin GIP} = \frac{\sin pPQ}{\sin qPQ} = \frac{Pq}{Qq} = \frac{v}{\omega \cdot GP};$$

therefore
$$GI = \frac{v}{\omega};$$

and since the length GI is independent of r, i.e., of the point P, we get the same point I, no matter what point P we choose. This proves the existence of an instantaneous centre, and, at the same time, finds its position.

29. Analytical Proof. Let Ox and Oy (fig. 19) be two rectangular axes fixed in space in the plane of motion; let a particle of the body at G have co-ordinates a and β, at any instant, with reference to Ox and Oy; let ω be the angular velocity of the body at this instant—i.e., the rate per unit of time at which the line joining any two particles is describing an angle with a fixed-space line, Ox; let P be a particle at a distance r from G; draw Gx' and Gy' parallel to Ox and Oy; and let PG make an angle θ with Gx', or Ox.

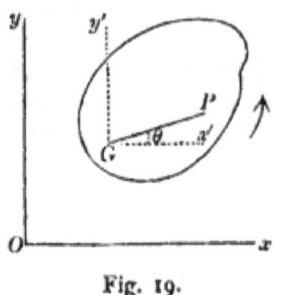

Fig. 19.

If x and y are the co-ordinates of P with reference to Ox and Oy,

$$x = a + r \cos \theta, \tag{1}$$

$$y = \beta + r \sin \theta. \tag{2}$$

Denote time-rates by dots—i.e., $\dfrac{dx}{dt}$ by $\dot{x}$, &c. Then

$$\dot{x} = \dot{a} - \omega r \sin \theta, \tag{3}$$

$$\dot{y} = \dot{\beta} + \omega r \cos \theta, \tag{4}$$

since $\dot{\theta}$ is ω; and observe that we have differentiated on the supposition that r is of invariable length.

If x' and y' are used for $r\cos\theta$ and $r\sin\theta$, the co-ordinates of P with reference to Gx' and Gy', we have

$$\dot{x} = \dot{a} - \omega y', \qquad (5)$$
$$\dot{y} = \dot{\beta} + \omega x'. \qquad (6)$$

Hence $\dot{x}$ and $\dot{y}$ will both be zero if we choose

$$x' = -\frac{\dot{\beta}}{\omega}, \quad y' = \frac{\dot{a}}{\omega}, \qquad (7)$$

that is we shall have chosen the particle which for the instant has no velocity—i.e., the particle which is at the instantaneous centre.

It is very easily seen that this analytical determination of the instantaneous centre from the known motion of a point G and the angular velocity of the body agrees with the geometrical method of last Article.

The co-ordinates, with reference to axes fixed in space, of the instantaneous centre are, therefore,

$$x = a - \frac{\dot{\beta}}{\omega}, \qquad (8)$$

$$y = \beta + \frac{\dot{a}}{\omega}. \qquad (9)$$

30. Notation for Space-Points and for Body-Points. Consider any one definite position of a moving body, and let the position of a point be determined according to any assigned law—such as, for example, that the point shall always be taken on the normal to the path of the body's centre of mass, at a distance from this point proportional to the time. At the position of this point we shall find a *particle* of the body (or of a rigid membrane imagined as prolonging the body). Denote the *particle* by P_b, and the *space-point* occupied by it by P_s. Now beyond the bare fact that they coincide—i.e., that their co-ordinates are the same—P_b and P_s have, in general, nothing in common.

Their velocities, and as we shall subsequently see, other properties also, are wholly different both in magnitude and in direction—notwithstanding that their positions coincide. In fact, their motions must be conceived to be as distinct as those of two moving points, describing wholly different curves, which happen to reach at the same instant a point at which their two paths cross each other.

Thus if the point determined is the Instantaneous Centre, the *particle* is at rest for the instant, while the space-point has a velocity. The two aspects—with respect to body and to space—of the Instantaneous Centre we shall distinguish by the notation I_b and I_s, respectively. Similarly, the body centrode will be denoted by C_b, and the space centrode by C_s; and, in general, the suffix b will be used to denote anything relating to the body, while s will denote the corresponding thing for space.

31. Velocities of I_b and I_s. The velocity of I_b is, of course, nothing. The velocity components of I_s are the time-rates of increase of x and y in equations (8) and (9) of Art. 29. Thus

$$\dot{x} = \dot{a} - \frac{d}{dt}\left(\frac{\beta}{\omega}\right) \qquad (1)$$

$$\dot{y} = \dot{\beta} + \frac{d}{dt}\left(\frac{\dot{a}}{\omega}\right) \qquad (2)$$

are the velocity components of I_s, expressed in terms of the motion of a definite particle, G, of the body, and the angular velocity of the body.

The velocity components of I_s can also be found by differentiating equations (1) and (2) of Art. 29, on the supposition that r is variable and that $\frac{d\theta}{dt}$ is not the angular velocity of the body—since it is not the rate at which the line joining two definite particles revolves.

Thus
$$\dot{x} = \dot{a} + \cos\theta\,\frac{dr}{dt} - r\sin\theta\,\frac{d\theta}{dt}.$$

But
$$r\cos\theta = -\frac{\dot{\beta}}{\omega}, \quad r\sin\theta = \frac{\dot{a}}{\omega},$$

so that
$$\dot{x} = \dot{a} - \frac{\dot{\beta}}{r\omega}\frac{dr}{dt} - \frac{\dot{a}}{\omega}\frac{d\theta}{dt},$$

which must, of course, be equal to the value in (1). Similarly for $\dot{y}$; and the student may show that both methods lead to consistent results.

32. Equations of the Body Centrode and Space Centrode. The equation of the Space Centrode is obtained by eliminating the variable (usually the *time*, t) from equations (8) and (9) of Article 29 and thus obtaining a relation between x, y, and constants.

To obtain the Body Centrode, i.e., the curve traced out in the body by those particles which have had, or will have no velocity at some time during the motion, draw two rectangular axes $G\xi$ and $G\eta$ at G, and let each of these axes always contain the same row of particles of the body—i.e., let them be fixed in the body. Let the line $G\xi$ make an angle ϕ with the fixed-space line Ox. Then if ξ and η are the co-ordinates of any point with reference to the axes $G\xi$ and $G\eta$, we have

$$\xi = (x-a)\cos\phi + (y-\beta)\sin\phi, \qquad (1)$$

$$\eta = (y-\beta)\cos\phi - (x-a)\sin\phi. \qquad (2)$$

Now $\dfrac{d\phi}{dt}$ obviously measures the angular velocity of the body at any instant; and if we give to x and y the values in (8) and (9), Art. 29, we have

$$\xi = -\frac{\dot{\beta}}{\omega}\cos\phi + \frac{\dot{a}}{\omega}\sin\phi, \qquad (3)$$

$$\eta = \frac{\dot{a}}{\omega}\cos\phi + \frac{\dot{\beta}}{\omega}\sin\phi, \qquad (4)$$

for the body-coordinates of the instantaneous centre. We must combine with these the equation

$$\frac{d\phi}{dt} = \omega, \qquad (5)$$

and then the relation between ξ and η—i.e., the equation of the Body Centrode—is to be obtained by eliminating the variable quantities from (3), (4), (5).

It is evident that if between the co-ordinates, referred to axes fixed in the body, of a point determined according to any law, an equation is found, by the elimination of variables, this equation will denote the *body-locus* of the point; and if an equation is similarly found for the co-ordinates of the point referred to axes fixed in space, it will denote the *space-locus*.

Thus, for example, the body centrode may be found by an equation between the radius vector drawn to the instantaneous centre from an invariable point (or particle) in the moving body, and the angle which this radius vector makes with fixed line (or invariable row of particles) in the body.

33. Point of Contact of a Line with its Envelope. Let AB (fig. 20) be a right line of which the points A and B are at any instant being displaced along two curves, AC and BD, fixed in space. Then since the absolute motions of the points A and B take place along the tangents to the curves at A and B, the instantaneous centre, I, is the point of intersection of the normals at A and B (Art. 26).

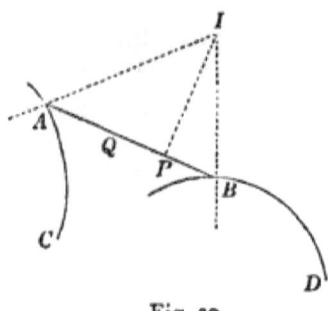

Fig. 20.

Then any point, Q, on the line is moving along the perpendicular at Q to the line QI. Draw IP perpendicular to AB; then the point P is for the moment moving along the line AB; P is, therefore, the point common to the line AB and to the consecutive position of AB; that is, it is the point of contact of AB with its envelope, since this point of contact is shown in the Differential Calculus to be the point of intersection of two consecutive positions.

Exactly the same construction holds for the point or points of contact of any moving curve whatever with its envelope.

In the same way, if a right line, AB, (fig. 21) pass continually through a fixed point, O, while one extremity, B, moves along a fixed curve, the instantaneous centre, I, is the

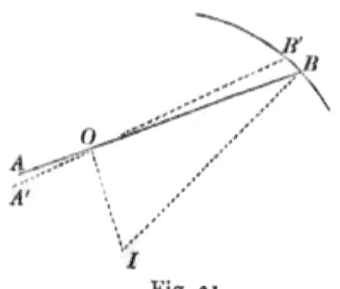

Fig. 21.

point of intersection of the normal to the curve at B with the perpendicular drawn to AB at O.

For, if $A'B'$ be a very close position of the moving line, the point of the line that was at O is now found close to O on OA'; the direction of the absolute motion of this point is therefore OA', or ultimately OA; while the point B moves ultimately along the tangent at B. Therefore, &c.

34. Oscillating Cylinder. The principle of the Cylinder used in oscillating engines is the same as that of the system of jointed bars represented in fig. 22. AB is a bar moving freely round a fixed axis at A; BD is a bar freely jointed to AB at B, and passing through a ring, C, which can rotate round its (fixed) axis to allow of the rotation of the bar BD.

We propose to find the relation between the angular velocities of AB and BD, and also the two Centrodes, C_a and C_b, of the motion of BD.

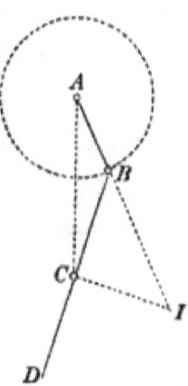

Fig. 22.

The point B of the bar BD is moving along a circle with A as centre; hence the instantaneous centre, I, for BD lies on AB produced; and by Art. 33 it also lies on the perpendicular at C to BD.

Hence if ω' is the angular velocity of BD at any instant, and ω that of AB, the velocity of B—a point on both bars—is, on the one hand, $\omega' \cdot IB$, and, on the other, $\omega \cdot AB$; therefore

$$\omega' = \omega \frac{AB}{BI}.$$

Thus $\omega' = 0$ when $BI = \infty$, i.e., when BD is a tangent to the circle described by B; in other words, the bar BD is in that position, receiving a motion of translation, which is along its own direction. Again, ω' is a maximum (if ω is constant) when BI is least, and this is so when the bars lie both in the position AC.

At any instant the component of velocity of any particle of the rod BD in the direction BD is $\omega' \cdot IC$, or $\omega \times$ perpendicular from A on BD, which is the same for all particles.

In practice, BD is a piston which works up and down in a

cylinder capable of revolving round a fixed axis perpendicular to the plane of motion.

To find the Centrodes of the motion of BD (those of AB both reduce, of course, to the point A).

Space Centrode, C_s. As explained in Art. 32, this curve will be known if we find a relation between the radius vector AI and the angle CAI.

Let $AB = a$, $AC = c$, $AI = \rho$, $\angle CAI = \theta$, $\angle ACB = \phi$.

Then in the triangle AIC we have
$$\rho \cos(\theta + \phi) = c \cos \phi,$$
and in the triangle ABC
$$a \sin(\theta + \phi) = c \sin \phi.$$

Each equation gives a value of $\tan \phi$; and equating these,
$$\frac{\rho \cos \theta - c}{\rho \sin \theta} = \frac{a \sin \theta}{c - a \cos \theta},$$
$$\therefore \rho + a = \frac{c^2 - a^2}{c \cos \theta - a}.$$

Now the polar equation of a hyperbola referred to one focus is
$$r = \frac{a(e^2 - 1)}{e \cos \theta - 1},$$
where a is its semi-axis and e its excentricity.

Hence if we construct a hyperbola having A for focus, its semi-axis in the direction AC and of length a, and its excentricity $\frac{c}{a}$, we obtain the Centrode C_s by diminishing each radius vector of the hyperbola by the length a, since we see that
$$\rho = r - a.$$

Body Centrode, C_b. As explained in Art. 32, this curve will be known if we find a relation between the radius vector BI and the angle CBI.

Now let $BI = \rho$, $\angle CBI = \theta$, $\angle ACB = \phi$.

Then
$$(\rho + a) \cos \theta = c \cos \phi,$$
$$a \sin \theta = c \sin \phi;$$
$$\therefore \sqrt{(\rho + a)^2 - a^2} = \frac{\sqrt{c^2 - a^2}}{\cos \theta}.$$

Examples. 47

We might practically construct this curve thus: take a point, K, in BD such that $BK = \sqrt{c^2 - a^2}$; at K draw a line perpendicular to BD; take any point, P, on this perpendicular; at P draw PQ perpendicular to BP and equal to a; draw BQ; take QR, equal to a, along QB towards B, and QR', equal to a, along BQ remote from B, and finally measure off BP' and BP'' along BP, equal to BR and BR', respectively. Then P' and P'' are points on the body Centrode. The curve consists of two infinite branches on opposite sides of the line KP, the branch towards B corresponding to positions of B on the lower portion of the circle round A between the points of contact of tangents to it from C; and the branch remote from B corresponding to the upper positions of B.

The curve C_s passes through the point C, which is the position of I when the directions of the bars coincide; and the curve C_b cuts the bar BD at two points distant $c-a$ and $c+a$ from B. These are evident *à priori*.

The machine whose principle we have here described is one of those whose object is *the conversion of continuous circular motion into alternating rectilinear motion*.

EXAMPLES.

1. A bar, AB, moves in one plane with given angular velocity round an axis fixed at A, while at B it is freely jointed to another bar, BC, whose extremity C is constrained to move along a fixed groove, AD; find the velocity of C in any position. [Crank and Connecting Rod.]

The instantaneous centre, I, for the bar BC is the point of intersection of AB with a perpendicular at C to AD. Now if ω is at any instant the angular velocity of BC, the velocity of B is $\omega \cdot BI$, and that of C is $\omega \cdot CI$; therefore
$$\frac{\text{velocity of } C}{\text{velocity of } B} = \frac{CI}{BI} = \frac{PA}{AB},$$
if CB is produced to meet AP, the perpendicular to AD at A, in P.

If Ω is the angular velocity of the bar AB, the velocity of B is $\Omega \cdot AB$; therefore the velocity of C is $\Omega \cdot PA$, and this velocity will be a maximum when PA is a maximum, if Ω is constant.

Let us find the position in which the ratio of the velocity of C to that of B is a maximum.

Now when AP is a maximum, it is equal to its next consecutive value (as far as the first order of small quantities), i.e., the point P is for a moment stationary, or in other words, P is at the moment the point of contact of BC with its envelope; i.e. (Art. 33) P is the foot of the perpendicular from I on BC. To determine this position, let $\angle BAC = \phi$, $\angle BCA = \theta$, $AB = a$, $BC = c$.

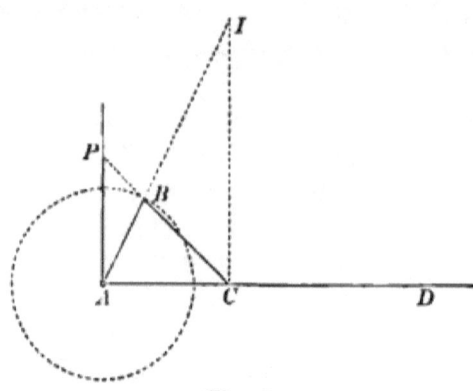

Fig. 23.

Then $PC = AC \sec \theta$; but, the angle at P being right, $PC = CI \cdot \sin \theta$, and $CI = AC \tan \phi$;

therefore $AC \sec \theta = AC \sin \theta \tan \phi$; or

$$\cot \phi = \sin \theta \cos \theta. \qquad (1)$$

Also $\qquad a \sin \phi = c \sin \theta, \qquad (2)$

from which equations ϕ and θ, corresponding to the position of maximum velocity ratio of C and B, must be found.

If we put $\dfrac{c}{a} = k$, and write z for $\sin^2 \phi$, we have, by eliminating θ,

$$z^3 - k^2 z^2 - k^4 z + k^4 = 0. \qquad (3)$$

Put $z = \dfrac{k^2}{3}(x + 1)$, and this equation becomes

$$x^3 - 12x - 11 + \frac{27}{k^2} = 0.$$

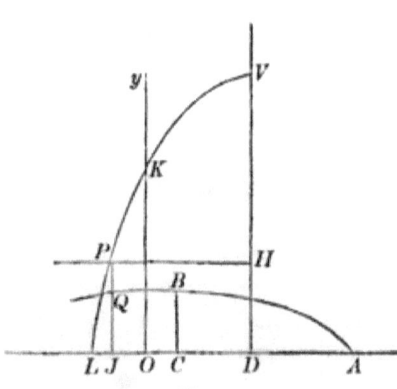

Fig. 24.

Practically, of course, c is much greater than a, so that k is always greater than unity.

A graphic construction for the position corresponding to the maximum ratio of the velocities of C and B may be given as follows:

Take any two rectangular axes, Ox and Oy; draw the curve, $LPKV$, whose equation is $y^2 = -x^3 + 12x + 11$; the point L is at a distance OL, equal to unity, on the negative side of O; the unit length is, of course, arbitrary. The point V is the extremity of an ordinate DV drawn at the point D whose distance from O is 2. The complete curve consists of an oval, of which $LPKV$ is a portion, together with a parabolic branch cutting Ox at the left side of O beyond L.

It is manifest that $y^2 = \dfrac{27}{k^2}$, so that the points V and L correspond to $k=1$ and $k=\infty$, respectively; that is to say, the portion of the curve taken will serve for all crank and connecting rod systems that can occur.

Now take the point C on Ox, such that $OC = \tfrac{1}{2}$, and construct an ellipse, ABQ, whose semi-axes, CA and CB, are $\tfrac{3}{2}\sqrt{5}$ and $\tfrac{1}{2}\sqrt{5}$, respectively.

Then for any given crank and connecting rod, draw $DH = \dfrac{\sqrt{27}}{k}$; draw HP parallel to Ox; from P draw the ordinate PJ, meeting the ellipse in Q, and the length QJ will be, on the scale adopted, the cosecant of ϕ, the angle corresponding to the maximum velocity ratio.

The proof of this may be left as an exercise to the student.

2. If any two points of a plane figure are guided, in any manner, along any two right lines in the plane of the figure, the Space Centrode and the Body Centrode are circles.

For, let F (fig. 25) be the moving figure, two points, P and Q, of which are guided along the fixed lines OA and OB. Then I, the instantaneous centre, is the point of intersection of two perpendiculars, PI and QI, to OA and OB. Describe a circle round the quadrilateral $IPOQ$. Then PQ is a given length, and the angle, POQ, which it sub-

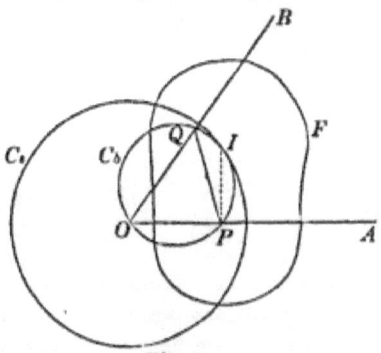

Fig. 25.

tends at the circumference of this circle is given; therefore the length of the diameter, OI, is given. That is, the point I is always at a given distance from the fixed-space point O. Hence C_s is a circle whose centre is O and radius $\dfrac{PQ}{\sin POQ}$.

To obtain C_b express a relation between I and some invariable particle, or line of particles, in the body. Now PQ is such a line of particles, and I is a point in the body such that PQ subtends a constant angle at it; i.e., the body locus of I is the circle $QIPO$, whose diameter is half that of C_s.

3. One point of a plane figure is guided along a fixed right line in its plane with constant velocity, while the body rotates with constant angular velocity; find the Centrodes.

Ans. C_s is a right line parallel to the given one, and C_b is a circle having the guided point for centre.

4. One point of a plane figure is carried round a fixed circle in its plane with constant velocity, while the body rotates with constant angular velocity; find the Centrodes.

Ans. If v is the constant velocity of the tracing point G, ω the angular velocity of the figure, a the radius of the given circle whose centre is O; then C_s is a circle with O for centre and radius $\frac{v}{\omega} - a$; while C_b is a circle with G for centre and radius $\frac{v}{\omega}$.

5. One point of a plane figure has harmonic motion in a right line in the plane of the figure, while the body rotates with constant angular velocity; find the Centrodes.

Ans. If G is the tracing point moving along the line OG, O being a fixed origin, and if $OG = a \sin \mu t$ while $\omega =$ the angular velocity of the body, C_s is the ellipse
$$x^2 + \omega^2 y^2 = a^2,$$
referred to OG and its perpendicular, Oy, at O as axes of x and y.

Taking the line in the body which originally coincided with Oy as initial line, and G as pole, the equation of C_b is
$$r = \frac{a}{\omega} \cos \frac{\mu \theta}{\omega}.$$

6. If one point of a body moves along a right line so that the distance, s, traversed in any time, t, is kt^2, where k is constant, while the body rotates with constant angular velocity about an axis perpendicular to the given line, prove that the Space Centrode is a parabola and the Body Centrode a spiral of Archimedes.

[This is the case of a heavy body set spinning about any axis through its centre of gravity and then dropped vertically.]

7. One point in a body moves so that its co-ordinates, x, y, referred to two fixed rectangular axes are given by the equations
$$x = at + kt^2, \quad y = bt,$$
where a, b, k are constants, while the body rotates with constant angular velocity about an axis perpendicular to the plane xy; prove that the Space Centrode is a parabola, and find the Body Centrode.

8. If a wheel moves so that its motion is a combination of slipping and rolling, find the instantaneous centre.

Ans. If at any instant ω is the angular velocity of the body and v the velocity of slipping of the point of the wheel in contact with the fixed surface, I is on the normal to this surface at a distance $\frac{v}{\omega}$ from the point of contact.

9. If a lamina moves so that two lines fixed in it always pass through two points fixed in space and in the plane of the lamina, prove that the Space and Body Centrodes are two circles, the diameter of the first being

half that of the second. (Wolstenholme's Book of *Mathematical Problems*, p. 417, second edition.)

10. Prove that the angular velocities of the connecting rod may be represented by the ordinates of an ellipse the abscissae of which represent the tangents of the inclinations of the connecting rod to the groove.

[With the notation of example 1, if $\dot{\phi}$ is the angular velocity of the crank, the angular velocity of the connecting rod is $\dfrac{\dot{\phi}}{k} \dfrac{\cos \phi}{\cos \theta}$.]

11. If a lamina moves so that two lines fixed in it always touch two circles fixed in space, prove that C_b and C_s are circles.

[This is at once reducible to example 9.]

12. If a lamina moves so that a right line fixed in it always touches a parabola fixed in space, while another right line fixed in the lamina at right angles to the first always passes through the focus of the parabola, prove that C_s is a parabola, and that C_b is the curve whose equation is $mr = p^2$, where r is the distance of any point on the curve from the point of intersection of the above two right lines, and p is the perpendicular from the point on the first right line, $4m$ being the latus rectum of the given parabola.

13. One point, A, of a lamina is carried with uniform velocity, v, along a right line Ox fixed in space and in the plane of the lamina, while the lamina performs small angular oscillations about A, defined by the equation $\omega = \epsilon\sqrt{a^2 - \theta^2}$, where θ is the (small) angle made at time t by a right line AG fixed in the body with a perpendicular to Ox, and a the (small) initial value of θ, ϵ being a constant; find C_s.

Ans. Its equation referred to O as origin is $y \sin \dfrac{\epsilon x}{v} = \dfrac{v}{a}$, the initial position of A being O.

CHAPTER II.

Acceleration of Velocity.

35. Definition of Acceleration. The term *acceleration of velocity*, or simply *acceleration*, means *rate of increase of velocity per unit of time*. A point moving in any manner has at any one instant different velocities (i.e., components of velocity) in different directions; and as these components have different time-rates of increase, the motion of the point is characterised by different accelerations in different directions.

Thus, the velocity of a moving point, in the position (x, y), parallel to the axis of x—a fixed line in fixed space—is $\frac{dx}{dt}$; and when the point reaches the position $(x+\Delta x, y+\Delta y)$, this component becomes $\frac{dx}{dt} + \Delta \frac{dx}{dt}$, the gain of velocity parallel to the axis of x being therefore $\Delta \frac{dx}{dt}$; and as this gain is made in the time Δt, it is made at the rate of $\frac{\Delta \frac{dx}{dt}}{\Delta t}$ units of velocity per unit of time. This in the limit is

$$\frac{d^2 x}{dt^2}.$$

Similarly the acceleration parallel to the axis of y is $\frac{d^2 y}{dt^2}$.

The time-rate of increase of any quantity whose value changes with the time may be found by the following simple rule which covers all possible cases—

Write down the value of the quantity at the point t; then write down the value of the quantity at the end of a very short interval of time, Δt; subtract the first value from the second, and divide the difference by Δt.

The limiting value of the fraction thus found, when Δt is indefinitely diminished, is the time-rate of increase of the quantity at the time t.

Thus, if the temperature is rising continuously at a certain place, and we wish to know its hour-rate of increase at 12 o'clock, we shall get a good idea of it by subtracting the temperature at 12 o'clock from the temperature at 5 minutes past 12, and multiplying the result by 12 (or dividing by $\frac{1}{12}$); but we should get a better approximation still by subtracting the temperature at 12 from the temperature at 1 second past 12, and multiplying the difference by 3600 (or dividing by $\frac{1}{3600}$).

36. Proper denomination of Acceleration. Acceleration is, then, velocity added (or subtracted) per unit of time; so that if length is measured in feet and time in seconds, velocity will be *feet per second*, and acceleration will be (feet per second) per second; or, as we shall write it, *feet per second per second*.

A minor inaccuracy in speaking of velocity and acceleration must be corrected. It is not at all uncommon to hear that 'velocity is space described in the unit of time,' and that 'acceleration is the amount of velocity accumulated in the unit of time.' This is quite an erroneous way of speaking. The unit of time may be of any magnitude—a day, a month, a year, or a century—and in the unit of time the motion considered may have passed through several positive and negative stages of velocity, and at the end of the unit of time the velocity may be exactly the same as at the beginning.

The truth is that velocity and acceleration have each reference to a particular instant, and each is time-rate of increase at that instant. Now if a train is moving past a station at a definite instant at the *rate of* 50 miles an hour, it by no means follows that 50 miles will be the distance travelled over actually in the next hour—for the train may stop in this time. If we regard velocity (as, of course, we must) as a *rate of* describing length, we shall get perfectly consistent and correct results in speaking of its velocity whether the unit of time is one hour, one second, or one century.

And similarly for acceleration—and, indeed, for every other rate.

Beginners are exceedingly prone to speak of acceleration as 'feet per second;' i.e., as velocity. The example of the above train will show the meaninglessness of such language. Thus, suppose the above train to be moving faster and faster. Now it is a perfectly intelligible question—at what rate per minute is its velocity increasing? Supposing that its velocity is measured in miles per hour, the question is simply—how many miles per hour are being added to its velocity per minute? And this is tantamount to saying that we are here measuring acceleration as *miles per hour per minute*, and not as miles per hour. Acceleration, then, must be spoken of as *feet per second per second*, or *miles per hour per hour*, or by some other equivalent expression.

No concise names are in common use for the *unit velocity* and the *unit acceleration*. Professor Lodge in his *Elementary Mechanics* proposes to call them *a speed* and *a hurry*, respectively; so that the acceleration g would be concisely described as '32.2 hurries,' (about).

37. **Acceleration of motion of a falling body.** The simplest acceleration is, of course, an acceleration of constant magnitude; and of such the simplest example is that afforded by the motion of a body falling near the earth's surface in vacuo. The force with which the earth attracts a small body (the *weight* of the body) does not sensibly vary if the body is carried some few hundreds of feet above the surface; so that if such a body be let fall in an exhausted tube, it will all through its fall be acted on by a constant force, there being no air to resist it with a force which would vary with the velocity. The effect of this constant force is to produce a constant acceleration, which is always denoted by the letter g, the acceleration being roughly at every place

$$32.2 \text{ feet per second per second;}$$

but the exact value depends, of course, on the position of the place on the earth's surface.

38. **Diagram of accumulated velocity.** If at any instant a is the acceleration of the velocity of a moving point in any

Acceleration along the Normal.

direction, the increment, Δv, of velocity acquired in the time Δt is given by the equation $\Delta v = a \Delta t$;

and the whole velocity accumulated in any time in the given direction is given by the equation

$$v = \int a\, dt,$$

the integral being taken throughout the given time.

Hence if we take two rectangular axes, Ox and Oy (fig. 1, Art. 2), and if the successive values, OM, &c., of t are laid down on Ox, and the corresponding values, MP, &c., of the acceleration a drawn perpendicularly to them, we shall have a curve, AB, of accelerations, and the area $aABb$ of any portion of it will represent the velocity accumulated in the given direction in the time interval represented by ab.

39. Acceleration along the Normal. When a point moves along any curve it has at each instant an acceleration along the normal, which is best found by the rule just given.

Let P be the position of the moving point at any time, t, Q the position which it occupies after the interval Δt, PN and QN the normals at P and Q, PT and QT' the tangents at P and Q, v the velocity at P, $v+\Delta v$ the velocity at Q, $\Delta\theta$ the angle between the tangents, or normals, at P and Q. We wish to find the acceleration along PN.

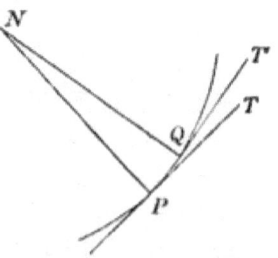

Fig. 26.

Then, velocity along PN at time $t = 0$;

velocity along PN at time $t + \Delta t$

$$= (v + \Delta v) \sin \Delta \theta = (v + \Delta v) \Delta \theta, \quad \text{nearly};$$

$$\therefore \frac{\text{gain of velocity}}{\Delta t} = \frac{(v + \Delta v)\Delta \theta}{\Delta t} = \frac{v \Delta \theta}{\Delta t}.$$

Now $PN \times \Delta\theta = PQ$, or $\rho\, \Delta\theta = \Delta s$, where $\rho = PN =$ radius of curvature at P, and $\Delta s = PQ$; therefore the *normal acceleration* is the limiting value of

$$\frac{v\, \Delta s}{\rho\, \Delta t}, \text{ i. e., } \frac{v^2}{\rho}, \text{ since } \frac{ds}{dt} = v.$$

Students sometimes feel a difficulty in understanding how a point P moving along any path can have an acceleration along the normal at P since at no point of its path has it ever any velocity along the normal. A familiar example ought to remove this difficulty, by pointing out the fact that—*it does not follow that because a moving point has at a particular moment no velocity along a certain direction, it has no* acceleration *in this direction*. Take the case of a stone thrown vertically up. When it is at its highest point, it has no velocity; yet, not only then, but at every instant during its upward (and, subsequently, downward) motion, is velocity being generated in it in the downward sense at about the rate of 32·2 feet per second per second. Hence even at the instant at which the velocity of the stone is zero, it has a downward acceleration, g. Similarly, when the moving point is at P (fig. 26) it has no velocity along PN; but when it gets to Q, it has acquired a component of velocity along PN—not along QN, of course.

40. Acceleration along the Tangent. The acceleration of a moving point along the tangent to its path at any instant is $\dfrac{dv}{dt}$, or $\dfrac{d^2s}{dt^2}$, where s is the length of the arc of its path measured from some fixed point on it up to the position, P, of the moving point (fig. 26).

For, velocity along PT at time $t + \Delta t$ is $(v + \Delta v)\cos \Delta \theta$, or $v + \Delta v$; and velocity along PT at time t is v;

$$\therefore \frac{\text{gain of velocity in time } \Delta t}{\Delta t} = \frac{\Delta v}{\Delta t} = \frac{dv}{dt}, \text{ in the limit.}$$

The acceleration along the tangent may also be written

$$v\frac{dv}{ds},$$

since it $= \dfrac{dv}{dt} = \dfrac{dv}{ds} \times \dfrac{ds}{dt} = v\dfrac{dv}{ds}$. Also it may be written

$$\frac{d^2s}{dt^2},$$

since $v = \dfrac{ds}{dt}$, and $\dfrac{dv}{dt} = \dfrac{d^2s}{dt^2}$.

The student must not imagine that the expression $\dfrac{d^2 s}{dt^2}$ for the acceleration along the tangent is evident without calculation because the acceleration along the axis of x is $\dfrac{d^2 x}{dt^2}$.

For, the latter expression holds only because the axis of x is a fixed line in space, whereas the tangent is a line whose direction is perpetually changing; hence an expression which holds for acceleration along the first would not necessarily hold for acceleration along the second. The application of the rule of Art. 35 shows that the expression for acceleration along the tangent happens to be the same as that for acceleration along a fixed direction; but, in general, it is not true that the acceleration along a changing direction is the second time-rate of increase of the co-ordinate of the moving point in the direction.

Cor. 1. It therefore appears that if the point travels along its path with constant velocity, the acceleration along the tangent is constantly zero, and the resultant acceleration is along the normal.

Cor. 2. In general, the resultant acceleration of the moving point is inclined to the normal at an angle whose tangent is $\dfrac{\rho}{v}\dfrac{dv}{ds}$.

41. Graphic construction for Resultant Acceleration. Through any point, P (fig. 27), draw a right line, PT, parallel to the direction of the velocity of the moving point P in fig. 26, Art. 39, at the time t; and let the length PT represent the velocity, v, of the moving point at this instant.

Fig. 27.

At P draw also PT' parallel to the direction, QT' (fig. 26), of velocity at the time $t+\Delta t$, and let the length PT' represent the velocity, $v+\Delta v$, at Q.

Then TT' is parallel to the direction of the resultant acceleration of the moving point.

For, the component of TT' in any direction is obviously the component of PT'—the component of PT in that direction;

but the difference of these components, divided by Δt, is, according to the rule of Art. 35, the component of the resultant acceleration. Hence TT' is the direction of this latter, and the magnitude of the resultant acceleration is the limiting value of the ratio
$$\frac{TT'}{\Delta t}.$$

The vector $\dfrac{TT'}{\Delta t}$, in its limiting value, is what Newton means by the term *change of velocity*.

42. Acceleration in a changing direction. Let KH (fig. 28) be a right line whose direction alters continuously according to any assigned law, and let it be required to find the expression for the component of the acceleration of the moving point P along KH.

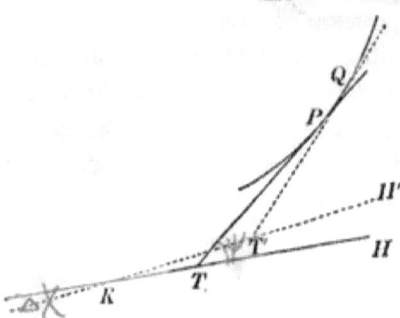

Fig. 28.

Let ψ be the angle, PTH, between the direction of motion at P and the line KH; let KH become KH' in the time Δt, while the direction of motion of the point has become QT'; v = velocity at P; $v + \Delta v$ = velocity at Q; $\psi + \Delta\psi = \angle QT'H'$; $\Delta\chi = \angle H'KH$, χ being the angle which KH makes with a fixed right line.

Adopting the rule of Art. 35, take the component of $v + \Delta v$ at Q along KH; this is $(v + \Delta v)\cos(\psi + \Delta\psi + \Delta\chi)$; and the component of v at P along $KH = v\cos\psi$. Hence the required component of acceleration is the limit of

$$\frac{(v + \Delta v)\cos(\psi + \Delta\psi + \Delta\chi) - v\cos\psi}{\Delta t}$$

$$= \frac{(v + \Delta v)\cos(\psi + \Delta\psi) - v\cos\psi - v\sin\psi\,\Delta\chi}{\Delta t}$$

$$= \frac{\Delta(v\cos\psi) - v\sin\psi\,\Delta\chi}{\Delta t}$$

$$= \frac{d}{dt}(v\cos\psi) - v\sin\psi\frac{d\chi}{dt}, \qquad\qquad (a)$$

in which reduction we have rejected the squares and products of the infinitesimals Δv, $\Delta \psi$, $\Delta \chi$.

43. Acceleration along and perpendicular to the Radius Vector. The expression (a) of last Article can be directly applied to find the components of acceleration of P along and perpendicular to the moving line OP (fig. 2, p. 3).

Identifying OP with the line KH in fig. 28, we have $\chi = \theta = $ the angle between OP and a fixed initial line; also $v \cos \psi = \dfrac{dr}{dt}$ (Art. 4), and $v \sin \psi = r \dfrac{d\theta}{dt}$.

Hence the acceleration along the radius vector OP is, by (a),

$$\frac{d^2 r}{dt^2} - r \left(\frac{d\theta}{dt}\right)^2. \qquad (\gamma)$$

Identifying the perpendicular to OP at P with the moving direction KH, we have $\chi = \dfrac{\pi}{2} + \theta$; $v \cos \psi = $ velocity perpendicular to $OP = r \dfrac{d\theta}{dt}$; $-v \sin \psi = $ velocity along $OP = \dfrac{dr}{dt}$. Therefore the acceleration perpendicular to OP is

$$\frac{d}{dt}\left(r \frac{d\theta}{dt}\right) + \frac{dr}{dt}\frac{d\theta}{dt}, \quad \text{or}$$

$$\frac{1}{r}\frac{d}{dt}\left(r^2 \frac{d\theta}{dt}\right). \qquad (\delta)$$

44. Central Acceleration. The particular case in which the resultant acceleration of a moving point is always directed towards a fixed point or centre is deserving of special notice on account of the part which it plays in kinetics—as, for instance, in the theory of planetary motions round the sun.

If the resultant acceleration is towards the fixed point O, there will be no component perpendicular to the radius vector; so that from (δ) of last Article we have

$$\frac{d}{dt}\left(r^2 \frac{d\theta}{dt}\right) = 0,$$

or $$r^2 \frac{d\theta}{dt} = h, \qquad (1)$$

where h is some constant. To give this equation a geometrical interpretation, observe that if P and Q are two close positions of the moving point, separated by the small time-interval Δt, and if $\Delta\theta$ is the circular measure of the angle QOP, the expression $\tfrac{1}{2}r^2\Delta\theta$ will be the area, QOP, traced out round O in the time Δt; so that $\tfrac{1}{2}r^2\dfrac{d\theta}{dt}$ is the area traced out *per unit of time* in the position P of the moving point.

Hence equation (1) gives as the essential feature of motion in which the resultant acceleration is always directed to a fixed centre the property that—*the time-rate of description of area round the fixed centre is constant in all positions of the moving point.*

The equation (1) can be put into another form which is useful. If Δs = length of the elementary arc PQ joining two very close positions of the moving point, and if p is the perpendicular from the pole O on the right line PQ (which is ultimately the tangent to the path of the moving point at P), the area QOP is
$$\tfrac{1}{2}p\Delta s;$$
so that the time-rate of description of area is $\tfrac{1}{2}p\dfrac{\Delta s}{\Delta t}$, or $\tfrac{1}{2}pv$, where v is the velocity along the tangent at P. Hence (1) is equivalent to
$$pv = h, \qquad (2)$$
where h is the constant expressing double the area traced out per unit of time about the pole.

This result can be deduced by another method. We know that the resultant acceleration of P can be broken up into $v\dfrac{dv}{ds}$ along the tangent and $\dfrac{v^2}{\rho}$ along the normal; and since the resultant is directed towards O, these components have equal and opposite moments about O. Hence
$$pv\frac{dv}{ds} - \frac{v^2}{\rho}p\cot\phi = 0,$$
where ϕ is the angle between OP and the tangent at P.

Now $\rho = r\dfrac{dr}{dp}$ (Williamson's Differential Calculus, chap. 17), and $\cot\phi = -\dfrac{dr}{rd\theta}$; therefore the last equation becomes

$$\frac{dv}{ds} + \frac{v\,dp}{r^2\,d\theta} = 0;$$

but $p\,ds = r^2\,d\theta$; therefore this equation gives $p\,dv + v\,dp = 0$, or $pv = $ constant.

Besides the expression (γ) of Art. 43, there is another very useful one for the acceleration of the moving point along OP (fig. 29) when the acceleration is central to O.

Let $OT = p = $ the perpendicular from the centre on the tangent at P; $\phi = \angle TPO$. Then if $v = $ velocity at P, $\rho = $ radius of curvature

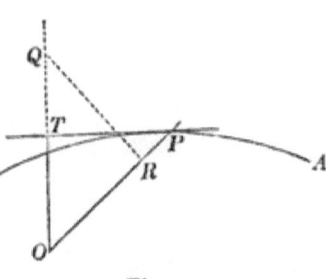

Fig. 29.

at P, the acceleration of P towards O is $\dfrac{v^2}{\rho}\operatorname{cosec}\phi$, since the resultant is along PO. But $\operatorname{cosec}\phi = \dfrac{r}{p}$, and $v = \dfrac{h}{p}$, therefore the resultant acceleration is

$$\frac{h^2 r}{p^3 \rho}, \text{ or } \frac{h^2}{p^3}\frac{dp}{dr}. \tag{3}$$

Of course the expression (γ) of last Article holds whether the acceleration is central or not.

Again, the expression (γ) of last Article may be put into a slightly different form which is useful in the Lunar Theory.

If we denote $\dfrac{1}{r}$ by u, we have $\dfrac{dr}{dt} = -\dfrac{1}{u^2}\dfrac{du}{dt}$; and if (as is the case in all central acceleration) $r^2\dfrac{d\theta}{dt} = h$, where h is a constant,

$$\frac{d}{dt} = hu^2\frac{d}{d\theta}; \text{ therefore } \frac{dr}{dt} = -h\frac{du}{d\theta}, \therefore \frac{d^2r}{dt^2} = -h^2u^2\frac{d^2u}{d\theta^2};$$

so that the expression (γ) becomes $-h^2u^2\left(u + \dfrac{d^2u}{d\theta^2}\right)$, which is the acceleration measured outward along the radius vector OP, i.e., from O to P. Hence

$$h^2u^2\left(u + \frac{d^2u}{d\theta^2}\right) \tag{4}$$

will express the acceleration of the moving point P measured from P towards the centre O.

If the acceleration is not central, $r^2 \dfrac{d\theta}{dt}$ will not be constant throughout the motion; but we may still put h for it, observing that h is variable, and the component acceleration along the line PO from P towards O is expressed as

$$h^2 u^2 \left(u + \dfrac{d^2 u}{d\theta^2}\right) + h u^2 \dfrac{dh}{d\theta}.$$

45. Case of Direct Distance. Let the central acceleration be directly proportional to the distance, PO, between the moving point, P, and a fixed point, O. Suppose the acceleration to be equal to μr; then if (x, y) are the co-ordinates of P referred to two rectangular axes through O, we have

$$\dfrac{d^2 x}{dt^2} = -\mu x, \quad \dfrac{d^2 y}{dt^2} = -\mu y.$$

Now each of these denotes a simple harmonic vibration the period being (see p. 26) $\dfrac{2\pi}{\sqrt{\mu}}$; and, as shown in p. 19, the resulting motion is motion in an ellipse having O for centre, the time of revolution being, of course, the same as that of each constituent harmonic vibration.

46. Hodograph. It appears from Art. 41 that if through any point O (fig. 30) we draw two right lines, Op and Oq, parallel to the tangents, PT and QT', to the path AB of a moving point, at two very close points, P and Q, the lengths of Op and Oq being proportional on any scale to the velocities at P and Q, the line pq will be parallel and proportional to the acceleration of the moving point at P. Suppose this process to be continued—in

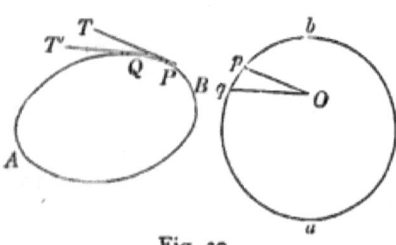

Fig. 30.

other words, suppose that at O we draw an infinite number of radii vectores, such as Op, each being parallel to a tangent to AB and proportional to the corresponding velocity of the moving point P; then we shall thus trace out a curve, ab, whose tangent at any point, p, is parallel to the direction of the resultant acceleration of the moving point P at the corresponding point on the curve AB. The curve ab thus obtained from AB is called (rather inappropriately) the *Hodograph* of the given motion. By Art. 41 it appears that $\frac{pq}{\Delta t}$, in the limit, is the magnitude of the acceleration of P, where Δt is the time of moving from P to Q. Hence *the velocity with which p describes the Hodograph is the acceleration of P.*

47. Hodograph of a Central Orbit. By a Central Orbit is meant an orbit described by a moving point whose resultant acceleration is in every position directed to a fixed point or centre, O. The characteristic property of such an orbit is (Art. 44) expressed by the equation

$$pv = h = \text{a constant,}$$

where p is the perpendicular from O on the tangent to the orbit at any point, P. Now about O as centre describe a circle whose radius is $\sqrt{h}$; let fall the perpendicular OT (fig. 29) on the tangent, PT, at P; and produce OT to Q so that $OQ \cdot OT = h$; then the locus of Q is the polar reciprocal of the given orbit with respect to the circle of radius $\sqrt{h}$. But $OT = p$, therefore $OQ = v$, so that the polar reciprocal constructs, by its radii vectores drawn from O, the velocities of the moving point P in directions always at right angles to those of the velocities. Hence if we turn the polar reciprocal round O through a right angle, it will be the Hodograph of the motion. In this particular case, therefore, of *central acceleration*, the Hodograph and the polar reciprocal of the orbit about O are virtually the same. Of course no such result holds when the motion of P takes place without acceleration towards a fixed centre.

If r, p, ρ denote the radius vector, perpendicular from O on tangent at P, and radius of curvature at P, and r', p', ρ' the corresponding things for the Hodograph (or polar reciprocal) of a central orbit, it is at once seen that

$$r' = \frac{h}{p}, \quad p' = \frac{h}{r}, \quad \rho' = \frac{hr^3}{\rho p^3}.$$

Observe that the tangent at Q to the polar reciprocal is obviously perpendicular to OP, and as by rotation this becomes the tangent to the Hodograph, we have the tangent to the Hodograph parallel to the radius vector OP, i. e., parallel to the direction of resultant acceleration of P—as we know (Art. 46) it must be in all cases, whether of central acceleration or not.

48. Case of Inverse Square of Distance. Let the acceleration of the moving point P be always directed to a fixed point, O (fig. 29), and let it be equal to $\frac{\mu}{r^2}$, where μ is a constant and $r = OP$. It is required to find the orbit. Describe the hodograph about O (fig. 30). Then $r^2 \frac{d\theta}{dt} = h$, therefore the acceleration $= \frac{\mu}{h} \frac{d\theta}{dt}$; but the acceleration of P equals the velocity of p, and it is directed along the tangent to the hodograph at p—which is parallel to PO; therefore $\frac{d\theta}{dt}$ is the angular velocity of tangent at p. Hence we have

$$\frac{\text{velocity of } p}{\text{angular velocity of tangent at } p} = \frac{\mu}{h} = \text{constant};$$

in other words, the radius of curvature of the hodograph at $p = \frac{\mu}{h}$ = constant, therefore the hodograph is a circle. The orbit of P is then (last Article) the polar reciprocal of the circle with respect to O—i. e., it is a conic having O for focus. (See Hamilton's *Elements of Quaternions*, p. 720.)

49. Accelerations of I_b and I_s. Let a, β be the coordinates, with reference to fixed axes, of any point, G, of a body moving with uniplanar motion. Then, as in Art. 29, the co-ordinates (x, y) of any other point, P, in the body will be given by the equations $x = a + r \cos \theta$,

$$y = \beta + r \sin \theta.$$

Accelerations of I_b and I_s.

Differentiating these twice with respect to the time, on the supposition that r is constant, we have

$$\ddot{x} = \ddot{a} - \omega^2 r \cos\theta - \dot{\omega} r \sin\theta, \qquad (1)$$

$$\ddot{y} = \ddot{\beta} - \omega^2 r \sin\theta + \dot{\omega} r \cos\theta, \qquad (2)$$

denoting, as in Art. 29, time-rates by dots.

Now if P is I_b, we know that $r\cos\theta = -\dfrac{\dot{\beta}}{\omega}$, $r\sin\theta = \dfrac{\dot{a}}{\omega}$ (Art. 29); and substituting these in the above expressions, we obtain

$$\ddot{x} = \ddot{a} - \frac{\dot{\omega}}{\omega}\dot{a} + \omega\dot{\beta}, \qquad (3)$$

$$\ddot{y} = \ddot{\beta} - \frac{\dot{\omega}}{\omega}\dot{\beta} - \omega\dot{a}, \qquad (4)$$

for the components, in fixed directions, of the acceleration of I_b.

The components of acceleration of I_s are obtained by differentiating the velocities of I_s in Art. 31. Thus

$$\ddot{x} = \ddot{a} - \frac{d^2}{dt^2}\left(\frac{\dot{\beta}}{\omega}\right), \qquad (5)$$

$$\ddot{y} = \ddot{\beta} + \frac{d^2}{dt^2}\left(\frac{\dot{a}}{\omega}\right). \qquad (6)$$

For clearness we collect in a table the co-ordinates, velocities, and accelerations of I_b and I_s, in terms of the motion of some one point, G, in the moving body.

	I_b	I_s
Co-ordinates.	$a - \dfrac{\dot{\beta}}{\omega},$ $\beta + \dfrac{\dot{a}}{\omega}.$	$a - \dfrac{\dot{\beta}}{\omega},$ $\beta + \dfrac{\dot{a}}{\omega}.$
Velocities.	0 0	$\dot{a} - \dfrac{d}{dt}\left(\dfrac{\dot{\beta}}{\omega}\right),$ $\dot{\beta} + \dfrac{d}{dt}\left(\dfrac{\dot{a}}{\omega}\right).$
Accelerations.	$\ddot{a} - \dfrac{\dot{\omega}}{\omega}\dot{a} + \omega\dot{\beta},$ $\ddot{\beta} - \dfrac{\dot{\omega}}{\omega}\dot{\beta} - \omega\dot{a}.$	$\ddot{a} - \dfrac{d^2}{dt^2}\left(\dfrac{\dot{\beta}}{\omega}\right),$ $\ddot{\beta} + \dfrac{d^2}{dt^2}\left(\dfrac{\dot{a}}{\omega}\right).$

The components of acceleration of I_b can also be written

$$\omega\left[\dot{\beta}+\frac{d}{dt}\left(\frac{\dot{\alpha}}{\omega}\right)\right] \text{ and } -\omega\left[\dot{\alpha}-\frac{d}{dt}\left(\frac{\dot{\beta}}{\omega}\right)\right],$$

and comparing these with the velocity components of I_a, we obtain the result that—

The direction of acceleration of I_b is at right angles to the direction of velocity of I_a, and the magnitude of the acceleration of I_b is equal to the velocity of I_a multiplied by the angular velocity of the body.

50. Instantaneous Acceleration Centre. We have seen that in one-plane motion there is at every instant a body-point, I_b, with no velocity, and that the velocity of every other point, P, takes place at right angles to the line PI_b and is proportional to the length of PI_b. We proceed to show that there is also at every instant a point, $\mathcal{J}$, of no acceleration, and to deduce analogous results for it. Putting $\ddot{x}=0$, $\ddot{y}=0$ in equations (1) and (2) of Art. 49, and using ξ and η for $r\cos\theta$ and $r\sin\theta$, we get

$$\xi = \frac{\omega^2 \ddot{\alpha} - \dot{\omega}\ddot{\beta}}{\omega^4 + \dot{\omega}^2}, \quad \eta = \frac{\dot{\omega}\ddot{\alpha} + \omega^2 \ddot{\beta}}{\omega^4 + \dot{\omega}^2},$$

for the co-ordinates, referred to the arbitrary point G, of the point, $\mathcal{J}$, of no acceleration. In uniplanar motion there is, therefore, always such a point, since we have obtained definite values for ξ and η.

The expression for the acceleration of any point, P, may now be simplified by choosing $\mathcal{J}$ as the point G of reference. Equations (1) and (2) of Art. 49 will become (if $\ddot{\alpha}=0$, $\ddot{\beta}=0$)

$$\ddot{x} = -\omega^2 \xi - \dot{\omega}\eta,$$
$$\ddot{y} = \dot{\omega}\xi - \omega^2 \eta.$$

Hence $\sqrt{\ddot{x}^2 + \ddot{y}^2} = \sqrt{\omega^4 + \dot{\omega}^2} \cdot P\mathcal{J},$

which proves that the magnitude of the acceleration of every point is proportional, at each instant, to its distance from the point $\mathcal{J}$, which is called the *Instantaneous Acceleration Centre*.

Again, if we put $\dfrac{\dot{\omega}}{\omega^2} = \tan\phi$, and $\sqrt{\omega^4 + \dot{\omega}^2} = k$, we have

$$\ddot{x} = -k(\xi\cos\phi + \eta\sin\phi),$$
$$\ddot{y} = k(\xi\sin\phi - \eta\cos\phi).$$

Instantaneous Acceleration Centre.

Denoting PJ by r and its direction angle by θ, these become

$$\ddot{x} = -kr \cos(\theta-\phi),$$
$$\ddot{y} = -kr \sin(\theta-\phi).$$

These values show that the direction of acceleration of the particle at P is the line PA drawn at the angle ϕ, whose tangent is $\dfrac{\dot{\omega}}{\omega^2}$, with JP. The sense in which ω is measured determines the side of the line JP at which we draw $\tan^{-1}\dfrac{\dot{\omega}}{\omega^2}$. If ω

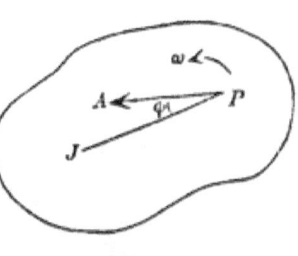

Fig. 31.

is measured in the sense denoted by the arrow, while $\dot{\omega}$ is negative, PA will be drawn at the under side of JP in the above figure. Now ϕ is at each instant the same for all particles. Hence—*in all uniplanar motion there is at each instant a point having no acceleration (the instantaneous acceleration centre), and the acceleration of every other point is in magnitude proportional to its distance from the acceleration centre, while its direction makes with the line joining the point to the acceleration centre an angle which, though varying with the time, is at each instant the same for all points in the moving body.*

If we are given, in magnitudes and directions, the accelerations of any two points, P and Q, the point J can be found. For, let the directions, PA and QB, of these accelerations meet in O, while the accelerations are p and q. Then the angles OPJ and OQJ are equal; therefore the circle round the triangle OPQ passes through J.

Again, $\dfrac{p}{q} = \dfrac{JP}{JQ}$; therefore J lies on the circle which is the locus of points whose distances from P and Q are in the given ratio $\dfrac{p}{q}$. One point of intersection of these two circles is J.

Cor. If the angular velocity of the body is constant, the accelerations of all points are directed to one point.

The whole system of accelerations of the various points of

the body at any one instant we shall call *an acceleration system* of the moving body.

There will, of course, be a fixed-space locus and a body-locus of J, which we shall call the Space Acceleration Centrode and the Body Acceleration Centrode, respectively.

51. Theorem. *Any two possible acceleration systems in uniplanar motion are superposable in a single acceleration system.*

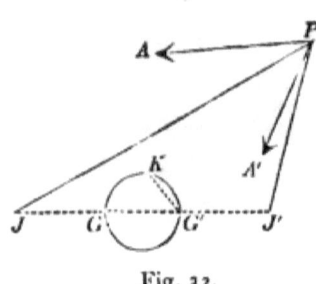

Fig. 32.

To prove this we shall show that if at any point, P, there act two forces, PA and PA', of magnitudes $\mu . PJ$ and $\mu' . PJ'$, respectively, and making angles, JPA and $J'PA'$, constantly equal to a and a', wherever P may be, J and J' being two fixed points, the resultant of these forces will, in all positions of P, be proportional to the distance of P from another fixed point, K, and will make a constant angle with PK.

Replace PA and PA' by their components along and perpendicular to PJ and PJ', respectively. Then we have two forces, $\mu \cos a . PJ$ and $\mu' \cos a' . PJ'$, along PJ and PJ', respectively; and also two forces, $\mu \sin a . PJ$ and $\mu' \sin a' . PJ'$, perpendicular to PJ and PJ', respectively. Now (see Statics, Art. 20) the first pair give a resultant equal to

$$(\mu \cos a + \mu' \cos a') . PG$$

directed from P to G, where G is the point dividing JJ' so that $\dfrac{JG}{J'G} = \dfrac{\mu' \cos a'}{\mu \cos a}$; and in the same way the second pair of forces give a resultant equal to

$$(\mu \sin a + \mu' \sin a') . PG'$$

acting perpendicularly to PG', where G' is such that

$$\frac{JG'}{JG'} = \frac{\mu' \sin a'}{\mu \sin a}.$$

Denote $\mu \cos a + \mu' \cos a$ by $\Sigma\mu \cos a$, and $\mu \sin a + \mu' \sin a'$ by $\Sigma\mu \sin a$. Now take G as origin and GG' as axis of x, and

let X and Y be the components of these forces along and perpendicular to GG'. Let x and y be the co-ordinates of P, and let GG' equal D, then
$$X = x\Sigma\mu \cos a - y\Sigma\mu \sin a,$$
$$Y = (x-D)\Sigma\mu \sin a + y\Sigma\mu \cos a.$$
Now a force consisting of components $m \cos \phi . PK$ and $m \sin \phi . PK$ along and perpendicular to PK would give components
$$m \cos \phi . (x - \xi) - m \sin \phi . (y - \eta),$$
$$m \sin \phi . (x - \xi) + m \cos \phi . (y - \eta),$$
in these directions, if the co-ordinates of K are ξ, η.

Identifying these with X and Y, we have
$$m \cos \phi = \Sigma\mu \cos a, \quad m \sin \phi = \Sigma\mu \sin a,$$
$$\xi = D \sin^2 \phi, \quad \eta = D \sin \phi \cos \phi.$$

Hence the reduction is possible and the point K is found. To construct K, we see that we must describe a circle on GG' as diameter, and from G' draw a line making the angle ϕ with the positive side of GG' (if $\tan \phi$ is positive); this line cuts the circle in K. Also we have for the constants m and ϕ
$$m^2 = (\Sigma\mu \cos a)^2 + (\Sigma\mu \sin a)^2,$$
$$\tan \phi = \frac{\Sigma\mu \sin a}{\Sigma\mu \cos a}.$$

Of course we can regard PA and PA' as two accelerations, and since K is not dependent on the point P, it follows that an acceleration system with constants μ and a whose centre is J and another with constants μ' and a' whose centre is J' compound a single acceleration system with constants m and ϕ (given above) whose centre is K.

In the general, or three-dimensional, motion of a rigid body the superposition of two acceleration systems, each of which is separately possible, is not possible.

52. Accelerations of different orders. The resultant of $\frac{d^2x}{dt^2}$ and $\frac{d^2y}{dt^2}$ has been called the acceleration of the point x, y. In the same way we may take the resultant of $\frac{d^3x}{dt^3}$ and $\frac{d^3y}{dt^3}$, and regard it as the acceleration of the acceleration, or an acceleration of higher order; and similarly the resultant of $\frac{d^4x}{dt^4}$ and $\frac{d^4y}{dt^4}$ is an acceleration of still higher order. For if through any point, P (fig. 27, p. 57), we draw PT and PT' parallel and proportional to the accelerations of a moving point at the times t and $t+\Delta t$, respectively, the line TT' will, for the reason explained in Art. 41, be parallel and proportional to the *change* of acceleration, and the direction cosines of TT' will be proportional to $\frac{d^3x}{dt^3}$ and $\frac{d^3y}{dt^3}$, since the co-ordinates of T (referred to axes through P) are proportional to $\frac{d^2x}{dt^2}$ and $\frac{d^2y}{dt^2}$, and those of T' are proportional to

$$\frac{d^2x}{dt^2} + \frac{d^3x}{dt^3}\Delta t, \text{ and } \frac{d^2y}{dt^2} + \frac{d^3y}{dt^3}\Delta t.$$

53. Theorem. *For acceleration of any order there is in uniplanar motion of a rigid body at each instant a centre or point of no acceleration of that order; and the accelerations of all points are related, in magnitudes and directions, to the centre exactly as the velocities and ordinary accelerations are related to the instantaneous centres of velocity and acceleration.*

For if we express the motion of every point in terms of the motion of some one point, we have, as in Art. 29,

$$x = a + r\cos\theta, \quad y = \beta + r\sin\theta.$$

Hence, denoting $\frac{d^n x}{dt^n}$ by $x^{(n)}$, &c.,

$$x^{(n)} = a^{(n)} + mr\cos\theta + m'r\sin\theta,$$
$$y^{(n)} = \beta^{(n)} + m'r\cos\theta - mr\sin\theta,$$

denoting $\frac{d^n(\cos\theta)}{dt^n}$ by $m\cos\theta + m'\sin\theta$, and observing that $\sin\theta = -\cos(\frac{\pi}{2} + \theta)$.

Now $x^{(n)}$ and $y^{(n)}$ will be zero for the values

$$-\frac{ma^{(n)} + m'\beta^{(n)}}{m^2 + m'^2}. \quad \text{and} \quad \frac{m\beta^{(n)} - m'a^{(n)}}{m^2 + m'^2}$$

of $r\cos\theta$ and $r\sin\theta$. Hence there is a centre of accelerations of this order, and by taking it as point of reference we may put, as in Art. 50, $a^{(n)} = \beta^{(n)} = 0$, and

Rolling on a Fixed Surface.

$$x^{(n)} = m\xi + m'\eta,$$
$$y^{(n)} = m'\xi - m\eta;$$

and it follows, exactly as in the case of the ordinary acceleration centre, that the resultant of $x^{(n)}$ and $y^{(n)}$ is proportional to the distance of the point x, y from the centre, and that it makes with this distance an angle whose tangent is $\dfrac{m'}{m}$, which is the same for all particles in the body.

54. Rolling on a Fixed Surface. When a body, M, rolls on a fixed surface, AB, the point, I, of the body which is at any instant in contact with the fixed surface has an acceleration which we proceed to find.

We suppose the rolling to be unaccompanied by slipping, so that at the end of the element of time Δt the point L of the body will be that which is in contact with AB, and it will occupy the position L' on AB, such that the length of the arc IL measured on the surface of the rolling body is equal to the length of the arc IL' measured on the fixed surface. The figure represents sections of the body and surface by the plane of motion.

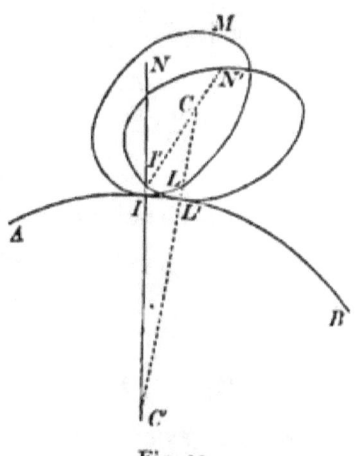

Fig. 33.

Consider the point I of the body which at the time t is the point of contact, and which at the time $t + \Delta t$ comes to I' such that $IL = I'L'$.

The point L' is the instantaneous centre at the end of the time Δt, so that I' is moving at right angles to $I'L'$, its velocity being $\omega \cdot I'L'$, where $\omega =$ the angular velocity of the body at the instant under consideration.

Let us find the acceleration of the point, I, of the body along the normal IN to the surface of contact. This normal becomes $I'N'$ by rotation. Draw the normal at L', and let it meet $I'N'$

in C and IN in C'. Then C is ultimately the centre of curvature of the contour of the rolling body at I, and C' is ultimately the centre of curvature of AB; so that if ρ is the radius of curvature of the body at I and ρ' that of the surface at I, we have $\rho = I'C$, $\rho' = IC'$. Denote the angles $I'CL'$ and $IC'L'$ by $\delta\theta$ and $\delta\theta'$.

Then the angle through which the body has turned in the time Δt is $\delta\theta + \delta\theta'$, since it is the angle between IN and $I'N'$.

Now, velocity of I along IN, at time t, $= 0$,

velocity of I' along IN, at time $t + \Delta t$,

$= \omega . I'L' . \cos(\delta\theta + \delta\theta') = \omega . I'L'$, nearly, $= \omega . \rho\delta\theta$;

$\therefore \dfrac{\text{gain of velocity}}{\Delta t} = \omega\rho \dfrac{\delta\theta}{\Delta t} = \omega\rho \dfrac{d\theta}{dt}$, in the limit.

But $\omega = $ limit of $\dfrac{\delta\theta + \delta\theta'}{\Delta t}$, and $\rho\delta\theta = \rho'\delta\theta'$, since the arc $I'L'$

$= $ arc IL';

$\therefore \omega = \left(1 + \dfrac{\rho}{\rho'}\right) \dfrac{d\theta}{dt}$, $\therefore \dfrac{d\theta}{dt} = \dfrac{\omega}{1 + \dfrac{\rho}{\rho'}}$;

therefore the acceleration of the point I of the body along IN is—

$$\dfrac{\omega^2}{\dfrac{1}{\rho} + \dfrac{1}{\rho'}},$$

or $h\omega^2$, if we put $\dfrac{1}{h}$ for $\dfrac{1}{\rho} + \dfrac{1}{\rho'}$.

If the concavity of the fixed surface is turned upwards, or in the same direction as that of the rolling body, the acceleration of the point of the body in contact is—

$$\dfrac{\omega^2}{\dfrac{1}{\rho} - \dfrac{1}{\rho'}},$$

for it will then be obvious that $\delta\theta - \delta\theta'$ is the angle turned through by the body in the time Δt.

There is no acceleration of the point I along the tangent, for

$\dfrac{\text{gain of velocity along tangent}}{\Delta t} = \dfrac{\omega . I'L' . \sin(\delta\theta + \delta\theta')}{\Delta t}$

$= \dfrac{\omega\rho\delta\theta (\delta\theta + \delta\theta')}{\Delta t} = 0$, in the limit.

Examples.

This expression for the acceleration of I holds only when the motion is *pure rolling*, i.e., when the point I of the body does not *slip* along the surface during the rotation. The student has already seen that the effect of slipping accompanying the rotation is to throw the instantaneous centre above or below the point of contact to a distance

$$\frac{v}{\omega}$$

along the normal, where v is the velocity of slipping (i.e., where $v \Delta t$ is the amount of slipping which takes place while the body turns through an angle $\omega \Delta t$).

EXAMPLES.

1. A lamina moves in its own plane so that two points in it describe two fixed right lines with given accelerations; find the acceleration of the instantaneous centre. (Wolstenholme's *Book of Mathematical Problems*, p. 416, 2nd ed.)

Let P and Q (fig. 34) be the two points in any position, AO and BO the lines along which they move.

At P and Q erect perpendiculars, PI and QI, to OA and OB; then I is the instantaneous centre for the lamina. Describe a circle round OPQ; it will pass through I. Also if J is the acceleration centre, J will be on the circle, since (Art. 50) the angles OPJ and OQJ are equal, and each is $\tan^{-1}\frac{\dot{\omega}}{\omega}$, where ω and $\dot{\omega}$ are

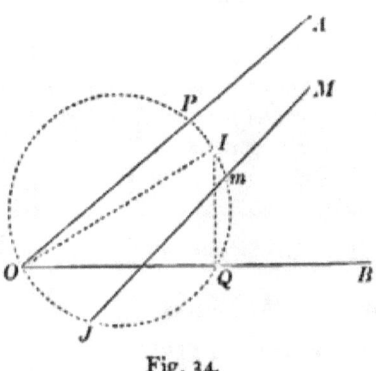

Fig. 34.

the angular velocity and angular acceleration of the lamina at the instant.

Again, the acceleration of the particle at I, i.e., the acceleration of I_b (Art. 50), is directed in the line IO, since $\angle OIJ$ is equal to OPJ; and if $k = \sqrt{\omega^4 + \dot{\omega}^2}$, the acceleration of I_b is (Art. 50) equal to $k \cdot IJ$.

Now if $PQ = a$, and $\angle AOB = a$, the diameter, OI, is $\frac{a}{\sin a}$, and $IJ = \frac{a}{\sin a} \cos OIJ = \frac{a\omega^2}{k \sin a}$, since $\tan OIJ = \frac{\dot{\omega}}{\omega^2}$. Therefore the acceleration of I_b is

$$\frac{a\omega^2}{\sin a}.$$

The acceleration of I_4 can be easily expressed in terms of the accelerations of P and Q. For, taking OA and OB as axes of x and y, respectively, denoting the co-ordinates of I by x, y, and the lengths OP and OQ by s and s', we have

$$x \sin^2 a = s - s' \cos a,$$
$$y \sin^2 a = s' - s \cos a.$$

Differentiating twice with respect to the time, and remembering that the square of the resultant of $\ddot{x}$ and $\ddot{y}$ is $\ddot{x}^2 + 2\ddot{x}\ddot{y} \cos a + \ddot{y}^2$, we get the square of the acceleration of I_4 to be

$$\frac{\ddot{s}^2 - 2\ddot{s}\ddot{s}' \cos a + \ddot{s}'^2}{\sin^2 a},$$

$\ddot{s}$ and $\ddot{s}'$ being the accelerations of P and Q.

Of course $\ddot{s}$ and $\ddot{s}'$ are not independent. We have

$$\ddot{s} = k \cdot PJ, \quad \ddot{s}' = k \cdot QJ,$$

where $k = \sqrt{\omega^2 + \dot{\omega}^2}$; and also $PJ^2 - 2PJ \cdot QJ \cos a + QJ^2 = a^2$,

$$\therefore \quad \ddot{s}^2 - 2\ddot{s}\ddot{s}' \cos a + \ddot{s}'^2 = k^2 a^2.$$

2. If two points in a lamina are guided along any two right lines fixed in space in the plane of the lamina, with accelerations, $\ddot{s}$, $\ddot{s}'$, satisfying any fixed relation of the form $f\left(\dfrac{\ddot{s}'}{\ddot{s}}\right) = 0$, the acceleration centre is an invariable point in the body, and also the acceleration of any one particle is always in the same direction in fixed space—the direction being different for different particles.

Firstly, J (fig. 34) is always the same particle of the body. For, the accelerations, $\ddot{s}$ and $\ddot{s}'$, of P and Q being connected by a fixed homogeneous relation of the form $f\left(\dfrac{\ddot{s}'}{\ddot{s}}\right) = 0$, we have the distances PJ and QJ connected by the equation $f\left(\dfrac{QJ}{PJ}\right) = 0$, which determines a body-locus of J. But the circle POQ is an invariable circle in the body, since it is one having a fixed body-line PQ for a chord, and this chord subtends a constant angle at the circumference. Hence, in the case supposed, J will be a point of intersection of this circle with the other body-locus; it is therefore always the same particle of the body.

Secondly, let M be any particle whatever. Draw MJ meeting the circle in m. Since M and J are two invariable particles, the line MJ is invariable in the body, and so therefore is the point m; so likewise is the line Pm; and therefore the angle POm which Om makes with a fixed-space line OP is constant because it is that subtended in the circle by the invariable length Pm. Now at any instant the angle which the direction of acceleration of M makes with MJ is equal to $\angle JPO$, which is equal to the angle JmO; therefore the direction of the acceleration of M is parallel to mO—

which, as we have just proved, makes a constant angle with a line fixed in space. Therefore, &c.

For the particular case in which $\ddot{s} = \ddot{s}'$ see Wolstenholme's *Book of Mathematical Problems*, ex. 2430, p. 417, second ed.

3. Two right lines in a lamina are kept constantly passing through two points fixed in space, while the displacement of the body is made with constant angular velocity; it is required to find at any instant the acceleration of any point of the lamina, and also the Acceleration Centrodes of the motion.

Let P and Q (fig. 25, p. 49) be the two fixed-space points (which may be two small swivel rings) through which two lines, OP and OQ, of the lamina F are guided.

Now the points P and Q being fixed, and the angle POQ being given, the fixed-space locus of I (or C_s) is the circle POQ; and since the diameter of this circle is $\dfrac{PQ}{\sin POQ}$, the distance OI is fixed; so that the centrodes C_s and C_b of the present problem are exactly the reverse of those in p. 49.

Now the rotation of the lamina F in a small time Δt is the angle between OP and $O'P$, where O' is a position of O on the circle POQ after the time Δt. The angular velocity of F being ω, it therefore appears that the angular velocity of O about the centre of the circle POQ is 2ω, and if r is the radius of this circle, the velocity of O is $2r\omega$, so that this velocity is constant. Hence the acceleration of O is (Cor., Art. 50) wholly along the diameter of this circle passing through O—i.e., along the line OI.

Also since $\dot{\omega} = 0$, the acceleration centre, J, must be on the line OI (Art. 50); and since the acceleration of O is (Art. 39) $4\omega^2 r$, we have

$$\omega^2 \cdot JO = 4\omega^2 r,$$
$$\therefore JO = 4r.$$

This shows that the Body Acceleration Centrode is a circle described round O as centre with radius equal to four times that of the circle POQ. Also the distance between J and the centre of the fixed-space circle POQ is $3r$, so that the Space Acceleration Centrode is a circle concentric with POQ.

The acceleration of every point in the body is directed towards J (since $\dot{\omega} = 0$), and equal to $\omega^2 \times$ distance from J.

In the motion considered in this problem we therefore have the result (see example 9, p. 50) that—the Space Centrode (i.e., Space *velocity* Centrode) is a circle of radius r; the Body Centrode is a circle of radius $2r$, the Space Acceleration Centrode is a circle of radius $3r$, and the Body Acceleration Centrode is a circle of radius $4r$.

4. A point P moves along a curve AB in such a manner that the line, Pp, of its resultant acceleration is always a tangent to another curve ab;

find a relation between the length of the tangent from P to the curve ab and the angular velocity of this tangent.

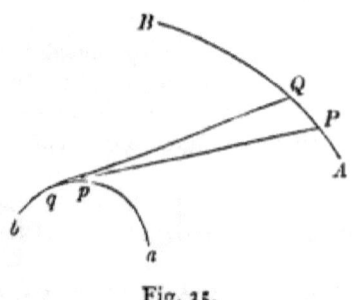

Fig. 35.

Let the line, Pp, of resultant acceleration at P touch ab in p; let Q be the position of P after the time Δt, and Qq the line of acceleration at Q touching ab at q; let $pq = d\sigma$; $\tau = Pp$; and let $d\psi =$ the angle between Pp and Qq. Then since Pp is the direction of resultant acceleration at P, the component acceleration perpendicular to Pp is zero. This component we proceed to calculate by the Rule of Art. 35.

Let $v =$ velocity at P; $v + dv =$ velocity at Q; $\phi =$ angle between Pp and tangent to AB at P; $\phi + d\phi =$ angle between Qq and tangent at Q; then the angle between Pp and the tangent at Q is $\phi + d\phi - d\psi$; so that

velocity perpendicular to Pp at time t is $v \sin \phi$,

velocity perpendicular to Pp at time $t + dt$ is $(v + dv) \sin (\phi + d\phi - d\psi)$.

Hence $\dfrac{\text{gain of velocity perp. to } Pp}{dt} = \dfrac{(v+dv)\sin(\phi+d\phi-d\psi) - v\sin\phi}{dt}$,

and since there is no acceleration in this direction, we must have

$$(v + dv) \sin (\phi + d\phi - d\psi) - v \sin \phi = 0,$$

or
$$d(v \sin \phi) - v \cos \phi \cdot d\psi = 0,$$

$$\therefore \quad \frac{d(v \sin \phi)}{v \sin \phi} = \cot \phi \, d\psi. \qquad (1)$$

Now by drawing from Q a perpendicular to Pp, we have

$$\cot \phi = \frac{d\sigma - d\tau}{\tau \, d\psi}.$$

Hence from (1) we have in the limit

$$\frac{d(v \sin \phi)}{v \sin \phi} + \frac{d\tau}{\tau} = \frac{d\sigma}{\tau},$$

$$\therefore \quad \tau v \sin \phi = h e^{\int \frac{d\sigma}{\tau}}, \qquad (2)$$

by integration. Now $v \sin \phi =$ velocity of P perpendicular to Pp; and this is also equal to $Pp \dfrac{d\psi}{dt}$, or $\tau \dfrac{d\psi}{dt}$; therefore from (2) we have

$$\tau^2 \frac{d\psi}{dt} = h e^{\int \frac{d\sigma}{\tau}}. \qquad (3)$$

The integral $\int \dfrac{d\sigma}{\tau}$ is, of course, performed over the second curve between points corresponding to the first and last point considered on the path of the moving point.

The particular case in which the resultant acceleration of the moving point is always directed to a fixed centre has been already discussed (Art. 44). In this case the curve pq above reduces to a point; Pp becomes r, the radius vector from this point to P; ψ becomes θ, the usual polar angle; and $d\sigma$ is always zero, so that the right side of (3) becomes h, and we have $r^2 \dfrac{d\theta}{dt} = h$, as in Art. 44.

5. If a body rolls, with uniplanar motion, on a fixed surface, find the acceleration of any particle on the normal at the point of contact.

Ans. If ω and $\dot{\omega}$ are the angular velocity and acceleration of the body at any instant, P the point on the normal IN (fig. 33) whose acceleration is required, and $\dfrac{1}{h} \equiv \dfrac{1}{\rho} + \dfrac{1}{\rho'}$, the components, $\ddot{x}$, $\ddot{y}$, of the acceleration of P parallel to the tangent and normal at I are given by the equations

$$\ddot{x} = -\dot{\omega} r; \quad \ddot{y} = (h-r)\omega^2.$$

6. In the last case find the position of the acceleration centre.

Ans. If H is the point on the normal IN distant h from I, and IK a line drawn at the left-hand side of IN making $\tan^{-1} \dfrac{\dot{\omega}}{\omega^2}$ with it, J is the foot of the perpendicular from H on IK. [The angular velocity ω is supposed to be measured counterclockwise in the figure. Also observe from last example that for H we have $\ddot{x} = -h\dot{\omega}$, $\ddot{y} = 0$.]

7. In all uniplanar motion the lines of resultant acceleration of all particles on the same right line envelop a parabola.

[The Acceleration Centre is its focus.]

8. If a rigid body is set rotating about a horizontal axis passing through its centre of gravity, and then allowed to fall, find the Body Acceleration Centrode.

Ans. A circle round the centre of gravity as centre with radius $\dfrac{g}{\omega^2}$, where ω is the (constant) angular velocity of the body.

9. A point P describes an orbit with acceleration f directed always to a fixed centre O; prove that the acceleration of the corresponding point, P', of the Hodograph is in general non-central, and makes with the radius vector OP' an angle χ given by the equation

$$\tan \chi = \dfrac{\dfrac{d}{d\theta} \log \dfrac{f}{r}}{1 - \cot \phi \dfrac{d}{d\theta} \log f},$$

where ϕ is the angle between OP and the tangent at P, θ is the polar angle of P, and $r = OP$.

10. Hence (and also independently) prove that the only law of central acceleration which makes the acceleration also central in the Hodograph is the law $f = \mu r$, where μ is a constant.

11. Prove that the radius of curvature at any point of the Hodograph of a central orbit $= \dfrac{\text{Force} \times (\text{distance})^2}{2 \,.\, \text{A real Velocity in orbit}}$.
(Hamilton's *Elements of Quaternions*, p. 720.)

12. Show how the components of acceleration of the third and higher orders of the point I of the rolling body (Art. 54), along the tangent and normal, are to be found.

13. If the crank (p. 48) rotate with constant angular velocity, find, in any position of the system, the acceleration centre for the connecting rod.

Ans. With the notation in p. 48, at C draw a line CJ making with the lower side of AC an angle, ACJ, whose tangent $= (k^2 - 1) \tan \theta \sec^2 \phi$, and meeting the circle round ABC in J. Then J is the required point.

14. A rod, AB, moves with its extremities, A and B, guided along two fixed rectangular lines, OA and OB, in such a way that if θ denotes the angle BAO at any instant, the angular acceleration, $\ddot{\theta}$, is given by the equation
$$\ddot{\theta} = n \sin \theta,$$
where n is constant; find the acceleration centre at any instant.

Ans. Let a be the initial value of θ; at A draw the right line AJ meeting the circle described round OAB in J, and such that
$$\tan JAO = \tfrac{1}{2} \frac{\sin \theta}{\cos a - \cos \theta};$$
then J is the required point. The motion is that of a heavy uniform rod sliding down between a smooth wall and a smooth horizontal plane.

CHAPTER III.

General Theorem of Epicycloidal Motion.

55. Theorem. *At every instant during the uniplanar motion of a rigid body the Space and Body Centrodes touch each other at the instantaneous centre.*

For if the co-ordinates and velocities of any point, G, in the body, with reference to axes fixed in space are at any instant, as in Art. 29, a, β, $\dot{a}$, $\dot{\beta}$, and if ω is the angular velocity of the body at that instant, the co-ordinates of I (ξ, η) referred to body-axes at G are (Art. 32)

$$\xi = -\frac{\dot{\beta}}{\omega}\cos\phi + \frac{\dot{a}}{\omega}\sin\phi,$$

$$\eta = \frac{\dot{\beta}}{\omega}\sin\phi + \frac{\dot{a}}{\omega}\cos\phi,$$

ϕ being the angle made with the fixed-space-axis of x by the body-axis of x at G, and $\dfrac{d\phi}{dt}$ being, of course, ω. Now differentiate these with respect to t, and put $\phi = 0$, so that we shall obtain the velocities parallel to the fixed-space-axes of x and y with which I_b travels over the Body Centrode, C_b. Then

$$\frac{d\xi}{dt} = -\frac{d}{dt}\frac{\dot{\beta}}{\omega} + \dot{a} = \frac{d}{dt}\left(a - \frac{\dot{\beta}}{\omega}\right),$$

$$\frac{d\eta}{dt} = \dot{\beta} + \frac{d}{dt}\frac{\dot{a}}{\omega} = \frac{d}{dt}\left(\beta + \frac{\dot{a}}{\omega}\right).$$

But if (x, y) are the fixed-space co-ordinates of I, or co-ordinates of I_s, we have $x = a - \dfrac{\dot{\beta}}{\omega}$, $y = \beta + \dfrac{\dot{a}}{\omega}$; therefore

$$\frac{d\xi}{dt} = \frac{dx}{dt}, \quad \frac{d\eta}{dt} = \frac{dy}{dt},$$

so that $\dfrac{dy}{dx} = \dfrac{d\eta}{d\xi}$, or the tangents to C_s and C_b are the same line. The lengths, $\sqrt{dx^2 + dy^2}$ and $\sqrt{d\xi^2 + d\eta^2}$, of their arcs between two successive instantaneous centres are also the same.

Hence C_b rolls without sliding on C_s; or—

If a rigid body moves, in any way whatever, parallel to one plane, the motion may be in all respects produced by causing the Body Centrode to roll, without sliding, on the Space Centrode.

Geometrical proofs of this proposition will be found in many works. (See Clifford's *Kinematic*, p. 137.)

This is the fundamental theorem of *Epicycloidal Motion*, the consequences of which it is proposed to develop in the present Chapter.

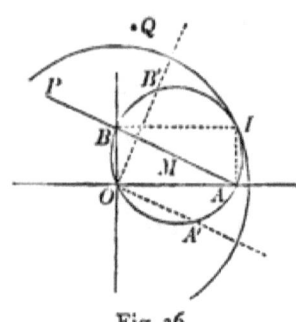

Fig. 36.

56. Elliptic Compass. If a right line, AB, of fixed length have its extremities carried along two fixed rectangular grooves, OA and OB, and carry a tracing pencil at any point P in its length, this pencil will trace out an ellipse of which the axes are PA and PB. This instrument is called an *Elliptic Compass*.

Let us now replace this motion by the epicycloidal motion of the Centrodes, C_b and C_s.

The instantaneous centre, I, is the point of intersection of perpendiculars to the grooves at A and B. Also $OI = AB = $ a constant; therefore I is always at a constant distance from O a fixed point in space; again, if M is the middle point of AB, $MI = $ a constant; therefore the distance of I from an invariable body-point is constant; and hence C_s is a circle with centre O and radius AB, while C_b is a circle of half the radius.

The groove motion may therefore be replaced by the rolling of the circle AOB on that of double the radius, the latter being fixed in space, and the point P (rigidly connected with the small circle) will trace out an ellipse.

It is not merely one point, P, rigidly connected with the

rolling circle which traces out an ellipse; *every* point, such as Q, also traces out an ellipse. For, draw QM, meeting the small circle in A' and B', and draw two fixed-space lines OA' and OB'. Then the particles (or body-points) at A' and B' are moving at right angles to $A'I$ and $B'I$, respectively, i. e., along OA' and OB'; hence the line $A'B'$ of invariable length is moving so that its extremities are describing two fixed rectangular lines, OA' and OB', therefore the invariable point Q on this right line traces out an ellipse.

If the extremities A and B of a moving line describe two non-rectangular right lines, the motion may be otherwise produced, as above, by the rolling of one circle inside another of double the diameter; for the student will easily see that C_b and C_s are in this case also circles, one round ABO, and the other with O as centre—as in the above case of two rectangular lines.

The theorem of the Elliptic Compass may, of course, be otherwise stated thus—if two points of a lamina are made to describe two fixed right lines, every point in the lamina describes an ellipse.

57. Oblique Elliptic Compass.

Given two right lines, OA' and OB' (fig. 37), it is required to trace out by continuous motion an ellipse of which OA' and OB' are semi-conjugate diameters in magnitudes as well as in directions.

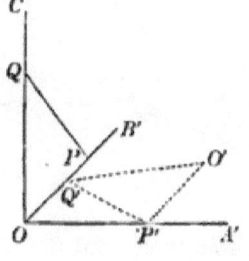

Fig. 37.

Suppose a triangle $P'Q'O'$ to move with the extremities, P' and Q', on the lines OA' and OB'; then the locus of its vertex, O', is an ellipse whose centre is O.

If we take OA' and OB' as axes of x and y; if $\angle B'OA' = a$; if the sides $Q'O'$ and $P'O'$ are m and n respectively; and if $a + Q'O'P' = \omega$, the equation of the ellipse is easily found to be

$$\frac{x^2}{m^2} + \frac{2xy}{mn}\cos\omega + \frac{y^2}{n^2} = \frac{\sin^2\omega}{\sin^2 a}.$$

G

Hence OA' and OB' will be conjugate diameters in *directions* if $\omega = \dfrac{\pi}{2}$; and the *lengths* of the semi-conjugate diameters will be

$$\frac{m}{\sin a} \quad \text{and} \quad \frac{n}{\sin a}.$$

Hence if we wish OA', OB' to be the semi-conjugates, we must take $\quad m = OA' \cdot \sin a, \quad n = OB' \cdot \sin a.$

At O erect OC perpendicular to OA' and equal to it. Take OP = perpendicular from B' on $OA' = OB' \sin a$; draw PQ parallel to $B'C$. Then the triangle PQO is the one whose motion traces out the required ellipse, as is very easily seen.

A piece of paper cut into the size of this triangle can be used with great ease for describing the curve.

The ordinary elliptic compass—that in which the lines OA' and OB' are the axes of the ellipse—is a particular case of the above; for then PQ becomes the difference of the axes and the tracing triangle becomes a right line.

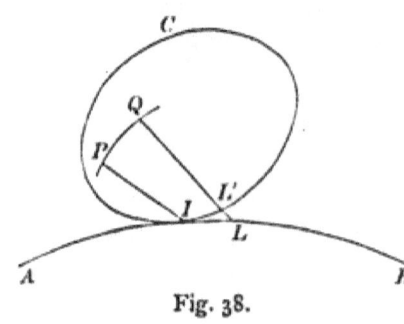

Fig. 38.

58. Area of any Roulette. If any curve roll without slipping on a fixed curve, the curves traced out by all points rigidly connected with the rolling curve are called *Roulettes*.

Thus, if a circle roll along a right line, any point on the circumference of the circle traces out the particular Roulette called a Cycloid.

Let any plane curve, C (fig. 38), roll on any fixed plane curve, AB, and let the rolling curve carry with it a point P which traces out a roulette in fixed space. Let I be the point of contact of the two curves at any instant, let L' be a point on C which after a small motion comes to L on the fixed curve, and let Q be the point to which P comes at the end of this motion. We shall regard the area $IPQL$ as the area of the portion of the roulette generated by this motion, and the area of any portion whatever of the roulette will, defined thus, be the area included between any two normals to the roulette and the corresponding portions of the fixed curve and the roulette. This elementary area = area LIP + area LPQ; and since the distance between L and L' is an infinitesimal of the second order, these points may be regarded as coincident, so that area LIP = area $L'IP = dC$, where dC signifies an elementary polar area of the curve C traced out round P as pole.

Again, if $d\omega$ is the angle between the tangents to C at I and L', and $d\omega'$ the angle between the tangents to AB at I and L, the angle between PL (or PL') and QL is $d\omega + d\omega'$, the angle of rotation of the rolling curve; so that if PI is denoted by r, the area $LPQ = \tfrac{1}{2}r^2(d\omega + d\omega') = \tfrac{1}{2}r^2 d\Omega$, say.

Hence, if $d\Sigma$ denotes the element of area of the roulette, we have
$$d\Sigma = dC + \tfrac{1}{2}r^2 d\Omega, \tag{1}$$
$$\therefore \Sigma = C + \tfrac{1}{2}\int r^2 d\Omega. \tag{2}$$

We may, indeed, put $dC = \tfrac{1}{2}r^2 d\theta$, where θ is the angle made by PI with a fixed line, so that we can write
$$\Sigma = \tfrac{1}{2}\int r^2 (d\theta + d\Omega),$$
and the integration may extend over the whole curve C, or between any two points on it.

It is necessary, however, to be more explicit as to what we mean by the area of a roulette, and to point out the fact that portions of the area may sometimes require to be understood as *negative*.

Supposing that the lines, PI and QL (or QL'), joining two consecutive positions of the tracing point to the corresponding positions of the instantaneous centre intersect each other at a point, O, in their *unproduced* lengths, the element of area which we consider is

area PQO − area OIL.

In fact, area $PQL = \tfrac{1}{2}r^2 d\Omega$, and area PIL (or dC) = $\tfrac{1}{2}r^2 d\theta$, and to avoid taking the portion POL twice

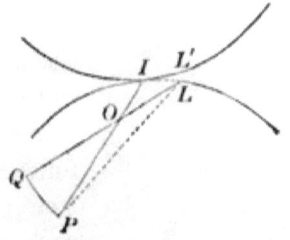

Fig. 39.

over, we must take the arithmetical difference between these areas; the *algebraic* sum of them would do, however, and our formula (1) may be regarded as applicable, since if we take any fixed point, P, in the plane of any curve, the radius vector from P to the curve must be regarded as tracing out triangular elements of area (such as PII', or dC) of one sign so long as it is approaching a tangent to the curve from P (by rotation in one direction), and as tracing out elements of area of the *opposite* sign so long as it is receding from the tangent (by an opposite rotation). In estimating thus the area of a plane curve it is not necessary that we should have a *tangent*, or *tangents*, in the ordinary sense of the word; it is enough that we should have positions of the tracing radius vector in which this line changes the direction of its rotation. The curve may, in fact, be a polygon with sides crossing each other in any complicated manner. The vectorial area of any curve (closed or not) must be understood with this reference. And the rule for estimating not only the element, $d\Sigma$, but any integrated portion, Σ, of the area of a roulette, as used above, is this—

Start from the first position, P_1, of the tracing point; follow the roulette round to any point, P_n; then move along the normal, $P_n I_n$, to the roulette as far as I_n, which is the point on the fixed curve corresponding to the point P_n; then move along the fixed curve through the points, IN INVERSE ORDER, which have been instantaneous centres, until the first position of the instantaneous centre, I_1, is reached; finally, move along the normal $I_1 P_1$ back to the original point, P_1. The vectorial area of the complex path thus traced out, estimated round any point as pole, is the area of the roulette.

This rule applies to every case that can arise; and special attention is directed to the fact that the contour of the fixed curve must be passed over in the inverse order in which its points have been instantaneous centres, because we may have to deal with cases in which the roulette is a closed curve (possibly with several loops), so that we have to move along it until we again reach the original point, P_1, before moving along the fixed curve at all.

In this case the area Σ may obviously be regarded as the area proper, A, of the roulette itself—without reference to the fixed curve, and taken with positive and negative portions according to the contrary rotations of a radius vector—diminished by the area proper, S, of the fixed curve estimated in like manner. That is, $\Sigma = A - S$.

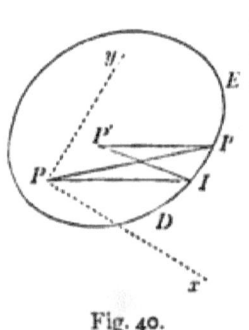

Fig. 40.

We may now regard the complex function $d\theta + d\Omega$ as a quantity varying from point to point on C, and having for each point on the curve a peculiar value associated with the point, so that the idea of rolling may be discarded, and the curve C may be considered fixed.

59. Areas of different Roulettes compared. Take

any other point P', (fig. 40), and consider the area of the corresponding portion of its roulette. We shall have for P'
$$d\Sigma' = d'C + \tfrac{1}{2}r'^2 d\Omega,$$
where $r' = P'I$, and $d'C$ is the polar element of area round P'.

Draw any two rectangular axes, Px, Py, at P; let x, y be the co-ordinates of P' with reference to them, and let PP' make an angle θ with Px. Then instead of dC (area IPI') we may use $\tfrac{1}{2}pds$, where p = perpendicular from P on tangent at I, and instead of $d'C$ (area $IP'I'$) we may use $\tfrac{1}{2}p'ds$, where p' = perpendicular from P' on tangent at I. But obviously, if ω = angle between p and axis of x,
$$p' = p - x \cos\omega - y \sin\omega,$$
$$\therefore\ d'C = dC - \tfrac{1}{2}x \cos\omega\, ds - \tfrac{1}{2}y \sin\omega\, ds,$$
so that if D and E are any two fixed points on the curve,
$$\text{area } DP'E = \text{area } DPE - \tfrac{1}{2}x\!\int\cos\omega\, ds - \tfrac{1}{2}y\!\int\sin\omega\, ds.$$

Now $\int\cos\omega\, ds$ and $\int\sin\omega\, ds$ are two constants, viz., the difference between the ordinates and the difference between the abscissae of D and E. Denote them by a and b; then
$$\text{area } DP'E = \text{area } DPE - \tfrac{1}{2}ax - \tfrac{1}{2}by.$$
Also $\quad r'^2 = r^2 + x^2 + y^2 - 2r(x \cos\theta + y \sin\theta),$
$$\therefore\ \int r'^2 d\Omega = \int r^2 d\Omega + \Omega(x^2+y^2) - 2x\!\int r\cos\theta\, d\Omega$$
$$- 2y\!\int r\sin\theta\, d\Omega,$$
Ω being $\int d\Omega$ between the points D and E.
Hence
$$\Sigma' = C - \tfrac{1}{2}ax - \tfrac{1}{2}by + \tfrac{1}{2}\!\int r^2 d\Omega + \tfrac{1}{2}\Omega(x^2+y^2) - gx - hy,$$
(if we denote $\int r\cos\theta\, d\Omega$ and $\int r\sin\theta\, d\Omega$ by g and h), or
$$\Sigma' = \Sigma + \tfrac{1}{2}\Omega(x^2+y^2) - (g+\tfrac{1}{2}a)x - (h+\tfrac{1}{2}b)y. \qquad (1)$$

From this expression we derive the following theorem—

All points which trace out Roulettes of the same area lie on a circle.

For if Σ' is constant while P' varies, the above equation (1) shows that the locus of P' is a circle.

Also, *by varying the area of the Roulette, we get a series of concentric circles.*

For, only the constant term of the above equation between x and y will change if Σ' is varied.

Precisely the same theorem holds for the areas of the pedals of a curve taken with respect to different poles, as has been proved by Steiner. (See Williamson's *Integral Calculus*, p. 202, third ed.)

A direct consequence of the above is Kempe's theorem—viz., *if one plane, sliding upon another, start from any position, move in any manner, and return to its original position after making one or more complete revolutions, every point in the moving plane describes a closed curve, and the locus, in the moving plane, of points which describe curves of equal area is a circle;* and by varying the area we get a series of concentric circles. (Williamson, *ibid.*, p. 210.)

According to the principle of epicycloidal motion, the motion of the plane may be conceived as produced by the rolling of C_b on C_a; every point describes a Roulette, and the areas of the Roulettes are connected by our theorem above.

It is not of course necessary for the truth of the theorem that the moving plane should be brought back to its original position; the only effect of bringing it back is to bring the points E and D (fig. 40) into coincidence, i.e., to make $a = b = 0$ in equation (1).

The centre of the system of circles is necessarily the point which traces out the Roulette of minimum area. If we use it as point of reference instead of P, and if K is the area of its Roulette, the area of any other is given by the equation

$$\Sigma = K + \tfrac{1}{2}\Omega(x^2 + y^2).$$

If the rolling takes place through the entire length of the curve C, a and b will both be zero, and the centre of the system of circles is such that

$$\int r \cos\theta\, d\Omega = 0, \quad \int r \sin\theta\, d\Omega = 0,$$

i.e., this point is the centre of mass of a distribution of matter over the curve C, the density at each point being proportional to $d\Omega$.

60. Theorem of Holditch. Two points, A_1 and A_2, at a constant distance apart describe two closed curves of areas (C_1) and (C_2), the line $A_1 A_2$ making a complete revolution so as to return to its original position; it is required to find the area of the curve described by any point P on the line $A_1 A_2$.

The motion may be regarded as produced by the rolling of the Body Centrode on the Space Centrode. Let the areas of these centrodes be denoted by (C_b) and (C_s), respectively, let $A_1 A_2 = l$, and let P divide $A_1 A_2$ so that $PA_1 = nl$.

Then the area between the roulette of A_1 (which is the curve C_1) and the Space Centrode is $(C_1)-(C_s)$, and if r_1 denote the distance between A_1 and any point, I, on the Body Centrode, we have by Art. 58,

$$(C_1)-(C_s) = (C_b) + \tfrac{1}{2}\int r_1^2 d\Omega. \tag{1}$$

Similarly
$$(C_2)-(C_s) = (C_b) + \tfrac{1}{2}\int r_2^2 d\Omega, \tag{2}$$
$$(X)-(C_s) = (C_b) + \tfrac{1}{2}\int r^2 d\Omega, \tag{3}$$

where X = area of curve described by P, $r_2 = A_2 I$, $r = PI$.

Now obviously $r^2 = (1-n)r_1^2 + n r_2^2 - n(1-n)l^2$. (4)

Multiply (1) and (2) by $1-n$ and n, respectively, and add; then

$$-(C_s) + (1-n)(C_1) + n(C_2) = (C_b) + \tfrac{1}{2}\int r^2 d\Omega + \tfrac{1}{2}\Omega n(1-n)l^2;$$

therefore from (3)

$$(X) = (1-n)(C_1) + n(C_2) - \tfrac{1}{2}\Omega n(1-n)l^2. \tag{5}$$

If the line $A_1 A_2$ returns to its original position, its total rotation, Ω, is 2π, and then

$$(X) = (1-n)(C_1) + n(C_2) - \pi n(1-n)l^2, \tag{6}$$

which is Holditch's Theorem.

It is to be observed that the area traced out by P is susceptible of a minimum value, the corresponding value of the ratio of the segments into which it divides the line $A_1 A_2$ being

$$\frac{\pi l^2 - \Delta}{\pi l^2 + \Delta},$$

where Δ stands for $(C_2)-(C_1)$.

The theorem of Holditch is proved otherwise thus by Mr. McCay. Let a consecutive position of the moving line $A_1 A_2$ be $A_1' A_2'$. From A_1' draw $A_1'D$ equal and parallel to $A_1 A_2$. Then the elementary area $A_1 A_2 A_2' A_1' = d[(C_2)-(C_1)]$ = area of parallelogram $A_1 D$ + area of triangle $DA_1' A_2 = //^m A_1 D + \tfrac{1}{2}l^2 d\Omega$.

Again, if P' is the position of P on $A_1'A_2'$, and if $A_1'Q$ is equal and parallel to A_1P, we have $d[(X)-(C_1)] = //^m A_1Q + \frac{1}{2}n^2 l^2 d\Omega$. But $n \times //^m . A_1D = //^m A_1Q$; therefore

$$nd(C_2) - nd(C_1) - \tfrac{1}{2}nl^2 d\Omega = d(X) - d(C_1) - \tfrac{1}{2}n^2 l^2 d\Omega.$$

Integrating this, we obtain equation (6).

61. Extension of Holditch's Theorem. If two vertices, A and B, of a triangle, ABC, of given magnitude are displaced along two given curves, the third vertex, C, will trace out a curve the relation of whose area to the areas of the two given curves we propose to find.

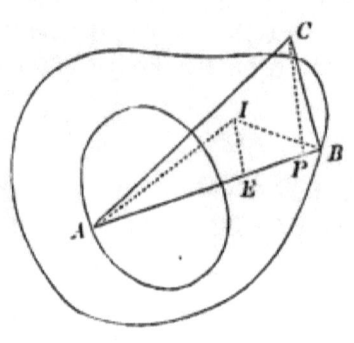

Fig. 41.

Denote the sides by a, b, c, and the areas of the curves by (A), (B), (C); let K denote the sum of the areas of the Space Centrode and the Body Centrode (what in our previous notation is $C_s + C_b$); let I be any position of the instantaneous centre; $AI = r$, $BI = r'$, $CI = r''$, $\angle IAB = \theta$.
Then

$$(C) = K + \tfrac{1}{2}\int r''^2 d\Omega$$

$$= K + \tfrac{1}{2}\int \{r^2 + b^2 - 2br \cos(A - \theta)\} d\Omega$$

$$= (A) + \pi b^2 - b \cos A \int r \cos \theta \, d\Omega - b \sin A \int r \sin \theta \, d\Omega, \quad (1)$$

assuming the rotation to be through 2π, so that the triangle returns to its original position. Now

$$(B) = K + \tfrac{1}{2}\int r'^2 d\Omega$$

$$= K + \tfrac{1}{2}\int (r^2 + c^2 - 2cr \cos \theta) d\Omega$$

$$= (A) + \pi c^2 - c \int r \cos \theta \, d\Omega,$$

$$\therefore \int r \cos \theta \, d\Omega = \frac{(A) - (B) + \pi c^2}{c}.$$

Substituting this in (1),

$$(C) = (A) + \pi b^2 - \frac{b \cos A}{c}\{(A) - (B) + \pi c^2\}$$
$$- b \sin A \int r \sin \theta \, d\Omega \, ;$$
$$\therefore \; c(C) = a \cos B (A) + b \cos A (B) + \pi abc \cos C$$
$$- bc \sin A \int r \sin \theta \, d\Omega.$$

Now observe that $r \sin \theta$ is the length of the perpendicular from I on AB, that the foot, E, of this perpendicular is the point of contact of AB with its envelope, and that if I' is a consecutive position of I and $I'E'$ the new perpendicular on AB from the instantaneous centre, the element of length of the envelope is $IPd\Omega + EE'$, or $r \sin \theta d\Omega + EE'$ (see Art. 68).

If, then, we denote the whole length of the envelope of AB by E_e, and by Δ the area of the triangle ABC, we have, since $\Sigma EE' = 0$ (the line AB returning to its original position)

$$c(C) = a \cos B(A) + b \cos A(B) + \pi abc \cos C - 2\Delta E_e, \quad (2)$$
or $\quad (C) = (P) + \pi p^2 - pE_e,$ $\qquad\qquad\qquad\qquad$ (3)

where (P) is the area traced out by the foot of the perpendicular from C on AB, and p is the length of this perpendicular.

Equation (3) is deduced in the following way more shortly by Mr. McCay. It is evidently a theorem relative to a line, CP, of constant length, p, whose extremities, C and P, trace out curves (C) and (P). Take then a consecutive position, $C'P'$, of this line, and at the points P, P' draw two normals to the curve (P), which of course meet in I, the instantaneous centre. Also at P and P' draw two lines PE and $P'E$ perpendicular to PC and $P'C'$, respectively, and meeting in a point E, which is obviously the foot of the perpendicular from I on AE. Finally from P' draw $P'D$ equal and parallel to PC. Then the element of area $CPP'C'$, which is denoted by $d[(C)-(P)]$, consists ultimately of the parallelogram $PCDP'$ and the triangle $P'DC'$, the area of the latter being $\frac{1}{2}p^2 d\Omega$, where, as before, $d\Omega = \angle$ between PC and $P'C'$; and the integral of this in a complete revolution is πp^2. The area of the parallelogram $= p \times PP' \sin PP'D = p \times IEd\Omega$, and, exactly as above, the integral of this in a complete revolution = length of envelope of PE, or E_e. Hence $(C) - (P) = \pi p^2 - pE_e$, as before.

Mr. McCay observes that equation (2) proves Kempe's Theorem.

For if we seek the locus of the point C so that (C) shall be constant, the equation is of the form

$$l(a^2+c^2-b^2)+m(b^2+c^2-a^2)+n(a^2+b^2-c^2)+kp = \text{const.}, \quad (4)$$

since (A), (B), c, E_c are constants. Referring the position of C to the line AB as axis of x and a perpendicular to it at its middle point as axis of y, we have

$$p=y, \quad a^2-b^2 = 2cx, \quad a^2+b^2 = 2(x^2+y^2)+\frac{c^2}{2},$$

therefore the locus of C as given by (4) is obviously a circle.

Equation (2) gives a result which serves as a verification—viz., if E_a, E_b, E_c are the lengths of the envelopes of the sides a, b, c,

$$aE_a + bE_b + cE_c = 4\pi\Delta, \quad (5)$$

which is, of course, evident so long as I is inside the triangle, since if p, q, r are the perpendiculars from I on the sides

$$ap\,d\Omega + bq\,d\Omega + cr\,d\Omega = 2\Delta\,d\Omega.$$

If I is outside the triangle, it is customary to regard $ap+bq+cr$ as still equal to 2Δ, with the convention that the perpendiculars are not all of the same sign; and the portion of the arc of the corresponding envelope must be regarded as negative.

In fact, though a right line may continuously revolve in the same sense (say clockwise), the radius vector, OP, from a fixed pole, O, to its point of contact with its envelope may have contrary rotations—as, for example, the tangent to a cusped curve. Using the polar formula

$$ds = \sqrt{r^2 + \left(\frac{dr}{d\theta}\right)^2} \cdot d\theta$$

for the element of arc, ds, we see that $d\theta$ will change sign when OP becomes the radius vector to the cusp, and so, therefore, will the element of arc.

62. Generalised Roulette. We may here investigate the area of a curve of which a roulette is a particular case. Suppose that a plane lamina of any shape is displaced in any manner in its plane; then a point fixed with reference to the lamina describes a roulette; but a point moving, according to any assigned law, with reference to the lamina, and also partaking of the absolute motion of the lamina, describes in fixed space a curve which, in the absence of a better name, may be called a *generalised roulette*.

To find an expression for the element of area of a generalised

roulette, suppose the motion of the displaced figure to be produced by the rolling of a curve C on a fixed curve AB; let
ab be the curve described in the moving figure by the tracing-point P; and during an indefinitely small rolling motion, $d\Omega$, which carries the point P to q, perpendicularly to PI, and L' to L, let the tracing-point travel over a length Pp of the curve

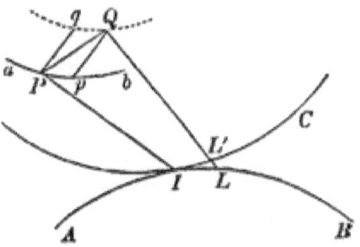

Fig. 42.

ab; then since the absolute motion of the tracing-point is obtained by compounding, according to the parallelogram law, the motions Pq and Pp, the element, PQ, of the space-path described by the tracing-point is the diagonal of the parallelogram whose adjacent sides are Pp and Pq; and the element of area of the generalised roulette is $PQLI$.

This is equal to $PqI + qQL'I$ (since L and L' are coincident to the first order of small quantities); and if $PI = r$, and we denote $qQL'I$ by dC_r, we have

$$d\Sigma = dC_r + \tfrac{1}{2} r^2 d\Omega,$$

where $d\Sigma$ stands for the required element of area.

The integral of dC_r will be the sum of the quadrilateral elements of area which are obtained by connecting the extremities of all elementary arcs of the curve C with the corresponding extremities of the corresponding elementary arcs of the curve ab. If the curves C and ab are closed, the integral of dC_r will be generally the area of the space included between them; and if the tracing-point merely oscillates along an arc, ab, of any length, the whole figure returning finally to its original position, the integral of dC_r will be generally the area of the curve C.

We may illustrate this by an example, due to Mr. E. B. Elliott, slightly more general than Holditch's theorem. (See Williamson, *Int. Cal.*, p. 209.) Suppose AB, fig. 41, p. 88, to be a right line, one extremity, A, of which travels round a curve (A), while the line itself is carried round in any manner and intersects another curve, (B), in a point, B, at a variable distance from A. It is required to find the area of the curve traced out by a point, P, taken always on AB and dividing AB in a constant ratio, the line returning exactly to its original position. Let the motion be produced by causing C_b to roll on C_s (Art. 59). Here the moving figure consists of a right line, AB, and P and B are two tracing points, each describing a generalised roulette, and each describing with reference to the line AB a curve of no area. Moreover, for each of the points B and P we have $\int dC_r = (C_b) =$ area of Body Centrode. Suppose I to be a position of the instantaneous centre; let $AI = r$, $BI = \rho$; $AB = l$, $AP = nl$, and denote by (X) the area of the path of P. Then, as in Art. 60,

$$(A) - (C_s) = (C_b) + \tfrac{1}{2}\int r^2 d\Omega,$$
$$(B) - (C_s) = (C_b) + \tfrac{1}{2}\int \rho^2 d\Omega,$$
$$(X) - (C_s) = (C_b) + \tfrac{1}{2}\int P I^2 d\Omega.$$

But $(1-n)r^2 + n\rho^2 = n(1-n)l^2 + PI^2$; therefore

$$(X) = (1-n)(A) + n(B) - \tfrac{1}{2}n(1-n)\int l^2 d\Omega,$$

and $\int l^2 d\Omega$ is simply the area of the relative path of A and B.

This result verifies in the particular case in which the line AB is always kept a tangent to the inner curve.

63. Amsler's Planimeter. The theory of some Planimeters is very simply and appropriately deducible from the preceding principles.

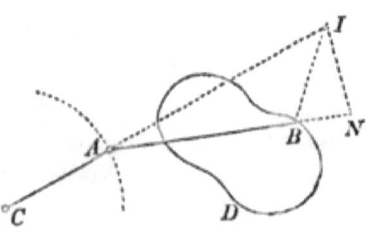

Fig. 43.

Let us, for example, consider Amsler's Planimeter. (Compare also Williamson's *Integral Calculus*, where different discussions of the instrument are found.)

Let CA and AB (fig. 43) be the two arms jointed at A; C the fixed end of one arm, and B the extremity of the other arm which traces out the contour D whose area is to be measured. Drawing the normal to the circle locus of A at A, and the normal at B to the curve D, we get I (their point of intersection) which is the point of contact of the Space and Body Centrodes.

If (A) denotes the area traced out by A, and (X) the area of D, we have, with the previous notation,

$$\left.\begin{array}{l}(X) = C_s + C_b + \tfrac{1}{2}\int BI^2 d\Omega, \\ (A) = C_s + C_b + \tfrac{1}{2}\int AI^2 d\Omega.\end{array}\right\} \quad (a)$$

But in the working of the instrument A describes no area about C, $\therefore (A) = 0$. Let fall IN perpendicular to AB; then since, if $AB = l$, $BI^2 - AI^2 = l^2 - 2l \cdot AN$, we have

$$(X) = \tfrac{1}{2} l^2 \Omega - l\int AN \cdot d\Omega. \quad (\beta)$$

But $\Omega = 0$, since, on the whole, the arm AB does not rotate.

Now take any point, P, on the arm AB, and at P let a graduated roller be fixed with its plane always perpendicular to AB. This roller, after any amount of motion, will indicate the total amount of motion of P perpendicular to AB. Now when I is the instantaneous centre, the elementary motion of P is

$$IP \cdot d\Omega,$$

and the component of this perpendicular to AB is

$$IP \cdot \cos IPN \cdot d\Omega, \quad \text{or} \quad PN \cdot d\Omega.$$

Hence if (R) stands for the reading of the roller,

$$(R) = \int PN \cdot d\Omega,$$

and, disregarding sign, equation (β) gives

$$X = l(R),$$

which is the result used in the working of the instrument.

We may observe that nothing depends on the curve along which A moves, since equation (β) will hold if $(A) = 0$, whatever the curve described by A may be. It is requisite only that A should *oscillate* back to its original position.

Let us obtain the result for this instrument on the supposition that the end A is allowed to trace out a curve of any area (A). Suppose the graduated roller fixed at a point P in AB at a distance c from A. Then if for a small motion $d(R)$ is the reading of the wheel [(R) being, of course, the amount of motion of P perpendicular to AB],

$$d(R) = IP \cdot \cos IPB \cdot d\Omega$$
$$= IA \cdot \cos IAB \cdot d\Omega - cd\Omega.$$

Now, as before,

$$(X)-(A) = \tfrac{1}{2} l^2 \Omega - l \int IA \cdot \cos IAB \cdot d\Omega$$
$$= \tfrac{1}{2} l^2 \Omega - l(R) - cl\Omega.$$

If AB is allowed to rotate through 2π, we have

$$(X) = (A) - l(R) + \pi l^2 - 2\pi cl.$$

By taking $c = \tfrac{1}{2} l$, we should obtain

$$(X) = (A) - l(R).$$

We may also observe that the graduated roller may be fixed anywhere on an arm attached rigidly to AB at any point, P, the axis of the roller being fixed parallel to AB; for if R is the point on the attached arm at which the roller is fixed, the elementary motion of R along RP is $IR \cdot \sin IRP \cdot d\Omega$, which is $PN \cdot d\Omega$ and is independent of the position of R.

Although not coming directly under the general theorem of epicycloidal motion, it may be well to give here the theory of another instrument, the principle of which has been employed by Amsler, and the purpose of which is to find the 'moment of inertia' of any plane figure *about a right line in its plane*.

Suppose Ox (fig. 44) to be the right line about which the moment of inertia of the area of the curve represented is required. From any point, A, on the line let an arm, AB, of constant length, a, be drawn to a point on the curve; let O be any fixed point on the line, and let θ denote the angle BAO.

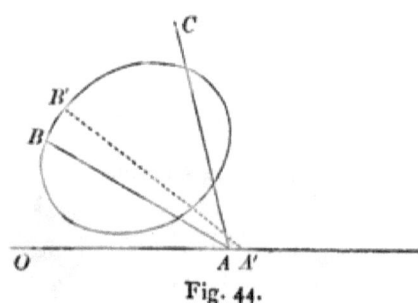

Fig. 44.

Let AB receive a slight displacement along the line and along the given curve, so as to come into the position $A'B'$; let $AA' = dx$, $\angle B'A'O = \theta + d\theta$, $OA = x$.

Now the moment of inertia of the quadrilateral $BAA'B'$ about Ox may be found by drawing from A' a line parallel and equal to AB, so that the quadrilateral is broken up into a parallelogram and a triangle. The moment of inertia of the parallelogram is

$$\tfrac{1}{3}a^3 \sin^3\theta \, . \, dx;$$

and the moment of inertia of the triangle about Ox is

$$\tfrac{1}{4}a^4 \sin^3\theta \, d\theta.$$

Hence if I is the required moment of inertia,

$$I = \tfrac{1}{3}a^3 \int \sin^3\theta \, dx + \tfrac{1}{4}a^4 \int \sin^3\theta \, d\theta \qquad (a)$$

$$= \tfrac{1}{4}a^3 \int \sin\theta \, dx - \tfrac{1}{12}a^3 \int \sin 3\theta \, dx + \tfrac{1}{4}a^4 \int \sin^3\theta \, d\theta. \qquad (\beta)$$

Now if a perpendicular, BM, is let fall from B on Ox, and A is the area of the given figure, $\quad A = \int BM d(OM)$

$$= \int a \sin\theta \, d(x - a\cos\theta)$$

$$= a \int \sin\theta \, dx + a^2 \int \sin^2\theta \, d\theta. \qquad (\gamma)$$

Again, draw AC equal to a, and making the angle $CAO = 3\theta$. Then if A' is the area of the curve traced out by C,

$$A' = \int a \sin 3\theta \, d(x - a\cos 3\theta)$$

$$= a \int \sin 3\theta \, dx + 3a^2 \int \sin^2 3\theta \, d\theta. \qquad (\delta)$$

Hence from (β), (γ), and (δ),

$$I - \tfrac{1}{4}a^2 A + \tfrac{1}{12}a^2 A' = \tfrac{1}{4}a^4 \int \sin^4 3\theta \, d\theta. \qquad (\epsilon)$$

AB and AC may then be simply two hands of a 'Watch' whose mechanism is at A, such that the angle turned through by the hand AC is always *three times* the angle turned through by the hand AB; the centre, A, of the Watch moves by a pivot along a groove coinciding with Ox, while a pencil at the extremity B of the hand AB is carried along the given curve.

We may (as in the ordinary use of Amsler's planimeter) work the instrument so that the hands come back exactly to their original positions without having, on the whole, rotated. In this case $\int \sin^2 3\theta \, d\theta$ will be zero, and we have simply
$$I = \frac{a^2}{12}(3A - A'). \qquad (\zeta)$$

Moreover, we may either place a tracing-pencil at C and *actually* trace out the locus of C, and then measure A', or place a roller on the hand AC (beyond C if necessary) and multiply the indication of this roller by a—as already explained—*without tracing out the locus of C at all*. Thus the instrument may be made its own planimeter.

According to the position of the line Ox, it may be convenient or not to let the hands perform a complete revolution. If they do, the right-hand side of (ϵ) is $\tfrac{1}{4}\pi a^4$, and (ζ) will be replaced by

$$I = \frac{a^2}{12}(3A - A') + \frac{\pi a^4}{4}, \qquad (\eta)$$

which is still a very simple result.

The same kind of instrument enables us to find very easily the position of the Centroid ('centre of gravity') of any plane area. For, assume any line, Ox, in its plane; then if $\bar{y}$ is the distance of the Centroid from Ox,

$$\bar{y} = \frac{\frac{1}{2}\int BM^2 d(OM)}{A}$$

$$= \frac{\frac{1}{2}a^2\int \sin^2\theta\, d(x - a\cos\theta)}{A} = \frac{\frac{1}{2}a^2\int \sin^2\theta\, dx}{A}$$

(if the whole rotation is zero)

$$= \frac{\frac{1}{4}a^2\int \cos 2\theta\, dx}{A} \text{ (disregarding sign)}$$

$$= \frac{\frac{1}{4}a^2\int \sin\left(\frac{\pi}{2} - 2\theta\right) dx}{A}.$$

Let, then, the hand AC always make the angle $\frac{\pi}{2} - 2\theta$ with Ox; that is, let one hand now move through *twice* the angle described by the other; and A'' be the area of the locus of C. Then

$$\bar{y} = \frac{a}{4} \cdot \frac{A''}{A}.$$

Instead of making AC the second hand, we may make the production of CA through A serve as such. This may be in some cases more convenient, since the hands in this case rotate in *opposite senses*.

Having found the distance of the Centroid from Ox, we may then find its distance from another arbitrary line, and these two will give the actual position of the point.

64. Line-Roulettes. If a curve, C (fig. 45), roll on a fixed curve, AB, every right line, such as PD, which is carried by the rolling curve traces out by envelopment a curve, which we shall designate as a *Line-Roulette*, in contradistinction to the ordinary or *Point-Roulette* traced out by every *point* carried by the rolling curve.

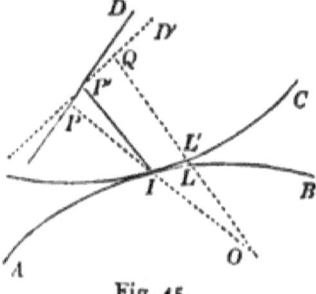

Fig. 45.

The point of contact of PD with its envelope is the foot, P, of the perpendicular on it from I, the instantaneous centre. Suppose the curve C to roll so that L', a point on it very near I, comes to L on the fixed curve, and let QD' be the corresponding displaced position of the line PD. The foot, Q, of the perpendicular from L on QD' is the point of contact of the line

QD' with the Line-Roulette; so that P and Q are two very close points on the curve.

To find the area of the Roulette, observe that the point I of the rolling curve is displaced to a position, I', coincident with I as far as the second order of small quantities; so that the line IP is, as far as small quantities of this order are concerned, displaced into the position IP', where P' is the foot of the perpendicular from I on QD'. Now the element of area $PILQ$ $= PIP' + P'ILQ$; and if $d\Omega$ (as in Art. 58) is the angle PIP' through which the curve C has rolled, and $y = IP$, we have

$$PIP' = \tfrac{1}{2} y^2 d\Omega,$$

$$P'ILQ = y dx,$$

where $dx = P'Q$, the distance between the feet of two successive ordinates, IP' and $L'Q$ (which may be taken as coincident with LQ), let fall on the carried line from two consecutive points, I and L', of the curve C.

If (as in Art. 60) (A) = the area of the Line-Roulette, (S) = area of the fixed curve AB, we have

$$(A) = (S) + \tfrac{1}{2} \int y^2 d\Omega + \int y dx.$$

Now $\int y dx$ is simply the area, (C), of the rolling curve. Hence

$$(A) = (S) + (C) + \tfrac{1}{2} \int y^2 d\Omega. \tag{a}$$

65. Analogue of Kempe's Theorem. *If a plane be moved about in any manner over a fixed plane, all those right lines in the moving plane which have traced out by envelopment Roulettes of the same area are tangents to the same conic, and by varying the area we get a series of Confocal Conics.*

This theorem follows easily from equation (a) of last Article. For, the motion may be regarded as produced by the rolling of the Body Centrode on the Space Centrode (Art. 55); and by (a) of last Article, we may discard the notion of rolling by performing the integration of $y^2 d\Omega$ over the Body Centrode, y being the perpendicular from any point of the Centrode on the axis of x, which we regard as tracing out a Roulette of area A. Now let the equation of any other line in the plane be

$$\lambda x + \mu y + \nu = 0; \tag{1}$$

then the perpendicular on it from the point (x, y) of the curve is
$$\frac{\lambda x + \mu y + \nu}{\sqrt{\lambda^2 + \mu^2}};$$
and if the area of the Line-Roulette of this line is (X), we have, by (a) of last Article,
$$[(X)-(C_s)-(C_b)](\lambda^2+\mu^2) = \tfrac{1}{2}\int(\lambda x+\mu y+\nu)^2 d\Omega$$
$$= a\lambda^2 + b\mu^2 + c\nu^2 + 2h\lambda\mu + 2g\lambda\nu + 2f\mu\nu, \quad (2)$$
if we observe that $a \equiv \tfrac{1}{2}\int x^2 d\Omega$ is a quantity simply depending on the nature of the Body Centrode and the Space Centrode, and in no way depending on the variable line (1).

Now if (X) is constant, equation (2) is the tangential equation of a conic (see Salmon's Conic Sections, chap. XVIII.); hence all lines which give Roulettes of equal area are tangents to the same conic. Moreover by varying (X), equation (2) denotes a series of confocal conics (Salmon, *ibid.*).

This theorem is proved in the following elegant manner by Mr. McCay. It is shown in Salmon's *Conic Sections* (p. 339, 6th ed.) that if $L_1 = 0$, $L_2 = 0$, ... are the equations of six right lines which all touch the same conic, we have
$$l_1 L_1^2 + l_2 L_2^2 + \ldots = 0,$$
where $l_1, l_2, \ldots$ are certain multipliers. Hence if $p_1, p_2, \ldots p_6$ are the perpendiculars from any point on six lines which all touch the same conic,
$$a_1 p_1^2 + a_2 p_2^2 + \ldots + a_6 p_6^2 = 0, \quad (3)$$
where $a_1, a_2, \ldots a_6$ are constant multipliers. Moreover these multipliers are connected by the relation
$$a_1 + a_2 + \ldots + a_6 = 0, \quad (4)$$
as may be seen by considering the perpendiculars as drawn from an infinitely distant point.

Now take any six lines (such as PD, fig. 45), which all touch a conic, and which envelop roulettes of areas $(A_1), (A_2), \ldots (A_6)$; then, calling the perpendiculars from I on them $p_1, p_2, \ldots$ and putting K for $(S)+(C)$ in equation (a) of last Article, we have
$$(A_1) = K + \tfrac{1}{2}\int p_1^2 d\Omega; \quad (A_2) = K + \tfrac{1}{2}\int p_2^2 d\Omega \ldots \ldots$$

Multiplying these by $a_1, a_2, \ldots$ adding, and attending to (3), we have
$$a_1(A_1) + a_2(A_2) + \ldots + a_6(A_6) = 0. \tag{5}$$

Now suppose that we have chosen five of the lines so that $(A_1) = (A_2) = \ldots (A_5)$; then, denoting the common value of these by (D), we have $(a_1 + a_2 + \ldots + a_5)(D) + a_6(A_6) = 0$, or by (4)
$$(A_6) = (D);$$

so that every line which envelops a roulette of the constant area (D) is a tangent to the conic determined by the five specially chosen lines.

66. McCay's Theorem. *The centre of the system of Confocal Conics coincides with the centre of Kempe's Circles.*

For, take any conic of the system and two rectangular tangents, PD, QD, to this conic, intersecting in the point D. Let the perpendiculars from I (fig. 45) on these tangents be p and q, and let (X) denote the area of the roulette enveloped by the tangents. Then $(X) - (C_a) - (C_b) = \frac{1}{2} \int p^2 d\Omega = \frac{1}{2} \int q^2 d\Omega$; therefore $(X) - (C_a) - (C_b) = \frac{1}{4} \int (p^2 + q^2) d\Omega = \frac{1}{4} \int ID^2 \cdot d\Omega$. But if (D) is the area of the roulette traced out by D, we have from Art. 60 $(D) - (C_a) - (C_b) = \frac{1}{2} \int ID^2 \cdot d\Omega$. Hence
$$(D) = 2(X) - (C_a) - (C_b);$$

and as all tangents to the conic chosen give (X) constant, it follows that (D) is constant. Hence all points on the director circle of the conic trace out roulettes of the same area; i.e., the director circle is a Kempe circle; but its centre is the same as that of the conic, and Kempe's circles are all concentric; therefore so are the conics. The common centre may be regarded as the centre of mass of a certain distribution (Art. 59).

67. Curvatures of Point-Roulettes and Line-Roulettes. Circles of Inflexions and Cusps. If in fig. 38, p. 82, the lines PI and QL are produced to meet in the point O, the radius of curvature at P of the roulette traced out by P is the ultimate value of PO.

Now if $d\chi = \angle POQ$, we have $PQ = PO \cdot d\chi$; but $PQ = PI \cdot d\Omega$, since PI is at right angles to PQ, and $d\Omega$, the angle through which C has revolved, is PIQ (see p. 83). Denote PI and PO by r and R, respectively. Then

$$Rd\chi = rd\Omega. \qquad (1)$$

But in the triangle ILO, we have $\dfrac{IL}{d\chi} = \dfrac{LO}{\sin LIO}$; or if the angle made by IP with the normal at I is ψ, we have

$$\frac{ds}{d\chi} = \frac{R-r}{\cos \psi}.$$

Again, $d\Omega = ds \left(\dfrac{1}{\rho} + \dfrac{1}{\rho'}\right) = \dfrac{ds}{h}$ (Art. 54), where h is put for $\dfrac{\rho \rho'}{\rho + \rho'}$. Hence from (1)

$$R = \frac{r^2}{r - h \cos \psi}. \qquad (2)$$

The value of R will be infinite if P is such that $r - h \cos \psi = 0$, i.e., if P is taken anywhere on a circle described at the upper side of AB, passing through I, and having its diameter (of length h) coincident with the normal at I. Points on this circle are therefore points of inflexion on the roulettes to which they give rise; and the circle is hence called the *Circle of Inflexions*.

Similarly for the Line-Roulette (fig. 45, p. 96). The lines PO and QO are two very close normals, therefore each is ultimately equal to the radius of curvature of the roulette at P, and the angle between them is $d\Omega$, the small angle of rotation corresponding to the displacement represented. If R_1 is the radius of curvature of the roulette,

$$R_1 d\Omega = PP' + P'Q = yd\Omega + dx,$$

$$\therefore R_1 = y + \frac{dx}{d\Omega} = y + \frac{dx}{ds} \cdot \frac{ds}{d\Omega},$$

where $ds = IL' =$ element of arc of the rolling curve. Now $\dfrac{dx}{ds} = \cos \psi$, where ψ is the angle between IP and the normal at I to C drawn towards its concave side, and $\dfrac{ds}{d\Omega} = h$ (as above). Hence $\qquad R_1 = y + h \cos \psi.$ \qquad (3)

Now R_1 will be zero if $y + h \cos \psi = 0$, i.e., if P lie on a circle of diameter, h, coincident with the normal at I drawn towards the *convex* (or downward) side of the curve C. So that if IH is the diameter, R_1 will be zero if the line PD passes through H, and for such a line P will therefore be a *cusp* on the roulette. This circle (which is the circle of inflexions exactly reversed in position) is therefore a *Circle of Cusps* on all roulettes corresponding to lines passing through H.

But, in addition to this, the circle is also the locus of the centres of curvature on all line-roulettes—as is evident from equation (3).

For another discussion of the curvature of roulettes, the student may consult Williamson's *Differential Calculus*, chap. xix.

68. Length of a Roulette. The element of arc, PQ (fig. 45, p. 96), of the roulette traced out by P is $PI \cdot d\Omega$, since PI is at right angles to PQ, and the angle QIP is (neglecting the displacement of I, which is infinitesimal compared with IL) $d\Omega$, the small angular displacement of the rolling curve; and it has been shown that $d\Omega = d\theta + d\theta'$ = the sum of the angles of contingence of the rolling and fixed curves (Art. 54). Hence the length of the roulette traced out by P is

$$\int r\,(d\theta + d\theta'), \qquad (1)$$

extended over the rolling curve.

In the case of the line-roulette, the element of arc, PQ (fig. 45, p. 96), is equal to $PP' + P'Q$, or $y\,d\Omega + dx$, taking the line PD as axis of x for the rolling curve. Hence the length of the roulette is

$$\int y\,d\Omega + \int dx, \qquad (2)$$

extended over the rolling curve. For the portion of the roulette corresponding to a complete revolution of a closed rolling curve the term $\int dx$ vanishes.

From (2), or from its equivalent given in equation (5), Art. 61, follows a theorem due to Mr. McCay—viz., that *if a plane area be moved about in any manner in its own plane, and return to its original position after any number of revolutions, the lines which have enveloped roulettes of the same length are tangents to the same circle; and by varying* the length, we get a series of circles concentric with Kempe's circles.

For, take any three lines forming a triangle ABC; then with the notation of Art. 61,

$$aE_a + bE_b + cE_c = 4\pi\Delta. \tag{3}$$

Now suppose that the lines a, b, c envelop roulettes of equal lengths, L; then (3) gives $(a+b+c)L = 4\pi\Delta$. But if r is the radius of the circle inscribed in the triangle, $(a+b+c)r = 2\Delta$; hence $L = 2\pi r$.

Take any fourth line which envelops a roulette of the same length as those enveloped by the lines AC and BC; then evidently the radius of the circle touching this line and the lines AC, BC must also be r; i.e., the fourth line must be a tangent to the circle inscribed in the previous triangle. Now taking CP (fig. 41, p. 88) as a moving line, and supposing that C and P are two points at a constant distance, p, apart on a Kempe circle, equation (3) of Art. 61 gives [since $(C) = (P)$] $E_c = \pi p$; so that the length of the envelope of a perpendicular to CP at P is equal to the length of the envelope of a perpendicular to CP at C; hence all lines drawn perpendicularly at the extremities of a chord of constant length in a Kempe circle envelop roulettes of the same length. But the circle, in the body, which they all envelop is obviously one concentric with the Kempe circle; therefore, etc. Hence the length of the envelope of a line-roulette depends simply on the perpendicular distance of the line from the centre of Kempe's circles.

69. Inverse Roulette Problem. Given a curve AB (fig. 38, p. 82) and a curve PQ, it is required to determine a curve, C, such that by its rolling on AB, a point carried by it may trace out the given curve PQ.

It is obvious that if P is any position of the tracing point, the corresponding point, I, at which the given base-curve, AB, and the required curve are in contact is the point in which the normal at P to the curve PQ meets AB. Now the curves PQ and AB being assigned, we shall be able to express a relation

Examples.

between the length PI and the angle, ϕ, which PI makes with the tangent at I to AB; but this tangent is also a tangent to the required rolling curve. Hence, if r is the radius vector PI from P to the required curve, we shall have some such equation as
$$r = f(\phi),$$
which defines the curve, and can be converted, if necessary, into the usual polar form of equation.

Examples.

1. Prove that the length of an epi- or hypo-trochoid is expressed in the same manner as the length of an elliptic arc.

The case is that of a circle rolling on or inside another circle. Let O be the centre of the rolling circle (whose motion we may entirely neglect), P the tracing point, I any point on the circumference, $PI = r$, $PO = c$; then (Art. 68) the length of the roulette is $\int r\,d\Omega$, where

$$d\Omega = d\theta \pm d\theta' = \left(\frac{1}{a} \pm \frac{1}{b}\right) ds,$$

where a and b are the radii of the circles, and ds the element of circular arc at I. If $\phi = \angle POI$, $ds = a\,d\phi$; therefore the length of the roulette is

$$\left(1 \pm \frac{a}{b}\right) \int r\,d\phi$$

$$= \left(1 \pm \frac{a}{b}\right) \int \sqrt{c^2 - 2ca\cos\phi + a^2} \, . \, d\phi,$$

which expresses the length of an elliptic arc. If put into the usual form of an elliptic function of the second kind, it becomes by using $\psi \equiv \pi - \phi$, instead of ϕ,

$$\frac{2(a+c)(b \pm a)}{b} \int \sqrt{1 - \frac{4ac}{(a+c)^2} \sin^2 \frac{\psi}{2}} \, . \, d\frac{\psi}{2}.$$

2. If a curve C slip continuously on a fixed curve AB (fig. 38, p. 82) without rotation at any instant, prove that—

(a) the increments of the co-ordinates, ξ, η, of any point P carried by the slipping curve corresponding to a motion over a small arc of the fixed curve are given by the equations

$$d\xi = \left(1 + \frac{\rho'}{\rho}\right) dx; \quad d\eta = \left(1 + \frac{\rho'}{\rho}\right) dy,$$

where ρ and ρ' are the radii of curvature of the fixed and slipping curve at I, and (x, y) are the co-ordinates of the point I considered as a point on the fixed curve;

(b) the tangent at I is parallel to the tangent to the locus of P;

(c) the radius of curvature of the locus of P is equal to $\rho + \rho'$.

[The locus of a point P carried in this way is properly called a *Glissette*. Thus the glissettes of all points P are essentially the same curve, placed in different positions.]

3. Prove that the lengths of the epi- or hypo-trochoids traced out by any two inverse points with respect to the rolling circle are to each other as the square roots of the distances of the tracing points from the centre.

(Observe that for two inverse points $\dfrac{r}{r'}$ is constant, where r and r' are the distances of any point on the circle from them. Then use Art. 68.)

4. If a curve (Cartesian oval) whose equation is $lr + mr' = k$, referred to two fixed points, A and B, roll on a right line, prove that for a complete revolution the lengths, L, L', of the roulettes traced out by A and B are connected by the equation $lL + mL' = \pi k$.

5. Two parallel right lines at a constant distance apart are moved about in any manner in their plane and brought back to their original position; prove that the difference of the areas traced out by two points on them which are on any common perpendicular is constant, whatever perpendicular be chosen.

(It is $\pi h^2 - hE$, where h = distance between them, and E = length of envelope of either.)

6. A system of points at constant distances apart is displaced in any manner in its plane; prove that the mean of the areas of the curves traced out by the points is equal to the area of the curve described by their centroid, increased by $\tfrac{1}{2} k^2 \Omega$, where k is their radius of gyration about the centroid, and Ω is the whole angle of displacement.

(Write down the equations (1), (2), and similar ones, of Art. 60 for the points and for their centroid, and add.)

The same result holds if the moving figure is any continuous plane area.

7. Prove that if a graduated roller whose plane is parallel to a moving bar has its centre rigidly attached to the bar by an arm of length h perpendicular to the bar, the reading of the roller after any motion which brings the system back to its original position will be

$$E - 2\pi h,$$

where E is the length of the envelope of the moving bar.

8. When two points of a lamina are guided along two fixed grooves, show that—

(a) The locus of points which describe right lines is the circle C_b;

(b) There is only one point which describes a circle (the centre of C_b);

(c) All points at the same distance from this point describe equal ellipses.

[Mr. McCay. See Art. 59.]

9. Prove that Holditch's Theorem follows from equation (3), Art. 61. [Mr. M°Cay.]

(Take any point, R, on the line PC; then the perpendiculars to PC at P and R envelop parallel curves the difference of whose lengths is

$$2\pi \cdot PR.)$$

10. If any plane figure be displaced in any manner in its plane and return to its original position after any number of complete revolutions, prove that all points which have described roulettes whose centroids ('centres of gravity') are on a given line lie on a conic.

CHAPTER IV.

Mass-Kinematics of Solid Bodies.

70. Definitions. The *momentum* of a moving particle in any direction is defined to be the product of the number of units of mass in the particle and the number of units of velocity in its component of velocity in that direction; so that if m and v are the mass and velocity component of the particle, its momentum in the assumed direction is mv. If instead of being a particle, the body is a solid of any magnitude moving so that all its particles are moving with the *same velocity*, v, in magnitude and in direction, its momentum is Mv, where M is the mass of the body. The *resultant* momentum of a moving particle is, of course, the product of its mass and *resultant velocity*.

The *moment of momentum* of a particle moving in any plane about any point in the plane is the product of its momentum and the perpendicular from the point on the line along which the particle is moving. If p (measured in the same units of length as those employed in measuring v) is the length of this perpendicular, the moment of momentum of the particle about the point is mpv. Just as in Statics when we speak of the moment of a force about a *point*, we mean, in reality, its moment about an *axis* through the point perpendicular to the direction of the force, so the moment of momentum of a particle is, in reality, the moment of the momentum about an *axis* through the point perpendicular to the direction of motion. Since the motions which we consider in this work are all in one plane, or parallel to one plane, we may adhere to the expression 'moment about a *point*,' which saves the circumlocution 'moment about an axis through the point perpendicular to the plane of motion.'

The *force of inertia* of a moving particle, in any direction, is the product of its mass and its component of acceleration in that direction. Thus, if $\frac{d^2x}{dt^2}$ is the acceleration of the particle in the direction of the axis of x, the force of inertia in this direction is $m\frac{d^2x}{dt^2}$.

The *resultant* force of inertia of the particle is the product of its mass and its resultant acceleration. Thus, with the notation of Chap. ii. p. 65, the resultant force of inertia of a particle of mass m moving in any path is $m\sqrt{(\ddot{s})^2 + \frac{v^4}{\rho^2}}$; and its component forces of inertia along the tangent and normal to its path are $m\ddot{s}$ and $m\frac{v^2}{\rho}$.

The quantity which we have here defined as force of inertia is one which is already known in this country by another name. It is spoken of as above by Newton (*Principia*, Definition iii, Book i).

Newton says: 'The *vis insita*, or innate force of matter, is a power of resisting by which every body, as much as in it lies, endeavours to persevere in its present state, whether it be of rest or of moving uniformly forward in a right line.' And in his remarks on the definition he says that 'this *vis insita* may by a most significant name be called *vis inertiæ*. But a body exerts this force only when another force impressed upon it endeavours to change its condition; and the exercise of this force may be considered both as resistance and impulse; it is resistance in so far as the body, for maintaining its present state, withstands the force impressed,' &c.

This terminology has been wholly ignored by English writers, and, as a result, the fact that a body exerts a *kick* (if we may use the expression for clearness of illustration) against any agent which acts on it by direct contact or through a medium for the purpose either of deviating its motion from a rectilinear course or of accelerating its velocity, has been lost sight of. The student must carefully observe that the force of inertia of a moving particle is not a force acting *on* the particle, but one exerted *by* it on some agent direct or indirect—a kick against change of motion.

D'Alembert, in enunciating the kinetical principle known by his name, speaks of force of inertia as *effective force*, and this deviation from Newton's definition and the physical idea contained in it has been followed by English writers.

Most (if not all) of the modern French writers have adhered to Newton's definition, and some (such as Delaunay) very clearly emphasise the nature of force of inertia.

As the momentum of a particle was defined to have a moment about any point, so the moment of its force of inertia may also be taken about any point.

The *energy* of a moving particle is half the product of its mass and the square of its velocity, i. e., energy $= \frac{1}{2} mv^2$.

Of the quantities which we have defined, it will be observed that the last alone is defined without reference to *direction*.

Any quantity which has direction as well as magnitude is called a *directed quantity*, or a *vector* quantity; and any quantity which has merely magnitude but not direction is called an *undirected*, or *scalar*, quantity. Thus a *volume* is, like energy, a scalar quantity. A force, a couple, a velocity, is a vector magnitude, the vector representing any couple being a right line drawn in the direction and *sense* of its axis, its length being, on some conventional scale, proportional to the moment of the couple. An element of *area* must also be regarded as a vector quantity, the vector being drawn parallel to its normal and proportional to its magnitude. A vector, however, as defined in general in Quaternions is not *localised*, i. e., so long as its magnitude, direction, and sense are not altered, it may be drawn at *any* point in space. We shall be concerned chiefly with *localised vectors*, such as those, for example, which are drawn at each point of a moving body to represent the corresponding velocities or momenta *in the actual lines in which these velocities or momenta take place*.

Finally, the student is assumed to be familiar with the theorem of *mass-moments*, which expresses the distance of the centre of mass of any body, or collection of separate particles, from a plane, in terms of the masses of the constituent particles and their several distances from the plane (see *Statics*, p. 252); and with the ordinary elementary facts concerning *moment of inertia* (see Routh's *Rigid Dynamics*, chap. i, or Williamson's *Integral Calculus*, chap. x).

71. Momentum-System of a Rigid Body. *To reduce to*

its simplest form the momentum system of a rigid body moving with uniplanar motion.

Consider first a plane lamina moving in any manner in its own plane. Let G (fig. 18, p. 39) be the centre of mass of the lamina; let ω be the angular velocity of the lamina at any instant; let P be any point of the lamina, dm the element of mass at P, and I the instantaneous centre. Then the velocity of the particle dm at P is at right angles to IP, and its momentum is $\omega . IP dm$. Imagine at each point P a vector drawn to represent the momentum of the particle at the point; and, exactly as in the reduction of a system of forces in Statics, at I introduce two equal and opposite vectors, $\omega . IP dm$, each parallel and equal to the vector at P. This will give a system of vectors at I whose type is $\omega . IP dm$, and a system of couples whose type is $\omega . IP^2 dm$. The sum of all the couples is $\omega \int IP^2 dm$, or $M\omega(k^2 + IG^2)$, where k is the radius of gyration of the lamina about an axis through G perpendicular to its plane. Also (see *Statics*, p. 16) the resultant of a system of vectors whose type is $\omega . IP dm$, if each were directed from I to P, would be a vector $\omega M . IG$ directed from I to G; hence the resultant of the vectors when (as here) each is perpendicular to IP is a vector $M . \omega . IG$ perpendicular to IG.

The momentum-system is therefore equivalent to a momentum vector $M . \omega . IG$ at I perpendicular to IG, together with a momentum couple $M . \omega (k^2 + IG^2)$.

We may combine, exactly as in the reduction of forces, the vector at I and the couple so as to give a single localised momentum vector, by altering the arm of the couple and making each vector in it equal to $M . \omega . IG$. If IG is produced through G to I' so that $M . \omega . IG \times II' = M . \omega (k^2 + IG^2)$, we have the value of II', the new arm of the couple, when the vectors in it are a vector $M . \omega . IG$ at I (directly opposed to the previous vector), and an equal vector at I', which is, then, the point of application of the single momentum-vector to which the whole system is equivalent. From the above we have

$$II' = \frac{k^2}{IG} + IG,$$
$$\therefore \quad GI \times GI' = k^2, \qquad (a)$$

which gives the relation between the points I, I' in a simple form.

For the purpose of resolution and of moments, therefore, we may imagine the lamina concentrated into a single particle at I' and moving in the same direction and with the same velocity ($\bar{v}$, or $\omega . IG$) as the centre of mass, so that the magnitude of the resultant momentum is $M\bar{v}$.

COR. 1. If p is the perpendicular from any point, O, in the plane of the lamina on the line of motion of G, the moment of momentum of the lamina about O—i. e., about an axis fixed in space, but not necessarily in the body, at O—is

$$M(k^2\omega + p\bar{v}). \qquad (2)$$

For the perpendicular from O on the line through I' perpendicular to II' is $p + GI'$, or $p + \dfrac{k^2\omega}{\bar{v}}$.

COR. 2. If a lamina rotates round an axis fixed in space and in the lamina, the moment of momentum of the lamina about the axis is $Mk^2\omega$, where k is the radius of gyration of the lamina about the axis.

Secondly, let the body be a solid of any shape; let G be its centre of mass; through G draw an axis GH perpendicular to the plane of motion; divide the body into an indefinitely great number of thin plates by planes perpendicular to GH; let the instantaneous axis, IK, pierce any one of these plates (whose mass $= \delta p$) in the point i; and let g be the centre of mass of this plate.

Reduce the momentum of the plate, exactly as above, to a couple $\omega . \delta p . k'^2$, and a vector $\omega . ig . \delta p$ at i perpendicular to ig, k' being the radius of gyration of the plate about the instantaneous axis.

If h is the distance between the axes GH and IK, the sum of all the couples $= \omega M(k^2 + h^2)$, where $M =$ mass of the whole body and k its radius of gyration about GH.

To get the equivalent of the system of vectors whose type is $\omega . ig . \delta p$, take IK as axis of z; let I be the foot of the perpendicular from G on IK, and let IG be the axis of x; then the components of $\omega . ig . \delta p$ along the axes of x, y, z are $(-\omega y \delta p,$

Momentum-System of a Rigid Body.

$\omega x \delta p$, o), if (x, y, z) are the co-ordinates of g. Hence the system of vectors in question is equivalent to a vector $\omega \int x \delta p$, or $\omega . Mh$, or $M\bar{v}$, at I perpendicular to IG, together with two couples in the planes of xz and yz whose moments are $-\omega \int xz \delta p$ and $-\omega \int yz \delta p$. But $\int xz \delta p$ and $\int yz \delta p$ are the products of inertia of the body with reference to the planes xz and yz. Denote them by l and m; then the whole momentum-system is equivalent to a momentum vector $M\bar{v}$ at I, a momentum couple $M(k^2 + h^2) \omega$ whose plane is perpendicular to the instantaneous axis, and two momentum couples, $-l\omega$, $-m\omega$, whose axes are perpendicular and parallel to IG.

Consequently if $l = m = o$, i.e., if the axis GH, through G perpendicular to the plane of motion, is a principal axis, the case becomes the same as if the body were a plane lamina, and the momentum-system reduces to a single momentum-vector through a point I' determined as in equation (a).

In Kinetics it is shown that the moment of momentum of a body about an axis, which is generated when the body is struck by a given system of impulses, is equal to the sum of the moments of the impulses about the axis; and, in general, at each instant in the motion of a body, its moment of momentum about any axis is equal to the sum of the moments, about the axis, of any impulse-system which would generate the actual motion of the body from rest. If an axis is suddenly fixed through any point, O, of a moving lamina, this principle and the expressions in Cors. 1 and 2 give us the angular velocity with which the body begins to rotate about the fixed point. If this angular velocity is Ω, and the radius of gyration about O is k', we have

$$\Omega = \frac{k^2 \omega + p\bar{v}}{k'^2}.$$

If the moving body is a lamina, or if, as explained above, the case is the same as if it were a lamina, *the expression for the moment of its momentum about an axis passing perpendicularly to the plane of motion through any point on the circle described in the plane of motion through G on the line GI as diameter is*

$$Mk'^2 \omega,$$

where k' is the radius of gyration of the body about the assumed axis.

For, let P be any point on the circle described on GI as diameter; then $p = \dfrac{PG^2}{GI}$, and $\bar{v} = GI \cdot \omega$, therefore $p\bar{v} = \omega \cdot PG^2$, and therefore by (2) the moment of momentum about P is $M(k^2 + PG^2) \cdot \omega$, or $Mk'^2\omega$.

Hence if P denote the sum of the moments, about an axis, of any system of impulses applied to a body, the equation

$$I\omega = P,$$

where I denotes the moment of inertia of the body about the axis and ω the angular velocity generated by the impulse-system, holds round any point on the circle described on GI as diameter.

72. System of Forces of Inertia. *To reduce to its simplest form the system of forces of inertia of all the elementary particles of a rigid body with uniplanar motion.*

Firstly, let the body be a plane lamina moving in any manner in its own plane. Let G be its centre of mass, J the acceleration centre at any instant, P any point in the lamina at which the element of mass is dm, ω and $\dot{\omega}$ the angular velocity and angular acceleration of the lamina at the instant considered.

Then if $\sqrt{\dot{\omega}^2 + \omega^4} = \epsilon$, the acceleration of the particle at $P = \epsilon \cdot JP$ and its direction makes $\tan^{-1} \dfrac{\dot{\omega}}{\omega^2}$ with JP. The resultant force of inertia of this particle $= \epsilon \cdot JP \cdot dm$. Let it be represented by a vector drawn at P; and pursuing the same system of reduction as in the last Article, introduce at J two vectors equal and parallel to this vector, and opposite to each other. We shall thus have the vector $\epsilon \cdot JP \cdot dm$ at J, and a couple whose moment $= \dot{\omega} \cdot JP^2 dm$. The resultant of the system of vectors at J is a vector $\epsilon \cdot JG \cdot M$ parallel to the direction of acceleration of G; and the sum of the couples $= \dot{\omega} \int JP^2 dm = M(k^2 + JG^2) \cdot \dot{\omega}$, where $M =$ the mass of the lamina and k is its radius of gyration about an axis through G perpendicular to its plane. Just as before, change the couple into one in which each vector is $\epsilon \cdot M \cdot JG$, and in which the arm $= \dfrac{\dot{\omega}}{\epsilon}\left(JG + \dfrac{k^2}{JG}\right)$. Produce the line JG through G to J' so that

$$GJ \cdot GJ' = k^2, \qquad (a)$$

and the single vector $\epsilon \cdot M \cdot JG$, or $M \cdot \bar{a}$ (if $\bar{a}$ denotes the acceleration of G), to which the whole system reduces, will be at $\mathcal{J}'$ parallel to the direction of acceleration at G.

Hence the whole system of forces of inertia is the same as if the lamina were concentrated into a single particle of mass M at the point $\mathcal{J}'$ having the acceleration, $\bar{a}$, of the centre of mass.

COR. 1. If p is the perpendicular from any point, O, on the line of acceleration of G, the sum of the moments of the forces of inertia about O is $\quad M(k^2\dot{\omega}+p\bar{a})$.

COR. 2. The sum of the moments of the forces of inertia about an axis fixed in the body as well as in space is

$$Mk'^2\dot{\omega},$$

where k' is the radius of gyration about the axis.

Secondly, let the body be a solid of any shape; let G be its centre of mass; let $\mathcal{J}K$ be the acceleration axis at any instant, $\mathcal{J}$ being the foot of the perpendicular from G on this axis, and, as before, divide the body into a number of thin plates perpendicular to this axis, any one of which is met in the point j by the axis, the centre of mass of this plate being g, and its mass δp. Take the line $\mathcal{J}K$ as axis of z, $\mathcal{J}G$ as axis of x, the length $\mathcal{J}G$ being denoted by h, and let (x, y, z) be the co-ordinates of g.

Then reduce the system of forces of inertia of the plate to a couple $\dot{\omega}k'^2\delta p$ about the axis $\mathcal{J}K$ and a vector $\epsilon.jg.\delta p$ at j, k' being the radius of gyration of the plate about $\mathcal{J}K$. The sum of the couples for all the plates is

$$M(k^2+h^2)\dot{\omega},$$

where k is the radius of gyration of the whole body about an axis through G perpendicular to the plane of motion. Also the vector $\epsilon.jg$ has for components parallel to the axes of co-ordinates $(-\omega^2 x - \dot{\omega}y, \dot{\omega}x - \omega^2 y, 0)$; so that the system of vectors of which $\epsilon.jg.\delta p$ is the type gives rise to a vector at $\mathcal{J}$ whose components are $(-\omega^2 Mh, \dot{\omega}Mh)$ along and perpendicular to $\mathcal{J}G$—i.e., to a vector $\epsilon.M.\mathcal{J}G$ at $\mathcal{J}$ parallel to the line of acceleration of G—together with two couples with axes parallel and perpendicular to $\mathcal{J}G$, their moments being $-l\dot{\omega}+m\omega^2$ and $-l\omega^2-m\dot{\omega}$.

The whole system, therefore, reduces to a vector $M\bar{a}$ at J, together with a couple $M(k^2+h^2)\dot{\omega}$ about the axis JK, and the two couples just mentioned.

Consequently if the axis through G perpendicular to the plane of motion is a principal axis, the case becomes the same as if the body were a plane lamina, and the system of forces of inertia reduces to a single force of inertia localised at a point J' determined by equation (a).

On the line JG describe a circle, the angle, JPG, in one segment being $\pi - \tan\frac{\dot{\omega}}{\omega^2}$, and the angle in the other segment $\tan^{-1}\frac{\dot{\omega}}{\omega^2}$; then, assuming the body to be a lamina, or that (as explained above) the case is the same as if it were a lamina, *the sum of the moments of the forces of inertia taken about any point, P, on the circumference of this circle is*

$$Mk'^2\dot{\omega},$$

where k' is the radius of gyration of the body about P.

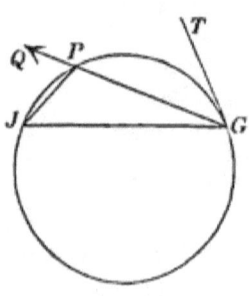

Fig. 46.

If $\phi = \tan^{-1}\frac{\dot{\omega}}{\omega^2}$, the angle $JPQ = \phi$, therefore the acceleration of P is along PQ. Also the acceleration of G is along the tangent GT to the circle, since $\angle JGT = \phi$. The diameter of the circle $= \frac{JG}{\sin\phi}$, therefore the perpendicular, p, from P on GT is $\frac{PG^2}{JG}\sin\phi = \frac{PG^2}{JG}\cdot\frac{\dot{\omega}}{\epsilon} = \frac{\dot{\omega}\cdot PG^2}{\bar{a}}$; hence substituting in (a) we have the moment of the forces of inertia $= M(k^2 + PG^2)\dot{\omega}$, or $Mk'^2\dot{\omega}$, as above.

It is an immediate consequence from Newton's Third Law that the system of forces of inertia is completely equivalent to the system of external forces acting on any body or any material system [1]. Hence the sum of the

---

[1] This principle was subsequently enunciated by D'Alembert, and since his time it held undisputed sway as *D'Alembert's Principle*, until Thomson and Tait pointed out the fact that it is clearly contained in Newton's Axiom. Indeed, assuming Newton's principle of the equality of action and reaction between every pair of mutually influencing bodies, it is surprising that the necessity for D'Alembert's Principle could have been felt.

moments of the forces of inertia about any axis is equal to the sum of the moments, L, of the external forces about the axis; and we have just proved that for an axis represented by any point on the circle JPG the expression for the moment of the forces of inertia is $I\dot{\omega}$, where I is the moment of inertia about the axis. Round all points on this circle we have therefore the equation
$$I\dot{\omega} = L.$$

73. Energy in Uniplanar Motion. Let dm be the element of mass at any point, P, in the body; from P let fall a perpendicular, r, on the instantaneous axis. Then if ω is the angular velocity of the body, the velocity of P is ωr, and the energy of the particle at P is $\frac{1}{2}\omega^2 r^2 dm$, so that the energy of the whole body is $\frac{1}{2}\omega^2 \int r^2 dm$, or $\frac{1}{2} Mk'^2 \omega^2$, where k' is the radius of gyration about the instantaneous axis. But since $\dfrac{\bar{v}}{\omega}$ is the distance between G and the instantaneous axis (p. 40), where $\bar{v}$ = velocity of G, we have $k'^2 = k^2 + \dfrac{\bar{v}^2}{\omega^2}$. Hence the energy is
$$\tfrac{1}{2} M(\bar{v}^2 + k^2 \omega^2). \tag{a}$$

Hence the whole energy may be described as consisting of two parts; one is $\frac{1}{2} M\bar{v}^2$, which is *the energy of the body on the supposition that it is condensed into a single particle moving with the velocity of the centre of mass;* and the other is $\frac{1}{2} Mk^2 \omega^2$, which is *the energy due to the rotation round the centre of mass.*

It is not true, in general, that the energy of the body is equal to the sum of that due to the rotation round *any* point and that due to the velocity of translation of the particle at this point; i.e., if v is the velocity of translation of any particle, P, and λ the radius of gyration about an axis through P perpendicular to the plane of motion, it is not in general true that the energy is $\frac{1}{2} M(v^2 + \lambda^2 \omega^2)$. Let us see whether there are any *particular* points for which it is true. Let I be the instantaneous centre, $GI = h = \dfrac{\bar{v}}{\omega}$, $PI = \rho$, $\angle PIG = \psi$; then v, the velocity of the particle at P, $= \omega \rho$, and $\lambda^2 = k^2 + PG^2 = k^2 + \rho^2 - 2h\rho \cos\psi + h^2$; therefore $v^2 + \lambda^2 \omega^2 = (k^2 + h^2 + 2\rho^2 - 2h\rho \cos\psi)\omega^2$; and if this $= \bar{v}^2 + k^2 \omega^2 = (k^2 + h^2)\omega^2$, we must have
$$\rho^2 - h\rho \cos\psi = 0, \text{ or } \rho = h \cos\psi;$$

i.e., the energy can be put into the form $\frac{1}{2}M(v^2+\lambda^2\omega^2)$ if P is any point on the circle described on GI as diameter.

Hence, in uniplanar motion—*the total energy of a body is equal to the sum of the energy due to the velocity of translation of any point on the circle described on GI as diameter and the energy due to the rotation round this point.*

There is an analogous property in the general, or three dimensional, motion of a rigid body, the proof of which is left to the student; viz., *the total energy of a rigid body is equal to the energy due to the velocity of a point P added to the energy due to the rotation about this point, if P is any point on the surface of the second degree whose equation, referred to the principal axes at G (the centre of mass), is*

$$\lambda^2 + \mu^2 + \nu^2 + a\lambda + b\mu + c\nu = 0,$$

where (a, b, c) are the component-velocities of G parallel to the axes, $\lambda \equiv \omega_2 z - \omega_3 y$, $\mu \equiv \omega_3 x - \omega_1 z$, $\nu \equiv \omega_1 y - \omega_2 x$, and $(\omega_1, \omega_2, \omega_3)$ are the angular velocities of the body about the axes at G.

74. Energy of Rapid Vibratory Motion. Suppose a very small particle of mass dm to have a velocity which at any instant is represented by $\epsilon \cos 2\pi \dfrac{t}{\tau}$, where τ is an extremely small period of time—say something like $\frac{1}{10000}$th part of a second. Then within even such a small period as one second the velocity has run through all its values 20,000 times, and the value of the energy of the particle will have varied between $\frac{1}{2}\epsilon^2 dm$ and zero 40,000 times. Hence we may look upon the energy of the particle as constant and equal to its mean value during the period τ, i.e., $\dfrac{\epsilon^2 dm}{2\tau}\displaystyle\int_0^\tau \cos^2 2\pi \dfrac{t}{\tau} dt$, or $\frac{1}{4}\epsilon^2 dm$.

We have in the previous Article spoken of the *total* energy of a body as being of the form $\frac{1}{2}M\bar{v}^2 + \frac{1}{2}Mk^2\omega^2$; and this would be correct if the body were *perfectly rigid*, or, in other words, if the velocity-components, u, v, of every point in it could be represented by the equations

$$u = a - \omega y, \quad v = b + \omega x, \qquad (a)$$

(as in p. 41). The energy thus expressed is often spoken of as the *energy of visible motion*—an expression which is, of course,

logically indefensible[1]—by which is signified the energy of the body considered as a strictly *rigid* body. The total energy of a *natural solid* in a state of motion may not be expressible in the form
$$\tfrac{1}{2} M \bar{v}^2 + \tfrac{1}{2} M k^2 \omega^2, \tag{β}$$
which results from the supposition that the velocity components of every particle are given by equations (a); for the molecules (using the term in its chemical sense) of the body may be executing rapid vibrations of very small amplitude about certain mean positions, while the whole system is moving through space. We do not know the precise forms of the vibratory velocity components of a molecule, but we may assume them to be of the type $\epsilon \cos 2\pi \dfrac{t}{\tau}$, so that the complete expressions for the velocity components of a molecule will be of the forms
$$u = a - \omega y + \epsilon \cos 2\pi \frac{t}{\tau}, \quad v = b + \omega x + \eta \sin 2\pi \frac{t}{\tau}, \tag{γ}$$
if all the motions are uniplanar; and in three-dimensional motion the expressions for the velocity components u, v, w, which hold for a *rigid* body (viz., in the usual notation, $u = a + \omega_2 z - \omega_3 y$, etc.), must be corrected by the addition of periodic terms of very short period, as in (γ).

Now, taking the expressions (γ)—the method of dealing with which is precisely the same for the case of three-dimensional motion—the total energy of the body will be
$$\tfrac{1}{2} \int [(a - \omega y + \epsilon \cos 2\pi \tfrac{t}{\tau})^2 + (b + \omega x + \eta \sin 2\pi \tfrac{t}{\tau})^2] dm, \tag{δ}$$
which will contain such terms as
$$a \int \epsilon \cos 2\pi \frac{t}{\tau} dm, \quad \int \epsilon y \cos 2\pi \frac{t}{\tau} dm, \quad \int \epsilon^2 \cos^2 2\pi \frac{t}{\tau} dm.$$
But, τ being extremely small, $\cos 2\pi \dfrac{t}{\tau}$ will, before the co-ordinates x and y have time to alter sensibly, run through all possible values, the mean of which will be zero; so that the

---

[1] Visible to whom? Instead of this expression, we shall use 'energy of the first order.'

first two of these three integral types vanish; and, as before explained, the third is practically equal to $\tfrac{1}{2}\int\epsilon^2 dm$.

Hence (δ) becomes

$$\tfrac{1}{2}\int[(a-\omega y)^2+(b+\omega x)^2]dm+\tfrac{1}{4}\int(\epsilon^2+\eta^2)dm, \qquad (\epsilon)$$

the first part of which—energy of the first order, or rigid-body energy—is equivalent to the form (β), and the second—energy of the second order—is evidently the *energy due to the molecular vibrations*, which we may denote by H; so that the total energy of a natural solid is, in general, of the type

$$\tfrac{1}{2}M\bar{v}^2+\tfrac{1}{2}Mk^2\omega^2+H. \qquad (\zeta)$$

The energy of molecular vibration, H, is exhibited, in general, in the form of *heat* or *sound*.

The subdivision of the whole energy of a natural solid is often carried farther than *two* terms—one expressing energy as if the body were *rigid*, and the other expressing energy of vibratory motion of *molecules* as if they were little rigid bodies—for it may happen that the *atoms* which constitute the molecules are disturbed in such a manner that the molecules cease to behave as rigid bodies.

Such is probably the case in a solid incandescent body moving in space and emitting rays from ultra-red to ultra-violet. In this case the terms introduced into the expressions for the velocity components (u, v) at any point would still be periodic, and we may regard H in the expression (ζ) as including the energy of *atomic* vibration.

If the x-displacement of a molecule from its natural position is of the form $a \sin 2\pi \dfrac{t}{T}$, the amplitude, a, is extremely small, but the *velocity* of this motion is $\dfrac{2\pi a}{T} \cos 2\pi \dfrac{t}{T}$, and the coefficient $\dfrac{2\pi a}{T}$, or ϵ, may, on account of the extreme smallness of T, be very great, so that $\int \epsilon^2 dm$ may in the case of a hot body moving in any way be several hundreds or thousands of times greater than the body's energy of the first order.

Thus, the very moderate amount of energy (of the second order) which is known as one *thermal unit* has been found by

Joule to be equivalent to 772 foot-pounds—the thermal unit being the quantity of heat necessary to raise 1 lb. of water at 60° F. one degree more in temperature. Adopting gravitation units, the energy of a body of weight w lbs. moving with a translational velocity v feet per second is $\dfrac{wv^2}{2g}$ foot-pounds; so that if $w = 1$, and this energy $= 772$, we have $v = 224$ feet per second (about); so that in order to possess the energy measured by one thermal unit in the form of energy of the first order, the pound of water should have a velocity of translation (common to all its particles) about three times that of the fastest express railway trains.

If the mean square of the resultant vibratory velocity of any molecule of a body is denoted by i^2, its energy of the second order $= \tfrac{1}{2}\int i^2 dm$; and, in analogy with moment of inertia, we may write this in the form $\tfrac{1}{2} M . \mu^2$,

M being the mass of the body, and μ^2 being obviously the mean square of molecular vibratory velocity. If θ is the temperature of the body, c its capacity for heat (referred to water), and J Joule's equivalent, this expression for the energy must be equal to $MJc\theta$; so that

$$\mu = \sqrt{2Jc\theta}.$$

In a material universe every body is directly in contact with some other body and in mediate communication with every other body (by means of air or ether), so that it is impossible for a body to move without communicating motion—or transferring some of its energy—to other bodies. Hence it is manifestly impossible to construct a material system which is *energy-tight*, i.e., which maintains undiminished a quantity of energy which it has once received. Even if we could construct such a system, it would be impossible to preserve its energy in the form of energy of the first order—i.e., rigid body energy—for, collisions or friction between its parts would inevitably set molecules in vibration; and since in such a system we have assumed the expression (ζ) to be constant, there would be a continual transformation of energy of the first order into energy of the second order (molecular energy).

Examples.

1. A lamina moves in any way in its own plane; show how to stop its motion completely by fixing one point.

2. A lamina moves in any way in its own plane; if it be required to fix a point in such a way that the angular velocity about it shall be given, find the locus of the point in the lamina.

Ans. A circle. If the new angular velocity $= \Omega$, $\omega =$ angular velocity of the lamina before the fixing of the point, $\bar{v} =$ velocity of centre of mass, G, before the fixing, the equation of the circle is

$$r^2 - \frac{\bar{v}}{\Omega} r \sin\theta + k^2\left(1 - \frac{\omega}{\Omega}\right) = 0,$$

G being the origin, and the line of motion of G before the fixing being the initial line, and k the radius of gyration about G.

3. Show that the fixing of a point entails a loss of energy (of the first order).

If u is the velocity of the point before fixing, h its distance from G, p the perpendicular from it on the line of motion of G before fixing, the loss of energy of the first order can be written

$$\tfrac{1}{2} M \cdot \frac{k^2 u^2 + (h^2 - p^2)\bar{v}^2}{k^2 + h^2}.$$

4. A lamina moves in any manner in its own plane; find the locus of a point the direction of whose velocity coincides with that of its acceleration. Find the locus also if these two directions are at right angles.

Ans. In each case a circle passing through the points I, J.

5. Prove that the surface locus in the theorem at the end of Art. 73 is a circular cylinder having the axis of resultant angular velocity at G and the axis of instantaneous screw motion for diametrically opposite generators. (Prof. Townsend.)

[The theorem of Art. 73 was published by the author in the *Educational Times* for July, 1882, and a discussion of it by Prof. Townsend appears in the number for August.]

6. If $\bar{v}$ is the resultant velocity of G, ψ the angle between the direction of $\bar{v}$ and the axis of resultant angular velocity (ω) in last question, show that the diameter of the cylinder $= \dfrac{\bar{v}}{\omega} \sin\psi$.

7. Find (from Arts. 71 and 72) the Virials, with respect to any point, of the momenta and forces of inertia of the particles of any lamina moving in its own plane.

CHAPTER V.

Analysis of Small Strains.

75. Nature of Strain. When a *natural solid* (such as a mass of iron or wood) is not acted upon by any external force, its molecules assume certain distances from each other, which are called their *natural distances;* but when some external force, or impulse, acts on the body—as, for instance, a pressure, whether continuous or sudden—some alteration of the molecular distances, however small, must result. These alterations may be so small as to be invisible to the eye, and yet they may produce effects which are otherwise very readily appreciated— e. g., when minute and rapid molecular alterations result in the production of sound, as in the case of a bell or a telephone.

Any alteration of the natural distances between particles of a body is called a *strain*. Every motion of a body is not accompanied by strain; for the motion may be one which is consistent with perfect rigidity. Thus, if a solid is slightly moved parallel to the plane xy so that the small changes (u, v) of the co-ordinates of every point (x, y) are given by the equations
$$u = a - \omega y, \quad v = b + \omega x, \tag{1}$$
where a, b, ω are the same for all points, no strain results since these equations express the displacement of a rigid body (Art. 29), a and b being the components of a motion of translation common to all points in the body, and ω a motion of rotation about the axis of z common to all points.

76. Changes in Relative Co-ordinates. Let fig. 47 represent a section of a body made by the plane (xy) of displacement; let Ox and Oy be two fixed axes of x and y; let

P be any point in the body; and Px, Py two lines parallel to Ox, Oy.

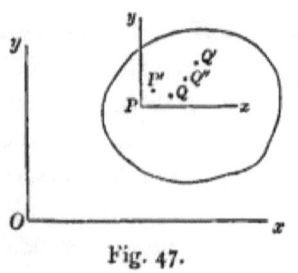

Fig. 47.

We confine our attention to *small* strains, and we suppose the displacements to be the same at corresponding points in all sections of the body parallel to the plane of the figure.

Suppose, then, that the body is strained in such a manner that P is brought to P', and Q a point very near P is brought to Q'. Let the components of PP' parallel to Ox and Oy be u and v. Then the strain is produced according to some law which assigns u and v as functions of the position of P.

Suppose that
$$u = f_1(x, y), \quad v = f_2(x, y). \tag{1}$$

Again, if the co-ordinates of Q with reference to Px and Py are (ξ, η), the displacements, u', v', of Q (i.e., the components of QQ') are

$$u' = f_1(x+\xi, y+\eta), \quad v' = f_2(x+\xi, y+\eta),$$

or
$$u' - u = \xi \frac{du}{dx} + \eta \frac{du}{dy}, \quad v' - v = \xi \frac{dv}{dx} + \eta \frac{dv}{dy}. \tag{2}$$

If the whole body receive a motion of translation equal and parallel to $P'P$, so that the point P' is brought back to P and Q' brought to Q'', no further strain results, and the expressions (2) are those for the co-ordinates of Q'' with reference to Q. Denote the changes $u'-u$ and $v'-v$ in the relative co-ordinates of Q with respect to P by $\Delta\xi$ and $\Delta\eta$, and we have

$$\Delta\xi = \xi \frac{du}{dx} + \eta \frac{du}{dy}; \quad \Delta\eta = \xi \frac{dv}{dx} + \eta \frac{dv}{dy}. \tag{3}$$

Denote $\frac{du}{dx}$ and $\frac{dv}{dy}$ by a and b; and denote $\frac{du}{dy}$ by $s - \omega$, and $\frac{dv}{dx}$ by $s + \omega$. Then these equations can be written

$$\Delta\xi = a\xi + (s-\omega)\eta; \quad \Delta\eta = b\eta + (s+\omega)\xi. \tag{4}$$

From these expressions we deduce the following results.

(1) *All points in the immediate neighbourhood of P which in the*

unstrained state of the body lay on a right line, continue after strain to lie on a right line (*with altered direction*).

For if (ξ', η') are the co-ordinates of any point Q'' with respect to Px and Py, we have $\xi' = \xi + \Delta\xi$, $\eta' = \eta + \Delta\eta$; therefore
$$\xi' = (1+a)\xi + (s-\omega)\eta,$$
$$\eta' = (s+\omega)\xi + (1+b)\eta, \qquad (5)$$

so that (ξ, η) are also linear functions of (ξ', η'); and if the first satisfy a linear equation, $\lambda\xi + \mu\eta + \nu = 0$, so must the second.

(2) *Two parallel lines in the immediate neighbourhood of P before strain become two parallel lines* (*with altered direction*) *after strain*. This follows by causing ν alone to vary in the equation of the above line. Hence a parallelogram becomes another parallelogram.

77. Transformation of Co-ordinates. Given the values of a, b, s, ω with reference to two rectangular axes, Ox, Oy, to find their values with reference to two other rectangular axes, Ox', Oy'.

Let Ox' (fig. 47) make the angle ϕ with Ox; let two axes, Px', Py', be drawn at P parallel to Ox', Oy'; let (x', y') be the co-ordinates of P with reference to Ox', Oy'; and let (u', v') be the values of (u, v) with reference to these directions.

Then
$$x = x'\cos\phi - y'\sin\phi,$$
$$y = x'\sin\phi + y'\cos\phi,$$
$$u' = u\cos\phi + v\sin\phi,$$
$$v' = -u\sin\phi + v\cos\phi.$$

Also $\dfrac{du'}{dx'} = \dfrac{du'}{dx}\cdot\dfrac{dx}{dx'} + \dfrac{du'}{dy}\cdot\dfrac{dy}{dx'}$, which from the above equations gives

$$\frac{du'}{dx'} = a\cos^2\phi + 2s\sin\phi\cos\phi + b\sin^2\phi. \qquad (a)$$

Similarly $\dfrac{dv'}{dy'} = a\sin^2\phi - 2s\sin\phi\cos\phi + b\cos^2\phi, \qquad (\beta)$

$$\frac{du'}{dy'} = -\omega + s\cdot\cos 2\phi + (b-a)\sin\phi\cos\phi, \qquad (\gamma)$$

$$\frac{dv'}{dx'} = \omega + s \cdot \cos 2\phi + (b-a)\sin\phi\cos\phi. \tag{δ}$$

Adding (γ) and (δ), and denoting $\frac{du'}{dy'} + \frac{dv'}{dx'}$ by $2s'$, we have

$$2s' = 2s\cos 2\phi + (b-a)\sin 2\phi. \tag{1}$$

Adding (α) and (β), and denoting $\frac{du'}{dx'}$, $\frac{dv'}{dy'}$ by a', b', we have

$$a' + b' = a + b. \tag{2}$$

Also if $\frac{dv'}{dx'} - \frac{du'}{dy'}$ is denoted by $2\omega'$, we have

$$\omega' = \omega. \tag{3}$$

The relations (2) and (3) being independent of ϕ, are *invariant* relations, showing that *the numbers $a+b$ and ω are the same whatever be the two rectangular lines, Px, Py, at P with respect to which they are calculated.*

Other invariant relations may be found by eliminating ϕ from the equations (α), ... (δ) in pairs.

From (1) we see that s' can be made zero by properly choosing the lines Px', Py'; and that the directions of these lines are given by the equation

$$\tan 2\phi = \frac{2s}{a-b}, \tag{4}$$

showing that the two lines in question are the axes of the conic discussed in the next Article.

Cor. 1. The equations (α), (β), and (1) expressing the components of the given strain with reference to a new set of axes, Px', Py', constitute *the resolution of strain*.

Cor. 2. Two strains, denoted for shortness by (a, b, s) and (a', b', s'), one expressed with reference to one set of axes, Px, Py, and the other expressed with reference to another set, Px', Py', are said to be equivalent when either produces the other. Hence we may replace one by the other. Thus we see that any strain (a, b, s) can always be replaced by one of the form (a', b', o).

78. Elongation in any Direction. The *elongation*, produced by the strain, in the direction PQ is defined to be *the* ratio which the change in the length of PQ bears to the original length PQ. Denote this elongation by ϵ; then

$$\epsilon = \frac{PQ'' - PQ}{PQ} = \frac{\Delta PQ}{PQ}.$$

Now $PQ^2 = \xi^2 + \eta^2$,

$\therefore PQ \cdot \Delta PQ = \xi \Delta \xi + \eta \Delta \eta$

$= a\xi^2 + 2s\xi\eta + b\eta^2$ by (4), Art. 76.

If $\angle QPx = \phi$, this gives

$$\epsilon = a\cos^2\phi + 2s\sin\phi\cos\phi + b\sin^2\phi, \qquad (1)$$

which expresses the elongation in any direction, ϕ.

Construct round P as centre the conic whose equation with reference to the axes Px, Py is

$$a\xi^2 + 2s\xi\eta + b\eta^2 = k^2, \qquad (2)$$

where k is any constant length. Then if r is the length of the line PQ intercepted by this conic, we have from (1)

$$\epsilon = \frac{k^2}{r^2}. \qquad (3)$$

If the above conic be called the *Elongation Conic*, we have the result that *the elongation in any direction is inversely proportional to the square of the radius vector of the Elongation Conic in that direction*.

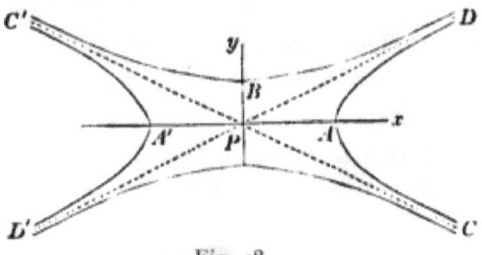

Fig. 48.

If the right-hand side of (1) is capable of changing sign, i. e., if we get real values for ϕ by putting the expression equal to zero, it follows that the elongation conic (2) is a hyperbola; that the elongation is zero in the directions of its asymptotes; and that these asymptotes separate the regions, DPC and $D'PC'$ (fig. 48), in which *elongation* takes place, from the regions, DPC'

and $D'PC$, in which *compression* takes place. To represent compression, we must put $-k^2$ for k^2 in (2); so that unless the right-hand side of (1) is incapable of changing sign—i.e., unless all lines through P are elongated or all contracted—there will be *two* conics required to represent the strain, these being conjugate hyperbolas whose common asymptotes are lines of no elongation or compression. When all lines through P are elongated or all compressed, the conic is an ellipse.

Cor. 1. *The elongations in the directions of the axes Px, Py are a, b;* and hence the elongation in any direction could have been inferred at once from equation (a), Art. 77.

Cor. 2. *The elongation is the same along all parallel lines in the neighbourhood of P.* For, the elongation in any direction at any near point, Q, is given by an expression of the form (1); and since a, b, s are small quantities of the first order, the values of these quantities at Q differ from them by small quantities of the second order; therefore, &c.

Cor. 3. *Any small area, A, near P is altered by strain into an area $(1+a+b) A$.* For, divide the area into small rectangles of the type $dx\,dy$, with sides parallel to the axes Px, Py. Then, by Cor. 2, the lengths dx and dy will become $(1+a) dx$ and $(1+b) dy$; and the small rectangle will become a parallelogram whose angle is $\frac{\pi}{2}-\gamma$, where γ is a small quantity of at most the first order. Hence the ratio of the strained to the unstrained area is $(1+a)(1+b) \cos\gamma$, i.e., $1+a+b$, rejecting quantities of the second and higher orders. And since all the elements of the area in question are altered in this ratio, the whole area is altered in this ratio, so that if ΔA is the change of area,

$$\frac{\Delta A}{A} = a+b.$$

The number $a+b$ is called the *areal dilatation* at P. A cylinder standing perpendicularly on the area has its *volume* altered in this ratio.

Cor. 4. From the result in Cor. 3 it follows that *the sum of the elongations along any two rectangular lines at P is constant—*

it must express the areal dilatation at P—a result already obtained, (2), Art. 77.

If the axes of reference Px, Py, at P are so chosen as to coincide with the axes of the elongation conic, s will be zero (see last Art.), and if we denote the values of a, b with reference to these axes by e_1, e_2, respectively, the equation of the elongation conic is
$$e_1\xi^2 + e_2\eta^2 = k^2. \tag{4}$$

In proving any general property of strain, simplicity is gained by taking the axes of co-ordinates at P in the directions of the axes of this conic.

79. Amount of Rotation of any Line. If ϕ' denote the angle $Q''Px$, the rotation of the line PQ will be $\phi'-\phi$. From equations (5) of Art. 76 we have $\tan\phi' = \dfrac{s+\omega+(1+b)\tan\phi}{1+a+(s-\omega)\tan\phi}$, or, neglecting the products of the small quantities,

$$\tan\phi' - \tan\phi = (b-a)\tan\phi + s + \omega - (s-\omega)\tan^2\phi.$$

But $\tan\phi' - \tan\phi$ is $\Delta\tan\phi$, or $\sec^2\phi \cdot \Delta\phi$, where $\Delta\phi = \phi' - \phi$. Hence

$$\Delta\phi = \tfrac{1}{2}(b-a)\sin 2\phi + s\cos 2\phi + \omega. \tag{1}$$

We may put $s = 0$, and then

$$\Delta\phi = \tfrac{1}{2}(e_2 - e_1)\sin 2\phi + \omega. \tag{2}$$

If $\phi = 0$ or $\dfrac{\pi}{2}$, i.e., *if PQ is either axis of the elongation conic, the amount of rotation is* ω—which may serve as a definition of the invariant ω.

It may happen that some line at P is not rotated. Putting $\Delta\phi = 0$, we have $\sin 2\phi = \dfrac{2\omega}{e_1 - e_2}$, which gives either two values of ϕ or none, according as $2\omega <$ or $> e_1 - e_2$.

80. Change of Inclination of Two Lines. If two lines in the unstrained state make angles ϕ, ϕ', with Px, the change in the angle between their directions after strain is $\Delta\phi - \Delta\phi'$, deduced from (1) of last Article. If $\Omega = \phi - \phi'$,

$$\Delta\Omega = (\overline{b-a}\cdot\cos\overline{\phi+\phi'} - 2s\cdot\sin\overline{\phi+\phi'})\sin\Omega. \tag{1}$$

Two rectangular lines with directions $\frac{\pi}{2}+\phi$ and ϕ will, after strain, include an angle $\frac{\pi}{2}-2\sigma$, where

$$2\sigma = (b-a)\sin 2\phi + 2s\cos 2\phi, \qquad (2)$$

or, in other words, 2σ is the cosine of the angle between the strained positions of the two rectangular lines. [Compare equation (1), Art. 77.]

Hence, in particular, putting $\phi = 0$, we see that *the number $2s$ is the cosine of the angle between the strained positions of the axes of reference Px, Py.*

81. Signification of the number s. The meaning of the quantity s can be presented in another way. Let Px, Py (fig. 49) be the unstrained positions of two rectangular axes at P. By strain these lines will become two right lines including an angle whose cosine is $2s$. Suppose that after strain P is brought back to its original position, and also that by a rigid-body rotation (which produces no strain) the line Px is brought back to its original position; then the line Py will occupy the position Py', where $\angle xPy' = \cos^{-1} 2s$. Also any line, IA, parallel to Px will become $I'A'$, also parallel to Px (Cor. 2, Art. 76).

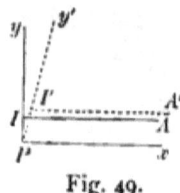

Fig. 49.

From I' let fall $I'p$ perpendicular to Px; then I' has advanced in front of P by the distance Pp, which $= 2s \cdot PI'$; but $PI' = (1+b)PI$,

$$\therefore \frac{Pp}{PI} = 2s(1+b) = 2s, \text{ approximately,} \qquad (a)$$

or in other words, every point in the line PA has advanced or slid forward parallel to PA through a distance which is $2s$ times the distance between PA and Px.

A practical example of such motion is afforded by a book with flexible binding. If one cover is placed on a table and kept fixed (this cover being represented by Px above), the successive pages above it may all be slid forward through distances proportional to their distances from the table (any page being represented by IA above). Such a strain is called *shearing strain*—which is formally defined thus—

When one plane is held fixed in a body, and all planes in the body parallel to it are slid parallel to it, those at one side in the same sense, and each through a distance proportional to its distance from the fixed plane, the strain so produced is called shearing strain.

The fractional amount of sliding, i.e., the ratio of the actual amount of sliding of any plane to the distance between this and the fixed plane, is called the *amount* of the shear.

Since we might equally well have brought the line Py back to its original direction, leaving Px rotated and strained, the number $2s$ expresses equally the amount of shear of lines parallel to Py—it may therefore be described as the shear of the axes Px and Py.

Similarly the number 2σ given in (2) of last Article is the amount of the shear of the lines $(\phi, \phi + \frac{\pi}{2})$.

82. The Strain Ellipse. Any small circle in the neighbourhood of P becomes an ellipse by strain. If we express (ξ, η) in terms of (ξ', η') from equations (5), Art. 76, we obtain approximately
$$\xi = (1-a)\xi' - (s-\omega)\eta',$$
$$\eta = -(s+\omega)\xi' + (1-b)\eta'. \qquad (1)$$

Consider now a small circle of radius r in the unstrained state, with P as centre. Its equation is $\xi^2 + \eta^2 = r^2$, which by (1) gives the relation
$$(\tfrac{1}{2}-a)\xi'^2 - 2s \cdot \xi'\eta' + (\tfrac{1}{2}-b)\eta'^2 - \tfrac{1}{2}r^2 = 0 \qquad (2)$$
between (ξ', η'). The ellipse denoted by this equation is called the *Strain Ellipse*. It is obviously co-axal with the elongation conic.

Every pair of equal rectangular lines through P, in the unstrained state, become conjugate semi-diameters of the Strain Ellipse.

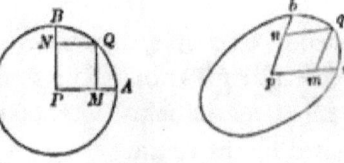

Fig. 50.

Let PA and PB (fig. 50) be two rectangular radii, each equal to r, of a circle described about P; let Q be any point on the circle; and let fall the perpendiculars QM, QN on PA, PB.

Let p, a, b, q, m, n be the displaced positions of $P, A, \ldots$. [The point p is in reality considered as coincident with P; but for clearness of figure they are represented apart.] Then the figure $pmqn$ is a parallelogram (Cor. 2, Art. 76). Also if ϵ is the elongation in the direction PA, we have $pa = (1+\epsilon)PA$, and $pm = (1+\epsilon)PM$; or $\dfrac{PM}{PA} = \dfrac{pm}{pa}$; and similarly $\dfrac{PN}{PB} = \dfrac{pn}{pb}$. But
$$\frac{PM^2}{PA^2} + \frac{PN^2}{PB^2} = 1, \quad \therefore \quad \frac{pm^2}{pa^2} + \frac{pn^2}{pb^2} = 1,$$
the second of which equations is that of an ellipse referred to two semi-conjugate diameters, pa, pb, as axes. [This, of course, furnishes another proof that a circle becomes an ellipse by strain.]

In particular, the semi-axes of the ellipse must be the strained positions of some two rectangular radii of the circle; hence *at every point, P, there are two particular rectangular lines which are strained into two rectangular lines, these latter being the axes of the strain ellipse at P;* and in general these latter are rotated from their original positions.

The *principal axes* of a strain at any point are those two rectangular lines which become the axes of the strain ellipse at the point; and the *principal elongations* at the point are the elongations along the principal axes.

If the axes of reference at P are taken in the principal directions of the strain, the equation (2) of the strain ellipse becomes
$$(\tfrac{1}{2}-e_1)x^2 + (\tfrac{1}{2}-e_2)y^2 - \tfrac{1}{2}r^2 = 0,$$
or
$$\frac{x^2}{(1+e_1)^2} + \frac{y^2}{(1+e_2)^2} = r^2,$$
which is, of course, directly evident.

83. Pure Strain. The strain at a point is said to be *pure strain* if the principal axes (axes of the strain ellipse) are not rotated by the strain.

Now the invariant ω is the amount of rotation of these particular axes (Art. 79). Hence the condition for a pure strain is
$$\omega = 0,$$

or expressed analytically with reference to any pair of axes

$$\frac{dv}{dx} - \frac{du}{dy} = 0,$$

which is the condition that the function $udx + vdy = 0$ should be a perfect differential.

Fig. 51 represents the nature of the displacements round P when the strain is pure. PA' and PB' are the semi-axes of the strain ellipse into which the circle of radius PA is converted, the extremities A and B of two rectangular radii moving along these lines to A' and B'.

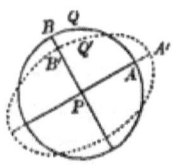

Fig. 51.

The general, or *rotational*, strain would be represented by turning the ellipse round P through an angle whose circular measure is ω, the lines PA' and PA being, of course, no longer coincident in direction. A pure strain is also called an *irrotational* strain.

The values of ξ' and η', equations (5), Art. 76, show that *every strain can be supposed to result from a pure strain followed by a rotation as of a rigid body;* for writing them in the forms

$$\Delta \xi = (a\xi + s\eta) - \omega \eta,$$
$$\Delta \eta = (s\xi + b\eta) + \omega \xi,$$

since ω belongs equally to all points near P, the changes $\omega \eta$ and $\omega \xi$ in ξ and η, respectively, indicate (p. 41) a motion of rotation of a rigid body about P—which is unaccompanied by strain; and the other terms indicate a strain in which the principal axes are not rotated.

If the axes of reference are the principal axes of the strain, $s = 0$, and the parts of $\Delta \xi$ and $\Delta \eta$ which are due to strain are

$$\Delta \xi = e_1 \xi, \quad \Delta \eta = e_2 \eta. \tag{a}$$

Hence the strain proper is produced by lengthening the co-ordinates of all points with respect to the principal axis of x in the ratio $1 + e_1 : 1$, and the co-ordinates parallel to the principal axis of y in the ratio $1 + e_2 : 1$.

Now a *simple elongation* of a body in a direction normal to a plane consists in the drawing out of each particle from the plane through a distance proportional to the natural distance of the particle from the plane. Hence equations (a) express two simple elongations parallel to the principal axes of the strain; so that—

Every small strain at a point can be produced by two successive simple elongations, followed by a rotation, as of a rigid body, about the point.

Considering only the strain proper, i. e., the pure portion, or neglecting the mere rotation, the strained position of any point, Q, on the circle (fig. 51) is easily found. For, if r is the radius of the circle, and ϕ the angle made by PQ with PA, the co-ordinates of Q are $\xi = r \cos \phi$, $\eta = r \sin \phi$; and if Q' is the point on the strain ellipse to which Q comes, the co-ordinates of Q' are $\xi' = (1 + e_1) r \cos \phi$, $\eta' = (1 + e_2) r \sin \phi$; and as the equation of the ellipse is $\dfrac{x^2}{(1+e_1)^2} + \dfrac{y^2}{(1+e_2)^2} = r^2$, we see that ϕ is the excentric angle of the point Q'. Hence producing PQ to R so that $PR = PA'$, a perpendicular from R on PA meets the ellipse in Q'.

The essential elements of a strain, therefore, are *elongation* (a, b) and *shear* (s); and they are not to be confounded with the other magnitude (ω) which enters into the expressions for the changes of relative co-ordinates (Art. 76), this last being mere rigid-body-rotation.

84. Case of no Expansion. When the invariant $a + b$ is zero, the areas of all small figures in the neighbourhood of P are unaltered by strain. The elongation conic becomes in this case a rectangular hyperbola, and it is, of course, accompanied by another rectangular hyperbola of compression (Art. 78).

Now we know that the equation $ax^2 + 2sxy - ay^2 = k^2$ becomes of the form $a(x^2 - y^2) = k^2$, if the axes of the curve are those of co-ordinates, and $a = \sqrt{a^2 + s^2}$. But this latter form of the equation of the conic corresponds to an elongation, a, along one axis, accompanied by a compression, a, along the other.

Again, if instead of the axes of the curve we take its *asymptotes* as the axes of co-ordinates, we know that its equation becomes $\qquad 2\sqrt{a^2 + s^2} \cdot xy = k^2,$

Shearing Strain.

which indicates a strain consisting wholly of *shear*, whose amount (Art. 81) is $2\sqrt{a^2+s^2}$.

Hence (see Cor. 2, Art. 77) any non-expansional strain $(a, -a, s)$ can be replaced by a non-expansional strain $(a, -a, 0)$ with no shear, with reference to axes properly chosen; or by a simple shearing strain $(0, 0, 2a)$, with reference to axes properly chosen.

Conversely, a simple shearing strain is not attended with expansion.

Let us consider geometrically the nature of the strain produced by an elongation of any amount, $a-1$ (no longer assumed to be small) along a line Px (fig. 52), and an equal compression along the perpendicular line Py.

Consider the line PA which is drawn making $\angle APx = \tan^{-1}a$.

Draw Pa at an angle of $\dfrac{\pi}{4}$ with Px.

Then if a parallel, Aa, to Px be drawn from any point, A, on the line PA, the point A will, by the elongation, viz., $a-1$, parallel to Px, be

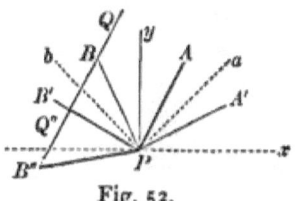

Fig. 52.

drawn out to a. Again, by the compression in the direction yP, the point a will be brought to A', such that aA' is parallel to yP and $PA' = PA$. For if (x, y) are the co-ordinates of A, (ax, y) are the co-ordinates of a, and $\left(ax, \dfrac{y}{a}\right)$ are those of A'; but $y = ax$, ∴ A is (x, ax) and A' is (ax, x), ∴ $PA' = PA$, and all lengths measured along PA remain unaltered. Evidently PA' and PA are equally inclined to Pa.

Again, considering the analogous line PB in the left hand quadrant, i.e., such that $\angle BPy = \angle APy$, the length PB is unaltered, and B comes to B', such that PB' is perpendicular to Pa. Introduce a rigid-body-rotation to bring PA' back to PA, and this will bring B' to B''; and it is easy to see that BB'' is parallel to PA, so that the point B is slid forward parallel to the plane PA through the distance BB''. Since all parallel lines are elongated or shortened in the same ratio, it follows that all lines parallel to PA remain unaltered in length. Hence if Q is any point on BB'', and Q'' the position of Q after strain, $B''Q'' = BQ$, so that Q and all points on the line BB'' slide through the same distance. The strain is therefore equivalent to a shear. If p is the perpendicular distance between BB'' and PA, the amount of the shear $= \dfrac{BB''}{p}$, which is easily proved to be equal to

$$a - \frac{1}{a}. \tag{1}$$

Neglecting the effect of pure rotation, the strain could otherwise be produced by holding fast the other unelongated line, PB, and sliding all lines parallel to it.

The lines bisecting the angle between the lines, PA and PB, of zero elongation are called the axes of the shear. They are Ox and Oy in the figure.

If the strain is small, $a = 1 + s$, where s is a small quantity, and $\dfrac{1}{a} = 1 - s$, so that the amount of the shear

$$= 1 + s - (1 - s) = 2s,$$

which agrees with previous results.

The expression for a shearing strain if the axis of x is taken coincident with one of the unelongated lines (held fixed) becomes $u = 2sy, \quad v = 0.$

85. Strain Potential. If at every point of the strained body the rotation, ω, is zero, the components, u, v, of the absolute displacement of every point are such that $udx + vdy$ is the differential of some function of the co-ordinates, $x, y,$ of the point (Art. 83). Denoting this function by $\phi(x, y)$, or simply by ϕ, we have at all points when the strain is pure

$$udx + vdy = d\phi.$$

The direction-cosines of the normal to the curve $\phi(x, y) =$ a constant being proportional to $\dfrac{d\phi}{dx}$ and $\dfrac{d\phi}{dy}$ are proportional to u and v; hence the absolute displacement of every point P takes place along the curve passing through the point and having for equation $\phi = C = $ a constant.

The function ϕ is called the *potential-function* of the strain, and the curves obtained by varying the constant in the equation $\phi = C$ are called curves of equal potential.

Into this subject we shall not further enter here, because it will be fully discussed in the next chapter.

EXAMPLES.

1. Prove analytically that the shear of any two rectangular lines intersecting at any point is equal to the difference between the elongations along the internal and external bisectors of the angle between them.

2. In any case of strain what is the line at any point which is most rotated by the strain? [The rigid-body-rotation is to be discarded.]

3. Represent graphically the shear of any two rectangular lines $(\phi, \phi + \frac{\pi}{2})$ at a point by the length of a radius vector drawn from the point along one of them.

Ans. Taking the principal axes of the strain as those of reference, the curve traced out by the extremity of the radius vector will be

$$r = \tfrac{1}{2}(e_2 - e_1)\sin 2\phi.$$

4. Prove that a simple elongation along any line is equivalent to a uniform areal expansion and a shear having the given line for one of its axes.

[Let the strain at P be a simple elongation ϵ along Px. This is equal to two superposed simple elongations each equal to $\tfrac{1}{2}\epsilon$. Introduce an elongation $\tfrac{1}{2}\epsilon$ along Py and a compression $\tfrac{1}{2}\epsilon$ in the sense yP. Then we shall have an elongation $\tfrac{1}{2}\epsilon$ along Px and an elongation $\tfrac{1}{2}\epsilon$ along Py (which make uniform areal expansion, i.e., their strain ellipse is a circle), together with an elongation $\tfrac{1}{2}\epsilon$ along Px and an equal compression along yP (which make a shear, Art. 84).]

5. Resolve a simple elongation, ϵ, along any line into its component (and therefore equivalent) elongations and shears with respect to two rectangular axes.

Ans. If ϕ is the angle made by the line with Px, the components are $\epsilon \cos^2\phi$, $\epsilon \sin^2\phi$, $\epsilon \sin 2\phi$—the last being the shear.

[Draw the line through P perpendicular to the given direction of elongation; then the perpendicular from Q on the first $= \xi \cos\phi + \eta \sin\phi$; and by the given elongation Q is drawn out along this perpendicular through a distance $\epsilon(\xi \cos\phi + \eta \sin\phi)$, whose projections on the axes of x and y are $\epsilon \cos\phi(\xi \cos\phi + \eta \sin\phi)$, or $\Delta\xi$, and $\epsilon \sin\phi(\xi \cos\phi + \eta \sin\phi)$, or $\Delta\eta$. Then comparing these values of $\Delta\xi$ and $\Delta\eta$ with those in Art. 76, we have the result.]

6. Find the strain ellipse resulting from the superposition of two small strains (a, b, s) and (a', b', s').

7. Given the components of a strain with reference to its principal axes, write down its components with reference to any two rectangular axes.

8. Prove the following construction for the displacement (due to pure strain) of any point—

At any point, Q', of the strain ellipse draw a tangent; from its centre, P, draw a perpendicular, PT, on the tangent; bisect $Q'T$ in M; then the line PM meets the circle from which the ellipse has come in the point Q whose strained position is Q'.

9. Find how much the moment of inertia of any small area at P about a line through P fixed in space is altered by strain.

[Take the fixed line as axis of x; let k be the original radius of gyration about Px; let the product of inertia of the area (S) about Px and Py be $S.p^2$; and let λ be the radius of gyration of the strained area about Px; then
$$\lambda^2 = (1 + 2b)k^2 + 2(s+\omega)p^2.]$$

10. Find the change in the moment of inertia about an axis through P perpendicular to the plane of the strain.

Ans. If Λ = radius of gyration after strain, K = radius of gyration before strain, k and k' radii of gyration round axes of x and y before strain, $\quad \Lambda^2 = K^2 + 2(bk^2 + ak'^2 + 2sp^2)$.

It follows that the quantity in brackets must be the same for all pairs of rectangular axes at P. It will be good exercise for the student to verify this. [See Art. 77.]

11. Prove that the centroid of any small area becomes by strain the centroid of the strained area.

12. If (r, θ) are the polar co-ordinates of any point P in a lamina, prove that $\phi \equiv Ar^n \sin n\theta$, where A is a constant, is a possible strain potential, and that the strain is at each point a shear.

[We easily find
$$a = -b = n(n-1)Ar^{n-2}\sin(n-2)\theta; \quad s = n(n-1)Ar^{n-2}\cos(n-2)\theta.]$$

13. In the last example prove that the equation of the curve of greatest elongation drawn through P is

$$r^{\frac{n}{2}}\sin\frac{n}{2}\left(\theta - \frac{\pi}{2n}\right) = C. \qquad \text{Prof. Townsend.}$$

[The elongation along a line at P making an angle λ with r is
$$n(n-1)Ar^{n-2}\sin(n\theta + 2\lambda).]$$

CHAPTER VI.

Kinematics of Fluids.

SECTION I.—**General Properties.**

86. Nature of a Fluid. The characteristics of bodies, or of classes of bodies, shade into each other by such gradual transitions, that it is impossible to mark off these classes (of *actually* existing bodies) by any strictly logical definitions. We may describe classes of bodies in a general way, or for such practical purposes as we may have in view at any instant, by seizing on certain well-marked characteristics, or properties which they possess in a more striking degree than other classes. But there is another and more definite course open to us. We may logically define certain *ideal* (non-existent) classes, to which the actually existing bodies approximate, and we may suppose these latter to be endowed with the ideal characters in our definition—subsequently correcting our theoretical results by the consideration of small departures from the properties assumed in the definition.

Every one easily recognises a broad distinction between a *Fluid* and a *Solid*. The former would be roughly described as a body such that its particles can be *very easily* moved about on each other; while the solid is such that its particles can be so displaced only with difficulty.

Very little reflection is sufficient to show how vague and unsatisfactory is such a definition of a fluid. Precision, therefore, compels us to define a body to which a fluid approximates. Such a body we call a *perfect fluid;* and it must be defined with reference to the nature of *stress* produced in it. The definition which we shall give is this—

A *Perfect Fluid* is a body such that, however it may be strained, the stress on every element plane throughout it is a normal pressure.

87. Compressibility. Consider any volume, v, of a body, and suppose that all over the surface of this volume we exert a uniform pressure of intensity p. For definiteness, we may imagine v to be measured in cubic centimètres, and p to be, therefore, estimated in force-units per square centimètre—say in kilogrammes per square centimètre. The effect of this pressure will be to diminish the volume to v', so that the absolute diminution of volume is $v-v'$, and the fractional amount per cubic centimètre is $\frac{v-v'}{v}$. Since this compression is produced by p units of force, the fractional compression per unit of force is
$$\frac{1}{v}\cdot\frac{v-v'}{p}.$$

Now for some fluids this quantity is extremely small, while for others it is very great. For example, if the body is water or mercury, it is found that even when p is extremely great, $v-v'$ is exceedingly small, so that the above expression for the compressibility is infinitesimal.

Again, if the body considered is air, oxygen, or hydrogen, it is found that for comparatively small values of p the fractional change of volume, $\frac{v-v'}{v}$, is great.

These two characteristics require us to divide fluids, for practical purposes, into two classes—*incompressible* and *compressible*, respectively—the first class being in strictness (like *perfect* fluids themselves) ideal, but closely approximated to, when the intensity of pressure applied to them is not enormous, by *liquids*. We shall therefore use the term *liquid* in the sequel to denote an *incompressible fluid*. To the compressible class belong *gases*.

In a liquid, therefore, the density is not altered by pressure, since by such means it is impossible to squeeze a given volume of it into a smaller space.

88. Uniplanar motion of a fluid. In accordance with

the plan of this work, we shall deal almost exclusively with fluid motion which is exactly the same in parallel planes, however much it may vary from point to point in any one of these planes. We shall therefore assume that the motions which we consider take place in the plane xy—there being, therefore, no displacements in the direction perpendicular to this plane.

89. Statistical and Historical points of view. Consider a fluid of infinite extent, i. e., occupying the whole of the plane xy. Then there are two ways of regarding the motion of the fluid—(1) we may fix our attention permanently on a *fixed point*, P (fig. 53), in space, and

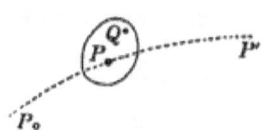

Fig. 53.

keep a continuous record of the velocities, directions of velocity, etc., which take place at this point, these velocities etc. being those of fluid particles as they pass through P; but of the series of velocities which any one of these particles will subsequently have we keep no record; or (2) we may fix our attention on an individual fluid particle, P_0, trace this particle all through its path, $P_0 P P'$, and keep a record of its successive velocities.

In the first case what we have done for the fixed-space point P we imagine to be done for all fixed-space points; and in the second case we imagine our record to be similarly kept for every individual fluid particle.

The first method is sometimes described as the *Statistical* method, and the second as the *Historical* method. They are also often called the Eulerian and the Lagrangian methods, respectively[1].

If we adopt the first method, it will be proper to speak of the velocity *at* the point P; so that if v is the velocity at P at the time t, v will be some function of x, y, t (where x, y are

---

[1] A very familiar instance of the nature of each of these methods is furnished by what is often seen at a great railway station. A policeman stationed at the place of exit noting the numbers of the cabs as they pass him serves to illustrate the first of the above methods; while the cabman, who remains with the cab throughout its motion, illustrates the second.

the co-ordinates of the fixed point P) of the form
$$v = f(x, y, t).$$
If we adopt the historical method of treating the motion, the velocity of the particle in the position P will be some function of t and the co-ordinates, x_0, y_0, of this fluid particle in some previous position, P_0; so that in this case we should have
$$v = F(x_0, y_0, t).$$
In the first method there is no such thing, for example, as $\frac{dx}{dt}$, because x is an invariable quantity. In the second method, on the contrary, $\frac{dx}{dt}$ is the component velocity of the particle at P along the axis of x.

90. Steady Motion. If after any time all the particles which pass through P successively pass through it with the same velocities and accelerations (both in magnitude and in direction), the record of the motion at P becomes constant, and there is said to be *steady motion* at P. If the same thing occurs at every other point of space, the whole motion is steady—there will then be different velocities at different points, but the velocity at each point will remain constant through time. In this case we have simply
$$v = \chi(x, y),$$
a function independent of t.

91. Line-Integrals and Surface-Integrals. Let Q be any quantity whose value depends on x, y, the co-ordinates of a point, i.e., of the form $f(x, y)$; then if we take the values of Q at all points of any curve, AB, multiply the value of Q at any point by ds, the element of arc at the point, and add these products together, the sum is called the *line-integral* of Q along the curve. The analytical expression is
$$\int Q \, ds.$$
If the curve is *closed*, so that the extreme points A and B coincide, this line-integral may or may not be zero, according to the form of the function Q. Thus, for example, let the closed curve be a circle, the axes being at its centre, and

suppose Q to be $\frac{y}{x}$; then $\int \frac{y}{x} ds = r \int_0^{2\pi} \tan\theta \, d\theta = 0$.

But if Q is $\tan^{-1}\frac{y}{x}$, the line-integral is $r\int_0^{2\pi} \theta \, d\theta$, which is $2\pi r^2$.

Again, take any area, multiply the value of Q at each point by the element of area, dS, at the point, and integrate the product over the whole area. The value of this integral is called the *surface-integral* of Q over the area; and the expression for it is

$$\int Q \, dS.$$

In the sequel we shall use dS to represent an element of *area* and ds to represent an element of *length*.

92. Flux across any curve. Let AB (fig. 54) be any curve in the plane xy. We propose to find the mass of fluid, per unit of time, which flows at any instant across this curve.

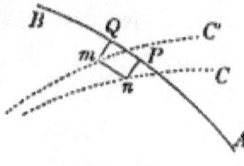

Fig. 54.

Take any point P on the curve; let v be the component of the velocity at P along the normal, nP, to the curve; take the length nP equal to $v \cdot \Delta t$, where Δt is an indefinitely small element of time; take a point, Q, along the curve at a very small distance, ds, from P; and construct the small rectangle $PQmn$. The velocity along nP of the particle at n is $v + \epsilon$, where ϵ is infinitesimal.

Now in the time Δt, the particle at n will have travelled in the direction nP through a distance equal to $(v+\epsilon)\Delta t$, or $v \cdot \Delta t$. [This particle may be moving along a curve nC, very obliquely to the normal nP at the instant considered; we say nothing about the *total* distance through which it goes in the time Δt; but the *component* of this distance along nP is $v \cdot \Delta t$.] But $v \cdot \Delta t = nP$, therefore the particle at n will have just reached the surface at the end of the time Δt; so will the particle at m, and all particles in the line mn, and all the other particles in the area $mnPQ$ will have *passed over* the curve; but those at a greater distance than Pn will not have reached the curve AB in the time Δt.

If we draw a plane parallel to the plane of the figure at a

142   *Kinematics of Fluids.*

unit distance above it, and imagine a cylinder constructed on AB with sides perpendicular to the plane of motion, and if ρ is the density of the fluid at P at the instant considered, the mass of fluid standing on the rectangle $mnPQ$ is

$$\rho \times nP \times ds, \text{ or } \rho v ds \cdot \Delta t,$$

and this is the mass which has passed over the length ds of the curve AB at P in the time Δt; therefore the mass of fluid which crosses the whole curve AB in this time is

$$\Delta t \int \rho v ds,$$

and therefore the mass which crosses it per unit of time is

$$\int \rho v ds$$

at the instant considered, the integration extending from one end of the curve to the other. If the motion is steady, v will be the same at all times at any point, P, and the flux will be constant.

93. Equation of Continuity. The change in density which takes place in any small interval of time, Δt, inside any element of volume (i.e., space-element) is due to the fluid which has passed in through and that which has passed out through the boundary of the element of volume, unless in this interval some fluid has been *created* inside the element of volume or unless some has been *annihilated*. The expression of this condition of non-creation and non-annihilation furnishes a relation holding universally between the density and the velocity at a point. Choose as the element of volume considered a small rectangle, $PQP'Q'$ (fig. 55), whose sides are parallel to the axes.

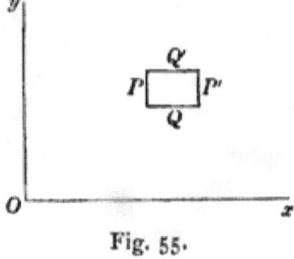

Fig. 55.

Let P be the middle point of the side parallel to Oy; P' the middle point of the opposite side; Q the middle point of the side parallel to Ox; Q' the middle point of the opposite side; ρ the density at P, where the component velocities parallel to the axes are u, v; ρ' the density at Q, where the components are u', v'.

Equation of Continuity.

Then the mass which flows *into* the area through the side P per unit of time is, by last Article,

$$\rho u\, dy;$$

and since the value of ρu at P' is $\rho u + \dfrac{d(\rho u)}{dx} dx$, the flux *out* through the side P' is

$$\left[\rho u + \frac{d(\rho u)}{dx} dx\right] dy.$$

Hence the time-rate of loss of fluid from the x-motion is $\dfrac{d(\rho u)}{dx} dx\, dy.$

Similarly the time-rate of loss from the y-motion is

$$\frac{d(\rho' v')}{dy} dx\, dy.$$

But if $\rho v = f(x, y)$, we have evidently

$$\rho' v' = f(x + \tfrac{1}{2} dx,\ y - \tfrac{1}{2} dy);$$

so that neglecting quantities like $dx^2 dy$ in comparison with $dx\, dy$, we have $\rho' v' = \rho v$, and the total flux of mass outwards is

$$\left[\frac{d(\rho u)}{dx} + \frac{d(\rho v)}{dy}\right] dx\, dy,$$

and the actual loss of fluid in the time Δt is this quantity multiplied by Δt.

But at the time t the mass standing on the little area (and having unit height) is $\rho\, dx\, dy$, neglecting infinitesimals as above; and at the end of the interval Δt, the mass standing on the area is $\left(\rho + \dfrac{d\rho}{dt} \Delta t\right) dx\, dy$; therefore in the interval Δt there is a *gain* of mass equal to $\dfrac{d\rho}{dt} dx\, dy\, \Delta t$, so that the *loss* per unit of time has the analytical expression $-\dfrac{d\rho}{dt} dx\, dy$.

Assuming, then, that this loss is due to the fluid which has passed through the boundary of the element, or in other words that no creation and no annihilation has taken place inside the element, we have

$$\frac{d\rho}{dt} + \frac{d(\rho u)}{dx} + \frac{d(\rho v)}{dy} = 0. \qquad (a)$$

This equation is invariably called the *equation of continuity*—

a term which is scarcely the best that could be chosen to express the idea.

94. Local and Total Time-Rates of Change. It is very important to understand the difference between the time-rate of increase of any quantity related to the moving fluid, according as we confine our attention to the successive states of affairs at one and the *same point fixed in space* or to one and the *same fluid particle*. We shall illustrate this by considering the density.

The fluid density at a fixed point P will, unless the fluid is homogeneous and of density not altering with the time, be a function of the position of P and of the time t; so that we may write
$$\rho = f(x, y, t). \tag{a}$$

The density at this same point at the end of the time $t+dt$ will be
$$f(x, y, t+dt),$$
so that there is a change of density at P of amount $\frac{d\rho}{dt}.dt$, or at the rate $\frac{d\rho}{dt}$ per unit of time.

This quantity $\frac{d\rho}{dt}$ is, therefore, the local *time-rate of increase of density at P*.

The moving fluid may be an incompressible one of different densities in different parts, and these different parts may pass over P as time goes on. Hence there is at each instant a local time-rate of increase of density at P; and we cannot express the incompressibility of the fluid by putting $\frac{d\rho}{dt}$ equal to zero.

Consider now the time-rate of increase of density *of the fluid particle* which is at P at the time t.

At the end of the time $t+dt$, the co-ordinates of this particle will be $x+udt$, $y+vdt$; and the density at this point will be by (a)
$$f(x+udt, y+vdt, t+dt),$$
which will be the (perhaps altered) density of our particle. Denote this new density by $\rho + D\rho$; then
$$\rho + D\rho = f(x+udt, y+vdt, t+dt);$$
$$= \rho + (u\frac{d\rho}{dx} + v\frac{d\rho}{dy} + \frac{d\rho}{dt})dt;$$

$$\therefore \frac{D\rho}{dt} = u\frac{d\rho}{dx} + v\frac{d\rho}{dy} + \frac{d\rho}{dt}, \quad (\beta)$$

which is the rate of change of density of the particle. We shall denote $\frac{D\rho}{dt}$, for conciseness, by $D_t\rho$. In the case of a liquid, therefore, the fact of its incompressibility is expressed by the equation $D_t\rho = 0$.

95. Equation of Continuity for a Liquid. As we have just said, for an incompressible fluid $D_t\rho = 0$. Now the equation of continuity, (a), Art. 93, is obviously the same as

$$\frac{d\rho}{dt} + u\frac{d\rho}{dx} + v\frac{d\rho}{dy} + \rho\left(\frac{du}{dx} + \frac{dv}{dy}\right) = 0, \text{ or}$$

$$\frac{1}{\rho}D_t\rho + \frac{du}{dx} + \frac{dv}{dy} = 0. \quad (1)$$

Hence the equation of continuity becomes for a liquid

$$D_t\rho = 0, \quad (a)$$

$$\frac{du}{dx} + \frac{dv}{dy} = 0. \quad (\beta)$$

96. Components of Relative Velocity of two close points. Let the components, parallel to the axes of x and y, of the velocity at P (fig. 53, p. 139) be u and v at the time t, and let ξ and η be the co-ordinates of a very close point Q with reference to axes drawn at P parallel to the fixed axes. Then u and v will be functions of x, y, t of the forms

$$u = f_1(x, y, t), \quad v = f_2(x, y, t); \quad (1)$$

and if at the *same time*, t, u' and v' are the components of the velocity at Q,

$$u' = f_1(x + \xi, y + \eta, t), \quad v' = f_2(x + \xi, y + \eta, t).$$

Hence
$$u' - u = \xi\frac{du}{dx} + \eta\frac{du}{dy},$$

$$v' - v = \xi\frac{dv}{dx} + \eta\frac{dv}{dy}.$$

It is to be observed that u and v are not now (as they were in the theory of small strains) small *displacements* of a particle, but *velocities* of any magnitudes. The equations just written are to be put into the forms (4), p. 122 by assuming, as previously,

$$\frac{du}{dx} = a, \quad \frac{dv}{dy} = b, \quad \frac{dv}{dx} + \frac{du}{dy} = 2s, \quad \frac{dv}{dx} - \frac{du}{dy} = 2\omega,$$

so that denoting the components of *relative* velocity of Q with respect to P by $\dot{a}$, $\dot{\beta}$, we have

$$\dot{a} = a\xi + s\eta - \omega\eta, \quad (2)$$
$$\dot{\beta} = s\xi + b\eta + \omega\xi. \quad (3)$$

Now (Art. 29) the terms $-\omega\eta$ and $\omega\xi$ in these equations are the components of velocity due to a rigid-body rotation, with angular velocity ω, round an axis through P perpendicular to the plane of motion; and the terms $a\xi + s\eta$ and $s\xi + b\eta$ are the components of velocity due to a pure strain at P (Art. 83).

Hence if we consider what happens in any *small element* of the fluid surrounding P, we see that, no matter *how* the motion of the fluid is produced,—

The relative velocity of any particle in the element, with respect to P, at any instant, is compounded of velocity resulting from two causes, viz.—

 (a) Rotation *of the element, as a rigid body, about an axis through P,* and

 (b) Pure Strain *of the element about P as a fixed point.*

Observe that this is stated with respect to the *relative* velocity of every particle in the element with respect to P. To get the *absolute* velocity of each point, Q, in the element, we must combine with this relative velocity the absolute velocity of P; so that the components of absolute velocity of Q are $u + \dot{a}$, $v + \dot{\beta}$—as is obvious from the definitions of $\dot{a}$ and $\dot{\beta}$.

97. Expansion. The *expansion* at any point of a fluid in motion is defined to be *the ratio which the time-rate of increase of any small volume at the point bears to that volume.* If the small volume selected is V, the expansion is $\frac{1}{V}\frac{dV}{dt}$, which will vary from point to point, and also (unless the motion is steady) from time to time.

Taking the small volume at P (fig. 55, p. 142) to be one

standing on the rectangle $dxdy$ with any height, it is easily seen that

$$V + \Delta V = V\left[1 + \left(\frac{du}{dx} + \frac{dv}{dy}\right)\Delta t\right],$$

$$\therefore \frac{1}{V}\frac{dV}{dt} = \frac{du}{dx} + \frac{dv}{dy}. \qquad (a)$$

The expansion is, then, $\frac{du}{dx} + \frac{dv}{dy}$, which we shall denote by θ.

The meaning of the equation of continuity, (β), Art. 95, for a liquid is, therefore, that at every point the expansion is nothing—which, of course, is necessarily the case since a liquid is incompressible.

The above equation (a) is exactly the same as (1), Art. 95; for since the mass of an element does not alter, $V\rho$ is constant, so that $\frac{1}{V}D_t V = -\frac{1}{\rho}D_t \rho$.

98. Rotation. Vortex Motion. If ω, the angular velocity, or, as we shall call it, the *rotation* or *spin*[1], of the fluid element at P is not zero, the motion is said to be *vortex motion* at P. It is also called *rotational* motion.

It may happen that all through certain limited regions of the fluid the rotation is not zero, while throughout the remainder of the fluid it is zero; then the regions in which ω exists are called vortex regions, i.e., the motion at each point of such a region is vortex motion. It is of fundamental importance that the student should clearly understand what it is that constitutes vortex motion. Such motion does not *necessarily* result from revolution about an axis. For example, a mass of water may be conceived to whirl round an axis in such a way that *though every particle of the water describes a circle round the axis, the rotation ω which constitutes vortex motion at each point is zero all through the liquid.* This will be proved a little farther on.

The essential thing to observe now is that the 'rotation' which constitutes vortex motion is not any angular velocity of the fluid particles about an axis fixed in space, but an *element-*

---

[1] The term *spin* is used by Clifford. With him the spin is a vector indicating at once the axis of the rotation and its magnitude.

rotation at each point, i.e., a rotation of a small element of the fluid about an axis drawn through a mean point inside the element.

Vorticity at any point does not depend on *velocities*, u, v, at the point, but on *rates of increase* of these velocities—its measure being $\dfrac{dv}{dx} - \dfrac{du}{dy}$—just as the sign or amount of the electrification at any one point of a conductor acted on by any inducing charges cannot be inferred from the value of the potential on the surface of the conductor, but depends on the *rate* at which this potential changes (both in magnitude and in sign) as we move outwards from the conductor along the normal at the point.

It will be observed that we are taking positive rotation (positive vorticity) in the sense from $+x$ to $+y$; and this sense we shall, for shortness, frequently denote by the notation $\overset{y}{\underset{x}{\uparrow}}$.

99. Irrotational Motion. At every point in the fluid at which $\omega = 0$, the motion is called *irrotational motion*.

At such points since $\dfrac{du}{dy} - \dfrac{dv}{dx} = 0$, the expression

$$u\,dx + v\,dy$$

is the differential of some function of x, y, and t. We shall denote this function by ϕ, so that in the non-vortical regions we have $\quad u\,dx + v\,dy \equiv d\phi(x, y, t),\ $ or $\ \equiv d\phi$, simply.

Of course if the motion has become steady, ϕ will not involve t.

The function ϕ is called the *velocity potential* of the motion; and we see that it always exists in those regions of the fluid where the vortical spin is zero.

If ϕ_1 is the value of ϕ at P, we may construct the curve, PA, whose equation is $\qquad \phi = \phi_1$.

This is called an *equipotential curve*.

We may at each instant map out the whole irrotational region by drawing a series of very close equipotential curves,

PA, $P'B$, $P''C$, ... and if the motion is steady this map will not alter with time.

100. Theorem. *The component of velocity at any point of an irrotational region along any direction is the line-rate of increase of the velocity potential in this direction.*

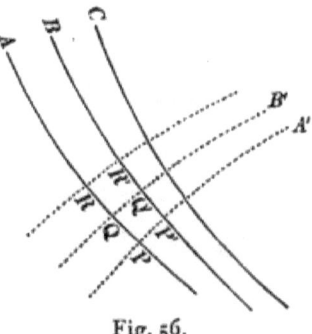

Fig. 56.

This is at once evident by taking the axis of x parallel to the direction; or it may be seen thus. Let P be the given point, the direction being PP', and P' very close to P. Let $PP' = ds$, and let the projections of PP' along the axes be dx and dy.

Then the velocity at P resolved along PP' is

$$u\frac{dx}{ds} + v\frac{dy}{ds};$$

but $u = \dfrac{d\phi}{dx}$, $v = \dfrac{d\phi}{dy}$, therefore this component is

$$\frac{d\phi}{dx}\frac{dx}{ds} + \frac{d\phi}{dy}\frac{dy}{ds}, \quad \text{or} \quad \frac{d\phi}{ds},$$

which is the line-rate of increase of ϕ as we go from P to P'.

Observe also that the *sense* of the component velocity along any line at a point *is the sense in which the velocity potential* **increases** *along the line*.

COR. 1. *The resultant velocity at any point in an irrotational region is in the direction of the normal at the point to the equipotential curve passing through it.*

For if P is the point (fig. 56), and PA the equipotential curve through it, the component velocity in the direction PQ is zero, since ϕ is the same at Q as at P. Hence the resultant velocity at P takes place along the normal, PP', to the curve PA.

If ϕ_P denotes the value of ϕ along the curve PA and $\phi_{P'}$ the value along $P'B$, the resultant velocity at P is approximately

$$\frac{\phi_{P'} - \phi_P}{PP'},$$

and the approximation will be closer the nearer the curves PA and $P'B$ are drawn to each other. If dn denotes the element of normal to the equipotential curve at P, the resultant velocity is
$$\frac{d\phi}{dn},$$
and the sense of its vector is that in which ϕ *increases*.

COR. 2. The component of velocity normal to *any* curve drawn through P is $\frac{d\phi}{dv}$, where dv is an elementary length measured along this normal.

COR. 3. *The fluid moves from regions of lower to regions of higher velocity potential.* Consequently, as we follow the course of a *stream-line* (see next Article) in the direction of motion the velocity potential continually increases; and if it should happen, as it may in some cases, that the stream-line returns to the point from which it started, forming a closed curve, it will follow of necessity that the potential at this point has more values than one. [This always happens when there are vortices present.]

101. Lines of Flow and Stream-lines. If at P we draw an element of length PP' in the direction of the resultant velocity at P, and then at P' continue the line in the direction, $P'P''$, of the resultant velocity at P', and so on, we get a continuous curve, PA', such that at every point on it the resultant velocity at the point is directed along the tangent to the curve. Such a curve is called a *Line of Flow*. Similarly at any other point, Q, we can draw a line of flow, QB'; and we can in this way map out the whole region by drawing lines of flow.

Every line of flow cuts every equipotential curve which it meets at right angles; for at each point the resultant velocity is along the tangent to the line of flow and along the normal to the equipotential curve.

Lines of flow exist in all parts of the fluid, whether vortex regions or irrotational regions; for every particle of the fluid must at any instant be moving in some definite direction.

When the motion becomes steady, each line of flow becomes

the *actual path of a fluid particle*, which is called a *stream-line*. If the motion is not steady, the map of lines of flow will change from instant to instant, so that the actual path of any particle which lies at the time *t* on a line of flow merely *touches* the line of flow at the point without coinciding with it[1].

If at any instant u and v are the components, parallel to the axes, of the velocity at the point x, y, the differential equation of the line of flow is

$$\frac{dy}{dx} = \frac{v}{u}, \text{ or}$$

$$u\,dy - v\,dx = 0. \tag{a}$$

In the case of a liquid the left-hand side of this equation is the exact differential of a function, $\psi(x, y)$, of x and y; for the condition that it should be is $\frac{du}{dx} + \frac{dv}{dy} = 0$, which is the equation (β) of Art. 95.

The function ψ is called the *function of flow*.

Thus, then, if the moving body is a liquid, and if its motion is steady, there exists at every point, whether in a vortex region or not, a flow function, ψ; and in the irrotational regions there exists at each point a velocity potential function, ϕ, in addition; and the equations

$$\phi = C,$$
$$\psi = C',$$

C and C' being any constants, denote a system of orthogonal curves, i. e., *every* curve of one system cuts orthogonally *every* curve of the other system which it meets.

102. Differential Equations for ϕ and ψ. In the case of a moving liquid, whether homogeneous or not, ϕ exists in irrotational regions and ψ throughout the whole. Now putting $\frac{d\phi}{dx}$ and $\frac{d\phi}{dy}$ for u and v in equation (β) of Art. 95, it becomes

---

[1] Clifford (*Kinematic*, p. 199) gives a good illustration of the relation and distinction between lines of flow and stream-lines in a case of unsteady motion: 'If a rigid body move about a fixed point, we know that its velocity-system at every instant is that of a spin about some axis through the fixed point, and consequently the lines of flow are circles about that axis. But in general the axis changes as the motion goes on, and the path of a particle of the body is not any of these circles.'

$$\frac{d^2\phi}{dx^2} + \frac{d^2\phi}{dy^2} = 0, \text{ or}$$

$$\nabla^2 \phi = 0, \tag{a}$$

where ∇^2 stands for $\frac{d^2}{dx^2} + \frac{d^2}{dy^2}$.

Again, since $u = \frac{d\psi}{dy}$ and $v = -\frac{d\psi}{dx}$, we have

$$\frac{d^2\psi}{dx^2} + \frac{d^2\psi}{dy^2} = \frac{du}{dy} - \frac{dv}{dx} = -2\omega \text{ (Art. 96)},$$

$$\therefore \quad \nabla^2 \psi = -2\omega, \tag{b}$$

where ω is the rotation at the point to which ψ belongs; so that in the regions in which the velocity potential, ϕ, exists (i.e., where $\omega = 0$) we have

$$\nabla^2 \psi = 0, \tag{c}$$

that is, both ϕ and ψ satisfy the same equation (Laplace's equation).

If the fluid is compressible, (a) does not hold in the irrotational regions; but by putting $\frac{d\phi}{dx}$ and $\frac{d\phi}{dy}$ for u and v in (1) of Art. 95, we have

$$\nabla^2 \phi = -\frac{1}{\rho} D_t \rho, \text{ or} \tag{d}$$

$$\nabla^2 \phi = \theta, \tag{e}$$

where θ is the expansion.

In this case (of compressibility) $u\,dy - v\,dx$ is not the differential of any function, so that the function ψ does not exist—although stream-lines, of course, exist. [See example 20, following.]

All cases of fluid (i.e., liquid or gas) motion may, however, be represented by the following[1] values of u and v:—

$$u = \frac{dP}{dx} + \frac{dN}{dy},$$

$$v = \frac{dP}{dy} - \frac{dN}{dx},$$

where P and N are functions of x, y. We have, in all cases,

---

[1] See a paper by Helmholtz (translated by Tait) in the *Phil. Mag.*, 1867.

Theorem.

$$\nabla^2 P = \theta \quad \text{and} \quad \nabla^2 N = -2\omega,$$

so that $u\,dx + v\,dy = dP + \dfrac{dN}{dy}\,dx - \dfrac{dN}{dx}\,dy$, which will be a perfect differential at all points where $\dfrac{d}{dy}\left(\dfrac{dN}{dy}\right) = -\dfrac{d}{dx}\left(\dfrac{dN}{dx}\right)$, i.e., where $\nabla^2 N = 0$, or $\omega = 0$. On the other hand, $u\,dy - v\,dx = dN + \dfrac{dP}{dx}\,dy - \dfrac{dP}{dy}\,dx$, which will not be a perfect differential unless $\nabla^2 P = 0$, i.e., unless $\theta = 0$, i.e., unless there is no expansion. We shall return to these expressions in a subsequent Article.

103. Flux of a Liquid across any Curve. In Art. 92 an expression for the time-rate of flow of any fluid across a given curve AB has been given. We shall suppose now the fluid to be a homogeneous liquid, so that the function ψ exists at all points. With the notation of that Article the flux is

$$\rho \int v\,ds\,;$$

but v = component of velocity at P along the normal

$$= u\frac{dy}{ds} - v\frac{dx}{ds} = \frac{d\psi}{ds},$$

where ψ is the flow-function at P; therefore the time-rate of flow of liquid across AB is

$$\rho(\psi_B - \psi_A),$$

the suffixes denoting the values of ψ at A and B.

104. Theorem. *The time-rate of flow of liquid across any direction at any point is the line-rate of increase of the flow-function in this direction.*

For, the time-rate of flow across PQ (fig. 54, Art. 92) is v, which is $\dfrac{d\psi}{ds}$.

This gives us a definition of the flow-function exactly analogous to that of the potential-function in Art. 100.

The following is also evident—*the line-rate of increase of ψ along any curve at any point is equal to the line-rate of increase of ϕ along an orthogonal curve at the point.*

For (see fig. 56, p. 149), assuming now that PQ is any curve at P, and PP' at right angles to it, the time-rate of flow across PQ is $\frac{d\psi}{ds}$; but it is also v, the normal velocity at P, which is $\frac{d\phi}{ds'}$, where ds' is the element of arc of PP', (Cor. 2, Art. 100). Hence at every point

$$\frac{d\phi}{ds'} = \frac{d\psi}{ds},$$

where ds and ds' are elements of any curve and an orthogonal curve respectively; and each of these expressions is the time-rate of flow across the first.

From this we see that the velocity at any point in any direction may be determined from the flow-function ψ instead of from the velocity potential ϕ; and its determination from the former is always possible, whereas the latter function may not exist at the point. [It will not exist when the motion at the point is vortical.]

The velocity in any direction (ds') at a point is the line-rate of increase of the flow-function in the perpendicular direction (ds)— the *sense* in which the differential coefficient of the ow-function is taken (sense in which ds is measured) is determined by the sense in which the rotation is measured. Thus, if P is the point considered, and the rotation, ω, at P is measured in the sense of the arrow (Fig. 57), the velocity in the direction Pa is $\frac{d\psi}{Pb}$ (when Pb, which is perpendicular to Pa, is diminished indefinitely); and the velocity in the direction Pb is $\frac{d\psi}{Pc}$, i.e., $-\frac{d\psi}{Pa}$ (Pa and Pc being diminished indefinitely). We have had at the outset a particular example in the equations $u = \frac{d\psi}{dy}$, $v = -\frac{d\psi}{dx}$, the sense of ω being from the axis of x to that of y.

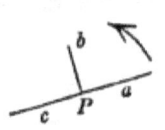

Fig. 57.

105. Flow and Circulation. DEF. The Flow along any curve, AB (fig. 54, Art. 92), at any given moment of time, is defined to be *the line-integral of the velocity along the curve between the extremities of the curve.*

Hence the flow from A to B is $\int (u\frac{dx}{ds} + v\frac{dy}{ds}) ds$, which is
$$\phi_B - \phi_A,$$
if the curve lies in an irrotational region. As explained in Art. 91, this *may* be a constant quantity for all curves whatever drawn from A to B, or it may not, according to the form of ϕ. When the curve is closed (or B coincides with A) the flow round it is called the *circulation* in it (Sir W. Thomson, *Vortex Motion*, Edinburgh Trans., 1869).

We shall calculate the value of the circulation round any small closed curve surrounding any point P (Art. 89) in the liquid. Let Q be a point on the curve; then the circulation round the curve is, with the notation of Art. 96,
$$\int (u' d\xi + v' d\eta)$$
taken over the same curve. Now observe that u, v, $\frac{du}{dx}$, $\frac{du}{dy}$ belong to the point P, and do not vary with Q; they may therefore be taken outside the integral sign; and since in the closed curve $\int \xi d\xi = 0$, $\int \eta d\eta = 0$, the circulation is
$$\frac{dv}{dx}\int \xi d\eta + \frac{du}{dy}\int \eta d\xi.$$

But, observing that we have taken the circulation round the curve in the same sense as that of the *rotation* at P, i.e., from x to y, it is clear that $\int \xi d\eta = -\int \eta d\xi =$ area of small curve. Denoting this area by δS, the circulation is
$$2\omega . \delta S,$$
i. e., twice the product of the area of the curve and the vortical spin inside it.

We can now prove that *the circulation round any curve A is equal to twice the surface-integral of the rotation taken over its area* (fig. 58).

For, inside A draw any other closed curve, A', very close to A at all points. Draw arbitrary lines aa', bb', cc', ... very close to each other across the curves, and apply the last result to each of the little areas $abb'a'$, $bcc'b'$, ... ; and we have

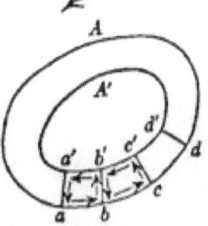

Fig. 58.

circulation round $abb'a' = 2$. area $abb'a' \times$ rotation within it,
,, ,, $bcc'b' =$,, $bcc'b \times$,, ,,
. . . .

If we add the left-hand sides together, a glance at the figure shows that the portions of circulation in all sides, bb', cc', ... which are common to two little areas destroy each other (as shown by the arrows); the total result of the summation is, therefore,

circulation in $A-$circulation in $A' = 2 \times$ surface-integral of rotation in the space between them.

Similarly the circulation in A' is equal to the circulation in any internal curve close to it, plus twice the surface-integral of the rotation over the space between them.

We may thus go on contracting the internal curve until it disappears, and we have the required result,

circulation round $A = 2\int\omega dS$,

where dS is any element of area within the curve, and ω the vortical spin at the position of the element.

This Article and Art. 103 enable us, therefore, to attach physical meanings to the functions

$$\phi_B - \phi_A \text{ and } \psi_B - \psi_A$$

which appertain to any two points A and B of the liquid.

From the theorem of this Article we can prove also that if a liquid is enclosed within a boundary every point of which is at rest, either there is no motion whatever of the liquid inside, or, if there is, the motion must be vortical. For the liquid particles at the boundary (if they are in motion) must move along the boundary, which is therefore a closed stream-line. And if there is no vortex inside, such a stream-line cannot exist, since the circulation along it would have to be zero by this Article, whereas the flux across every section of a tube of flow is the same, so that the circulation along it cannot vanish.

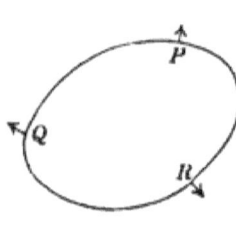

Fig. 59.

Non-vortical motion is therefore impossible inside a boundary which is kept at rest.

106. Theorem. If any closed curve, PQR, in the plane of motion be described in an irrotational region of a liquid; and if ϕ is the value of the velocity potential at any point, $\dfrac{d\phi}{dv}$ the normal rate of increase of ϕ measured constantly *outwards* from the curve, and ds the element of arc at the point, we have

$$\int \frac{d\phi}{dv} ds = 0, \qquad (a)$$

assuming that the velocity is not infinite anywhere inside the area.

For, (a) asserts that the line-integral of the normal velocity taken over the curve is zero. Now if A (fig. 58) is the closed curve, its area may be broken up into indefinitely small closed curves, such as $abb'a'$, ...; and the line-integral of normal velocity taken over each of these is zero. For, if any small closed curve be described round P (fig. 53), the line-integral of normal velocity round it is

$$\int \left[\left(u + \xi \frac{du}{dx} + \eta \frac{du}{dy} \right) d\eta - \left(v + \xi \frac{dv}{dx} + \eta \frac{dv}{dy} \right) d\xi \right];$$

and, as in Art. 105, this is $\left(\dfrac{du}{dx} + \dfrac{dv}{dy} \right). \delta S$, or $\theta \delta S$; but $\theta = 0$ since the fluid is a liquid. Hence (a) follows exactly as in Art. 105. This assumes that neither u nor v is infinite at any point; for if u is infinite, we must not assume $u \int d\eta = 0$; i.e., the closed curve A must not include a *source* or a *sink*.

Equation (a) holds whether the curve includes vortical regions or not; for all that we have assumed is that at every point inside $\theta = 0$, and the velocity is finite.

But if the curve cuts through a vortical region, the function ϕ does not exist at any point of the portion of the curve included in this region, so that we cannot represent the normal velocity at such a point by $\dfrac{d\phi}{dv}$; this normal component is

$$u \cos \theta + v \sin \theta,$$

where θ is the angle which the normal makes with the axis of

x; and hence for the portion of the curve included in the vortical region we must write the corresponding portion of the line-integral in the form

$$\int (u \cos \theta + v \sin \theta)\, ds.$$

107. Tubes of Flow. Assuming the motion to be steady, the flow-lines are stream-lines, and the particles at P and Q (fig. 56, p. 149) actually travel along the paths PA' and QB'. No liquid ever crosses any of the stream-lines, PA', QB', ... Hence the space between the stream-lines PA' and QB' is a channel through which liquid is flowing, none of the liquid within the channel ever leaving it.

Hence also it is obvious that the mass which flows across the section PQ of this channel in any time is the same as that which flows across any other section, $P'Q'$, etc., in the same time; and this quantity is measured by the product of the time and
$$\rho(\psi_P - \psi_Q), \text{ or } \rho d\psi.$$

Another way of expressing this fact is this—if ds_1 and ds_2 are any two normal sections of a channel,

$$\left(\frac{d\phi}{dn}\right)_1 ds_1 = \left(\frac{d\phi}{dn}\right)_2 ds_2,$$

where $\left(\frac{d\phi}{dn}\right)_1$ means the value of $\frac{d\phi}{dn}$ at the first section.

In the case which alone we consider (uniplanar motion) we imagine the channel $PQB'A'$ to be a section of a column of indefinite height, which is obtained by erecting perpendiculars to the plane xy at all points along PA' and QB', so that we are really considering a mass of liquid moving between two indefinitely high walls standing on the stream-lines.

In the general case (in which motion is not all parallel to one plane) the channels of never varying liquid are tubes—not columns between indefinitely high walls—and such a *tube of flow* may be constructed by describing any closed curve in the fluid and at each point of its contour describing the line of flow.

Although treating only of uniplanar motion, we shall use the term *tube of flow* as equivalent to the channel included between two lines of flow, such as the lines PA' and QB'.

108. Closure of Equipotential Curves and Stream-lines. We now proceed to inquire whether the equipotential velocity curves, $\phi = C$, or the stream-lines, $\psi = C$, can be closed curves, in the case of a liquid.

To begin with the latter, it is clear that *no tube of flow can begin at one place and end at a different place within the liquid.* For the flux across each end of the tube would be zero, and we know that the flux across all sections is the same; therefore there is no motion at all inside the tube. Hence a tube of flow, if it terminates, must do so at the boundary of the liquid, and the boundary will form part of it; but a tube of flow may obviously be a closed tube, i.e., one returning into itself. The stream-lines may, therefore, be closed curves within the liquid; but they cannot be so unless the liquid included contains a vortex. In a fluid which is devoid of all vortex motion no stream-line could be closed; for the circulation round a stream-line can never be zero (see Art. 100). Next, assume an equipotential curve, $\phi = C$, to be a closed curve. Then it must be possible to find immediately *inside* this curve another (necessarily) closed curve on which ϕ has a constant value, C', differing infinitely little from C, so that if dn is the normal distance between these curves at any point, the value of $\dfrac{d\phi}{dn}$ is $\dfrac{C'-C}{dn}$, which, though it has different values at all points on the curve, has everywhere the same sign; and it is easy to see that this will be impossible unless a source or a sink is included within the area of the curve. For if there is no included source or sink, we must have $\int \dfrac{d\phi}{dn} ds = 0$, the integral being taken over the contour of the curve; and this integral cannot vanish unless C' is the same as C. If there is a source or a sink inside, the integral is not required to vanish and the equipotential curve may be closed; but if there is not, C' and C must be the same; and continuing the same process, we see that ϕ must be constant throughout the space included by the curve, i.e., no motion is taking place inside. Hence if an equipotential curve is closed, there must be inside it either a source or a sink. (In illustration see example 5 following.)

109. Velocity System. By a *velocity system* within any region of a fluid we shall understand a diagram in which at every point of the region there is drawn a right line representing both in magnitude and in direction the velocity at the point. Unless the motion is steady, this diagram of the simultaneous velocities at all points will change from time to time.

It will be the same thing if we understand by the velocity system analytical expressions for the components, u, v, of velocity at each point, as functions of the co-ordinates of the point (and of the *time*, if the motion is not steady). For conciseness denote a velocity system by the notation

$$[u, v].$$

110. Theorem. *If on any closed curve enclosing, in the plane of motion, a portion of a fluid (liquid or gas) the velocity potential has at each point of the curve an assigned value, and if also at each point of the enclosed fluid the values of the expansion and vortical spin are assigned, there cannot be two distinct velocity-systems satisfying the assigned conditions.*

For, let ϕ denote the assigned value of the velocity potential at any point of the curve (the value of ϕ being, of course, different for different points on the curve); and, if possible, let there be two velocity-systems,

$$[u, v] \quad \text{and} \quad [u', v'], \tag{a}$$

each of which will give the same value of ϕ at the same point of the curve, and the same values of θ and ω at the same point inside. The exact reverse of any given motion is, of course, possible. Reverse everything in the second system, and then superpose the first on it. We shall thus have a constant (zero) velocity potential at each point on the given curve, and no expansion or vortical spin at any point inside. But (Art. 108) this requires complete rest at every point of the enclosed fluid (which with no *expansion* may, of course, be regarded as a *liquid*). Hence at every point inside the velocity-system $[u-u', v-v']$ is null, i.e., the two systems (a) are identical—which proves the proposition.

This theorem is of fundamental importance; for if we can find out any *one* velocity-system producing a certain distribution of velocity potential over a closed curve and of expansion and vortical spin inside it, we are now enabled to assume that this is the only system fulfilling the conditions.

We now proceed to determine (for the particular case of uniplanar motion) the velocity components, u, v, at each point when the values of θ and ω are assigned.

Assume, as in Art. 102,

$$u = \frac{dP}{dx} + \frac{dN}{dy}, \qquad (\beta)$$

$$v = \frac{dP}{dy} - \frac{dN}{dx}, \qquad (\gamma)$$

where P and N are (as yet) unknown functions of x, y. Then

$$\nabla^2 P = \theta, \text{ and } \nabla^2 N = -2\omega. \qquad (\delta)$$

But in the theory of attractions it is proved that if V is the potential, at any point, of matter attracting according to the law of the inverse square of distance, and ρ the volume-density of the matter at the point, $\nabla^2 V = -4\pi\rho$, where

$$\nabla^2 \equiv \frac{d^2}{dx^2} + \frac{d^2}{dy^2} + \frac{d^2}{dz^2}.$$

If the distribution of matter is the same at all points in space whose co-ordinates x, y are the same, differential coefficients with respect to z disappear, and in this case $\nabla^2 \equiv \frac{d^2}{dx^2} + \frac{d^2}{dy^2}$.

Hence equations (δ) show that—

(*a*) the value of P at any point, A, of the fluid may be regarded as the potential (per unit mass) produced at the point by a distribution of matter extending infinitely above and below the plane (x, y) of motion, the density of this matter at any point, Q, being $-\frac{\theta}{4\pi}$, where θ is the expansion at Q;

(*b*) the value of N at any point, A, of the fluid may be regarded as the potential (per unit mass) produced at the point by a distribution of matter extending infinitely above and below the plane (x, y) of motion, the density of this matter at any point, Q, being $\frac{\omega}{2\pi}$, where ω is the vortical spin at Q.

It may happen (as will be exemplified in a subsequent section) that there is expansion or spin at only certain isolated points of the fluid. In this case the distributions of matter which we imagine for the purpose of calculating P or N will be confined to infinitely long slender cylinders perpendicular to the plane of motion of the fluid.

Now, as we know that in the theory of attractions V can be expressed in the form $\int \dfrac{dm}{r}$, where dm is the element of mass at Q, and $r =$ the distance of Q from A, we see that

$$P = -\frac{1}{4\pi} \iiint \frac{\theta}{r} dx\, dy\, dz, \qquad (\epsilon)$$

$$N = \frac{1}{2\pi} \iiint \frac{\omega}{r} dx\, dy\, dz, \qquad (\xi)$$

which give the values of P and N, and therefore the values of u and v, when the values of θ and ω are assigned at each point.

For the determination of the velocity system resulting from a given distribution of expansion and vortical spin in the general (or three-dimensional) motion of a fluid, the student may consult Lamb's *Treatise on the Motion of Fluids*, p. 150.

111. Boundary Condition. The curve which bounds the field of motion may be absolutely fixed, or it may change from time to time. Now the problem of fluid motion may be presented to us in either of two forms—

(a) Given the expression for the velocity at each point, i.e., given the values of u and v at each point, determine the nature of the boundary of the field of motion which must correspond to this velocity system.

(b) The field of motion is bounded by a given fixed curve, determine the velocity system which must correspond to this bounded field.

The latter problem presents itself perpetually in the study of the flow of electricity.

Now a boundary of the field is, of course, in all cases a curve across which no flow takes place. Hence those fluid particles which are at the boundary either are not moving at all, or are moving tangentially to the boundary. This condition is easily expressed analytically.

Let the equation of the boundary at any instant be

$$F(x, y, t) = 0, \qquad (1)$$

so that we are including the case in which the boundary is changing with the time. Let the components of velocity of a particle at the point (x, y) on the boundary be U, V at the time t; then the co-ordinates of this particle at the time $t + \Delta t$ will be $x + U \Delta t,\ y + V \Delta t$; and whether the particle moves or is at rest, it must by supposition be still found on the boundary, so that

$$F(x + U\Delta t, y + V\Delta t, t + \Delta t) = 0. \qquad (2)$$

Combining this with (1), we have

$$U\frac{dF}{dx} + V\frac{dF}{dy} + \frac{dF}{dt} = 0, \qquad (3)$$

which is the condition to be satisfied in all cases (both in (a) and (b)) at the boundary.

If the boundary is invariable with the time, $\frac{dF}{dt} = 0$, and (3) becomes

$$U\frac{dF}{dx} + V\frac{dF}{dy} = 0. \qquad (4)$$

Since $\frac{dF}{dx}$ and $\frac{dF}{dy}$ are proportional to the direction-cosines of the normal to the boundary, this equation is obvious; for if $\overline{V}$ is the resultant velocity at the boundary and χ the angle which its direction makes with the normal, (4) is simply

$$\overline{V} \cos \chi = 0.$$

Any flow-line may obviously be taken as a boundary of the field since it satisfies the condition of having no flow across it.

Examples.

1. If the particles of a fluid describe a series of concentric circles, find the equation of continuity in its simplest form.

Let O be the common centre of the circles; P a point on any circle whose radius, OP, is r; θ the angle which OP makes with any fixed right line; Q a point on the circle indefinitely near P; $\theta + d\theta$ the angle of direction of OQ; describe with centre O a circle of radius $r+dr$, and produce OP and OQ to meet this circle in P' and Q'. Also let ρ be the density at P and $\dot{\theta}$ the angular velocity of the fluid particle at P about O, at the time t. [Observe we do not assume the fluid to be a liquid.] Then the normal velocity at the face PP' is $r\dot{\theta}$; therefore the flux into the area $PP'Q'Q$ in the time dt is $\rho r\dot{\theta}drdt$, and the flux out of it through QQ' is

$$\left[\rho r\dot{\theta}+\frac{d(\rho r\dot{\theta})}{d\theta}d\theta\right]drdt;$$

therefore the loss of fluid in time dt is

$$r\frac{d(\rho\dot{\theta})}{d\theta}drd\theta dt.$$

Again, the quantity in the area at time t is $\rho \times$ area, or $\rho r dr d\theta$; and the quantity in it at time $t+dt$ is $(\rho r+r\frac{d\rho}{dt}dt)drd\theta$; therefore the gain is $r\frac{d\rho}{dt}drd\theta dt$. Equating this to the former expression for the gain, we have

$$\frac{d\rho}{dt}+\frac{d(\rho\dot{\theta})}{d\theta}=0,$$

which is the required equation of continuity.

[Of course this equation could have been otherwise obtained by transforming the Cartesian equation of continuity to polar co-ordinates.]

2. Find, generally, the equation of continuity in polar co-ordinates.

At any point, P, whose polar co-ordinates are r, θ, let the fluid velocities, at the time t, along and perpendicular to OP be a and β, respectively, and at this time let ρ be the density at P. Draw a radius vector OQ equal in length to $r+dr$, and making an angle $d\theta$ with OP; from P draw Pp perpendicular to OQ, and from Q draw Qq perpendicular to OP. (Fig. 60.)

In the time dt the flux into the area by the face Pp is $\rho a \times Pp \times dt$, or $\rho a r d\theta dt$; and the flux out of it by the face Qq is

$$[\rho ar+\frac{d(\rho ar)}{dr}dr]d\theta dt;$$

therefore the loss of fluid by the motion along OP is

$$\frac{d(\rho ar)}{dr}drd\theta dt.$$

Again, the flux into the area by the face Pq is $\rho\beta drdt$; and the flux out by the face pQ is $[\rho\beta+\frac{d(\rho\beta)}{d\theta}d\theta]drdt$, therefore the loss by this motion is

$$\frac{d(\rho\beta)}{d\theta}drd\theta dt.$$

But the mass covering the area at the time t is $\rho r dr d\theta$; and the mass at time $t+dt$ is $(\rho+\frac{d\rho}{dt}dt)rdrd\theta$. Hence we have

$$r\frac{d\rho}{dt} + \frac{d(\rho a r)}{dr} + \frac{d(\rho\beta)}{d\theta} = 0. \tag{1}$$

Observe that ρ, the density at a fixed-space point P, is evidently in general a function of the time and the position of the point; so that

$$\rho = f(r, \theta, t).$$

At the end of the time $t + dt$, the fluid element which was at P will have polar co-ordinates $(r + adt, \theta + \frac{\beta}{r}dt)$, and will be at a point P' (which will not of course be Q unless PQ is so chosen as to be an element of the line of flow at P); and the density at P' is, by the above formula,

$$f(r + adt, \theta + \frac{\beta}{r}dt, t + dt).$$

But the particle at P' is the one which was at P, and if the density of this particle has changed in the time dt to $\rho + D_t\rho \cdot dt$, we have

$$D_t\rho = a\frac{d\rho}{dr} + \frac{\beta}{r}\frac{d\rho}{d\theta} + \frac{d\rho}{dt}. \tag{2}$$

Here of course $D_t\rho$ is the *total* time-rate of change of ρ (see Art. 94). Hence (1) can be written

$$\frac{1}{\rho}D_t\rho + \frac{1}{r}\left[\frac{d(ar)}{dr} + \frac{d\beta}{d\theta}\right] = 0, \tag{3}$$

so that the polar equation of continuity for a *liquid* is

$$\frac{d(ar)}{dr} + \frac{d\beta}{d\theta} = 0.$$

3. Transform the equations $\nabla^2\phi = 0$, $\nabla^2\phi = -\frac{1}{\rho}D_t\rho$ from Cartesian to polar co-ordinates.

If ϕ, the velocity potential, is expressed in polar co-ordinates, we have of course

$$a = \frac{d\phi}{dr}, \quad \beta = \frac{1}{r}\frac{d\phi}{d\theta}.$$

Hence equation (3) of last example becomes

$$\frac{1}{\rho}D_t\rho + \frac{d^2\phi}{dr^2} + \frac{1}{r}\frac{d\phi}{dr} + \frac{1}{r^2}\frac{d^2\phi}{d\theta^2} = 0.$$

[Of course it can be easily proved that $\nabla^2 \equiv \frac{1}{r}\frac{d}{dr} + \frac{d^2}{dr^2} + \frac{1}{r^2}\frac{d^2}{d\theta^2}$, by transformation.]

4. Transform, by elementary principles, the equation $\nabla^2\psi = -2\omega$ to polar co-ordinates.

Let P be the point at which the value of the flow function is ψ and the rotation ω, in the sense opposite to that of watch-hand rotation; let O be the origin of polar co-ordinates, $OP = r$, $OQ = r + dr$, $\angle QOP = d\theta$; let fall Pp and Qq perpendicular to OQ and OP; and express the fact that the circulation round the area $PqQp$ is $2\omega \times$ the area.

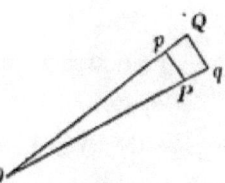

Fig. 60.

The velocity along Pq is

$$\frac{d\psi}{Pp} = \frac{d\psi}{rd\theta}, \quad \therefore \text{ flow along } Pq = \frac{d\psi}{rd\theta} \cdot dr;$$

the velocity along Pp is $-\frac{d\psi}{dr}$ (see Art. 104),

$\therefore$ flow along and in the sense of $pP = +\frac{d\psi}{dr} rd\theta;$

velocity along $qQ = -\frac{d\psi}{dr} - \frac{d^2\psi}{dr^2} dr,$

$\therefore$ flow along $qQ = -\left(\frac{d\psi}{dr} + \frac{d^2\psi}{dr^2} dr\right)(r+dr)d\theta;$

velocity along $pQ = \frac{d\psi}{rd\theta} + \frac{1}{r}\frac{d^2\psi}{d\theta^2} d\theta,$

$\therefore$ flow along $Qp = -\left(\frac{d\psi}{rd\theta} + \frac{1}{r}\frac{d^2\psi}{d\theta^2} d\theta\right) dr.$

Hence the total circulation counter-clockwise is

$$-\left(r\frac{d^2\psi}{dr^2} + \frac{d\psi}{dr} + \frac{1}{r}\frac{d^2\psi}{d\theta^2}\right) dr d\theta,$$

neglecting infinitesimals of the third order. Also the rotation multiplied by the area $= \omega \cdot r\, dr\, d\theta$; hence

$$\frac{d^2\psi}{dr^2} + \frac{1}{r}\frac{d\psi}{dr} + \frac{1}{r^2}\frac{d^2\psi}{d\theta^2} = -2\omega.$$

5. If the stream-lines of a liquid are right lines all passing through one point, find the law of irrotational motion.

Let O be the point of convergence (or divergence) of the lines of flow OP, OQ, ... (Fig. 60); then since the velocity at P perpendicular to OP is zero, we have

$$\frac{d\psi}{dr} = 0, \quad \frac{d\phi}{d\theta} = 0,$$

ψ being the flow function and ϕ the velocity potential (which by supposition exists at every point). The second equation gives, by example 3,

$$\frac{d^2\phi}{dr^2} + \frac{1}{r}\frac{d\phi}{dr} = 0, \quad \text{i.e., } \phi = A\log r + B,$$

where A and B are arbitrary constants. The first gives, by example 4,

$$\psi = A'\theta + B'.$$

The velocity at any distance $= \frac{d\phi}{dr} = \frac{A}{r}.$

The equipotential curves are circles having O for a common centre. If we take the integral $\int \frac{d\phi}{dn} ds$ over any one of them (of radius r) the result is $2\pi A$, and not zero (see Art. 106). The point O is then a source by which liquid is perpetually flowing out in all directions, or a sink down which it is perpetually flowing, according as the velocity is from or towards

O; and the system is what Clifford calls expressively a 'squirt' (*Kinematic*, p. 212).

Since (Art. 104) $\frac{d\phi}{dr} = \frac{d\psi}{rd\theta}$, we must have
$$A' = A.$$

[The same motion is supposed to take place in all planes parallel to the one chosen, so that the case is that of an *axial squirt*.]

The time rate of supply of liquid through the source is called the *strength* of the source. To find it, describe a circle of radius r round the source; then the velocity at each point on the circle is $\frac{A}{r}$, so that the time rate of supply is (Art. 92) $\frac{A}{r} \times 2\pi r$, or $2\pi A$. Hence if m is the strength of the source,
$$A = \frac{m}{2\pi}.$$

6. The stream-lines of a liquid being a series of concentric circles, find the law of motion, under the condition that it is irrotational.

In this case $\frac{d\phi}{dr} = 0$, $\frac{d\psi}{d\theta} = 0$, and we easily find
$$\phi = A\theta,$$
$$\psi = A \log r.$$

Here the velocity at distance r is $\frac{A}{r}$, and the circulation round any closed curve surrounding the origin is $2\pi A$.

We have, therefore, a mass of liquid *whirling* round a point (or, rather, round an axis) in concentric circles (or coaxal cylinders), and yet the 'rotation' is everywhere zero, except at O. Such a motion is called by Clifford 'a whirl' (*Kinematic*, p. 215). This example illustrates what has been already referred to in Art. 98, viz., the distinction between 'rotational' (that is *elementally* rotational) or vortical motion, and a motion in which there is bodily rotation of the particles round some external axis. This latter *may*, of course, be accompanied by vortical motion, but, as we see, the whirl can exist without the 'rotation.'

Observe here an instance—the simplest that could be presented—of a multiple-valued velocity potential. The potential difference between two points being defined as the flow along a definite path connecting them, if we start from a point P and go n times round the origin, returning finally to P, the circulation round this path is $2n\pi A$; so that there is an infinite number of values of ϕ for one and the same point, these values being a series in arithmetical progression with a common difference $2\pi A$.

7. Determine the form of the velocity value at every point so that the stream-lines may be a series of confocal ellipses, and the motion irrotational.

Any ellipse of the series may be defined by a single variable, namely, its semi-major axis, $\sqrt{\mu}$, suppose; and since all along a stream-line the flow-function, ψ, is constant we may regard ψ as a function of μ, so that

$$\psi = f(\mu), \tag{1}$$

and μ is at every point given by the equation

$$\frac{x^2}{\mu} + \frac{y^2}{\mu - c^2} - 1 = 0, \tag{2}$$

where $2c$ is the (given) distance between the foci of the system, and the axis of x has been taken along the major axes of the conics.

We have now to transform the equation for ψ, viz.,

$$\left(\frac{d^2}{dx^2} + \frac{d^2}{dy^2}\right)\psi = 0, \tag{3}$$

into a differential equation in which μ is the independent variable.

Evidently $\dfrac{d\psi}{dx} = \dfrac{d\psi}{d\mu} \cdot \dfrac{d\mu}{dx}$; $\dfrac{d^2\psi}{dx^2} = \dfrac{d^2\psi}{d\mu^2} \cdot \left(\dfrac{d\mu}{dx}\right)^2 + \dfrac{d\psi}{d\mu} \cdot \dfrac{d^2\mu}{dx^2}$;

hence (3) becomes

$$\left\{\overline{\left|\frac{d\mu}{dx}\right|}^2 + \overline{\left|\frac{d\mu}{dy}\right|}^2\right\}\frac{d^2\psi}{d\mu^2} + \nabla^2\mu \cdot \frac{d\psi}{d\mu} = 0. \tag{4}$$

Now the perpendicular, p, from the centre on the tangent to the ellipse (2) at the point x, y is given by the equation

$$\frac{1}{p^2} = \frac{x^2}{\mu^2} + \frac{y^2}{(\mu - c^2)^2};$$

therefore by differentiating (2) successively with respect to x and y, we have

$$\mu \frac{d\mu}{dx} = 2p^2 x, \quad (\mu - c^2)\frac{d\mu}{dy} = 2p^2 y. \tag{5}$$

Hence
$$\overline{\left|\frac{d\mu}{dx}\right|}^2 + \overline{\left|\frac{d\mu}{dy}\right|}^2 = 4p^2. \tag{6}$$

Also by differentiating equations (5), the first with respect to x and the second with respect to y, and substituting for $\dfrac{d\mu}{dx}$ and $\dfrac{d\mu}{dy}$, whenever they occur, their values given in (5), we have

$$\nabla^2 \mu = 2p^2 \left(\frac{1}{\mu} + \frac{1}{\mu + c^2}\right). \tag{7}$$

Hence (4) becomes, from (6) and (7),

$$2\frac{d^2\psi}{d\mu^2} + \left(\frac{1}{\mu} + \frac{1}{\mu - c^2}\right)\frac{d\psi}{d\mu} = 0,$$

which gives by integration

$$\psi = A \log(\sqrt{\mu} + \sqrt{\mu - c^2}) + B, \tag{8}$$

where A and B are arbitrary constants.

To find the velocity at any point. Since it takes place along the ellipse through the point, P, its value (estimated in the sense $\overset{y}{\underset{x}{\uparrow}}$) is $\dfrac{d\psi}{dn}$,

Examples.

dn being an element, PQ, of the normal at the point drawn inwards (Art. 104), and Q being a point on a consecutive ellipse of the series (on which the flow-function is $\psi + d\psi$); then, p being, as above, the perpendicular from the centre on the tangent at P, we have

$$p^2 = \mu - c^2 \cos^2 \theta,$$

where θ is the angle made by the tangent with the axis of x. The tangent at Q is parallel to that at P, so that $dn = -dp = -\dfrac{1}{2p} d\mu$ (since θ is the same for both tangents). Hence the velocity, V, at P is

$$-2p \frac{d\psi}{d\mu}, \text{ or}$$

$$V = -\frac{Ap}{\sqrt{\mu(\mu - c^2)}},$$

so that the velocity really takes place in the sense $\begin{smallmatrix}y\\x\end{smallmatrix}\downarrow$.

For points far removed from the centre the ellipses tend to become circles, and for such points p is approximately equal to $\sqrt{\mu}$. Hence at infinitely distant points $V = 0$, i.e., the infinitely distant particles are at rest.

At points taken on the same stream-line the velocities are directly proportional to the central perpendiculars on the tangents at the points.

The potential function, ϕ, may be calculated directly, as ψ was calculated, from the fact that the equipotential curves, being the orthogonal trajectories of the ellipses, must be a series of confocal hyperbolas; or it may be deduced from the velocity as follows.

If $ds =$ the element of arc of the ellipse at P in the sense of V, we have $\dfrac{d\phi}{ds} = V$, $\quad \therefore \dfrac{d\phi}{ds} = \dfrac{Ap}{\sqrt{\mu(\mu - c^2)}}.$ (9)

Now if $\sqrt{\nu}$ is the semi-major axis of the confocal hyperbola through P, we have

$$x^2 = \frac{\mu \nu}{c^2}; \quad y^2 = \frac{(\mu - c^2)(c^2 - \nu)}{c^2},$$

$$\therefore dx = \frac{1}{2c} \sqrt{\frac{\mu}{\nu}} d\nu; \quad ds = \tfrac{1}{2} \sqrt{\frac{\mu - \nu}{\nu(c^2 - \nu)}} d\nu,$$

($d\mu$ being zero along ds); hence

$$p^2 = \mu - c^2 \left(\frac{dx}{ds}\right)^2 = \mu - \mu \frac{c^2 - \nu}{\mu - \nu} = \mu \frac{\mu - c^2}{\mu - \nu}.$$

Substituting these values in (9), we have

$$\frac{d\phi}{d\nu} = \frac{A}{2} \cdot \frac{1}{\sqrt{\nu(c^2 - \nu)}},$$

$$\therefore \phi = A \sin^{-1} \frac{\sqrt{\nu}}{c} + B', \quad (10)$$

which gives ϕ as a function of the major axis of the equipotential curve.

The substitution of the above value of p in the expression previously given for the velocity gives

$$V = \frac{A}{\sqrt{\mu - \nu}}, \qquad (11)$$

estimated in the sense $\begin{smallmatrix} y \\ x \end{smallmatrix} \downarrow$.

For all points on the line joining the two foci, either μ or ν is constant. For all such points *between* the foci we have $\mu = c^2$, $\nu = x^2$, where x is the distance of the assumed point from the centre; for points on the line produced we have $\nu = c^2$, $\mu = x^2$. One ellipse of the series is the whole of the line *between* the foci, and one hyperbola consists of the two *external* portions of this line, each of which extends from a focus to infinity.

At a focus $\mu = \nu = c^2$, therefore at each focus $V = \infty$; yet the velocity is zero at all points *between* the foci, since, the ellipse having narrowed to a right line, there is a velocity from right to left and an equal velocity from left to right at all such points.

To investigate the motion if the hyperbolas are the stream-lines. If ψ is still used to denote the flow-function, and ϕ the velocity-function, they will simply interchange their previous values, so that we have

$$\phi = A \log(\sqrt{\mu} + \sqrt{\mu - c^2}),$$

$$\psi = A \sin^{-1}\frac{\sqrt{\nu}}{c},$$

neglecting the arbitrary constants added to them. If, as before, $ds =$ element of arc of ellipse, the velocity, U, along the hyperbola is $\frac{d\psi}{ds}$;

$$\therefore U = 2A\sqrt{\frac{\nu}{\mu - \nu}}.$$

At all points on the axis of y we have $\nu = 0$, $\mu = c^2 + y^2$; hence at all such points there is no velocity. At each focus the velocity is infinite as in the previous case. At points between a focus and the centre the velocities are perpendicular to the line FF', and they vary from ∞ at F to zero at the centre. At points on the line $F\infty$, the velocities (being along the hyperbola) are all along this line, and of magnitudes varying from ∞ at F to zero at infinity, since for all such points $\nu = c^2$ and $\mu = x^2$. The infinitely distant parts of the liquid are at rest, as in the previous case.

Since along the line $F\infty$ the velocities are along the line itself, this line may be considered as a fixed smooth wall round the edge, F, of which the liquid flows, those particles which have travelled along one side of the wall in the sense ∞F turning back, when they reach F, with infinite velocity and travelling off along the other side of the wall in the sense $F\infty$.

8. Determine the law of motion of a liquid in order that the stream-lines may be a series of confocal ellipses of Cassini, the motion being irrotational.

Let F, F' (fig. 61) be the foci, O the centre, f and f' the distances,

Examples. 171

PF, PF', of any point P in the plane of motion, from the foci, and let the equation of the Cassinian through P be

$$ff' = k^2, \qquad (1)$$

k being the constant belonging to this ellipse. Then if $OP=r$, $\angle POF=\theta$, the above equation is equivalent to

$$r^4 - 2c^2 r^2 \cos 2\theta + c^4 - k^4 = 0, \qquad (2)$$

where $c = OF = \tfrac{1}{2}FF'$.

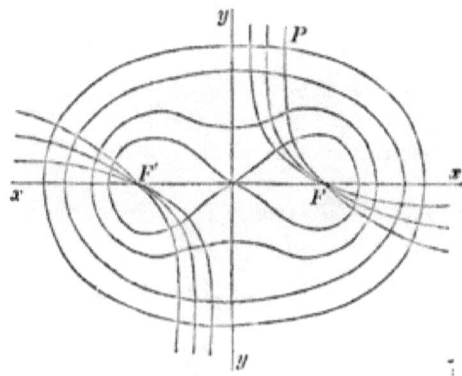

Fig. 61.

Now we have, by hypothesis,

$$\psi = f(k), \qquad (3)$$

some (as yet) unknown function of k; and by example 4,

$$\frac{d^2\psi}{dr^2} + \frac{1}{r}\frac{d\psi}{dr} + \frac{1}{r^2}\frac{d^2\psi}{d\theta^2} = 0. \qquad (4)$$

Transform this into an equation in which k is the independent variable. Now from (2) we have

$$\frac{dk}{dr} = \frac{r^3 - c^2 r \cos 2\theta}{k^3}, \quad (\theta \text{ being taken as constant}) \qquad (5)$$

$$\frac{dk}{d\theta} = \frac{c^2 r^2 \sin 2\theta}{k^3}; \quad (r \text{ being taken as constant}),$$

and (4) transforms into

$$\frac{d^2\psi}{dk^2}\left\{\overline{\frac{dk}{dr}}\Big|^2 + \frac{1}{r^2}\overline{\frac{dk}{d\theta}}\Big|^2\right\} + \frac{d\psi}{dk}\cdot\nabla^2 k = 0, \qquad (6)$$

or

$$k\frac{d^2\psi}{dk^2} + \frac{d\psi}{dk} = 0, \qquad (7)$$

$$\therefore \psi = A \log k + B, \qquad (8)$$

in which we may reject the arbitrary constant B.

Now the components of velocity along and perpendicular to OP are given by the equations

$$\frac{d\psi}{r d\theta} = \frac{A}{kr}\frac{dk}{d\theta} = \frac{Ac^2 r}{k^4} \sin 2\theta, \qquad (9)$$

$$-\frac{d\psi}{dr} = -\frac{A}{k}\frac{dk}{dr} = -\frac{Ar}{k^4}(r^2 - c^2 \cos 2\theta). \qquad (10)$$

Squaring these and adding, we have for the resultant velocity, V,

$$V = \frac{Ar}{k^2}, \qquad (11)$$

in virtue of (2).

On any one ellipse the velocity is directly proportional to the distance from the centre. Also at infinitely distant points $r = f = f' = k$, approximately, therefore at all such points $V = 0$, or the fluid is at rest at infinity.

To find the equipotential curves. Let ϕ denote the velocity potential at P. Then (ex. 3) we have

$$\frac{d\phi}{r d\theta} = -\frac{d\psi}{dr} = -\frac{Ar}{k^4}(r^2 - c^2 \cos 2\theta), \qquad (12)$$

$$\frac{d\phi}{dr} = \frac{d\psi}{r d\theta} = \frac{Ac^2 r}{k^4} \sin 2\theta. \qquad (13)$$

In these equations for ϕ we must get rid of the constant k, which defines a particular ellipse, by substituting its value in terms of r and θ from (2). Thus

$$\frac{d\phi}{dr} = \frac{Ac^2 r \sin 2\theta}{r^4 - 2c^2 r^2 \cos 2\theta + c^4};$$

and integrating with respect to r and adding an arbitrary function of θ, we have

$$\phi = \frac{A}{2} \tan^{-1}\left(\frac{r^2 - c^2 \cos 2\theta}{c^2 \sin 2\theta}\right) + f(\theta).$$

Differentiating this with respect to θ (taking r as a constant), and equating the result to the value of $\frac{d\phi}{d\theta}$ in (12), we have

$$f'(\theta) = -A,$$

$$\therefore f(\theta) = -A\theta,$$

rejecting an arbitrary constant. Hence

$$\phi = -A\theta + \tfrac{1}{2} A \tan^{-1}\left(\frac{r^2 - c^2 \cos 2\theta}{c^2 \sin 2\theta}\right). \qquad (14)$$

Multiply both sides of this equation by 2, divide it out by A, and take the tangents of both sides, and we have

$$r^2\left(\cos 2\theta - \sin 2\theta \tan\frac{2\phi}{A}\right) = c^2,$$

$$\therefore r^2 \cos 2\left(\theta + \frac{\phi}{A}\right) = c^2 \cos\frac{2\phi}{A}. \qquad (15)$$

Now observe that for any one equipotential curve ϕ is constant; hence the curve denoted by this last equation is a rectangular hyperbola, whose major axis is inclined at the angle $\frac{\phi}{A}$ to the line OF. By varying ϕ we get a series of rectangular hyperbolas having O for centre and all passing through the foci, F, F'.

The foci are points at which $V = \infty$, since for them $k = 0$. The centre, O, is a point at which there is no motion.

The value of ϕ given in (14) may be put into a simpler form. From it we have
$$\tan\frac{2\phi}{A} = \frac{r^2\cos 2\theta - c^2}{r^2\sin 2\theta}.$$

But if θ_1 and θ_2 denote the angles $PF\infty$ and $PF'\infty$, we find easily
$$\tan(\theta_1 + \theta_2) = \frac{r^2\sin 2\theta}{r^2\cos 2\theta - c^2}.$$

Hence
$$\frac{2\phi}{A} = \frac{\pi}{2} - (\theta_1 + \theta_2), \tag{16}$$

so that $\theta_1 + \theta_2$ is constant along any one equipotential curve (rectangular hyperbola). The circulation round any curve enclosing one focus is easily found. For, let the line yy form part of the curve, the remaining portion being one half of a very distant stream-line cut off by this line. Such a distant stream-line may obviously be considered as a circle, of radius R suppose, since the ovals ultimately become circles. Now at each point of the line yy the velocity is normal to the line, and the flow along this line is therefore zero. Also the velocity at each point of the semicircle is $\frac{A \cdot R}{k^2}$ by (11), or $\frac{A}{R}$ (since $k^2 = R^2$, nearly); therefore the flow round it is πA, which is therefore the circulation round every curve enclosing F; and the same value holds with reference to the other focus, F'.

These ellipses of Cassini and rectangular hyperbolas will be arrived at more simply from another point of view in a subsequent section.

9. Determine the circumstances of motion so that the stream-lines in the irrotational motion of a liquid may be a series of equiangular spirals having the same pole and the same constant angle.

The equation of any one of the spirals will be
$$r = ae^{k\theta}, \tag{1}$$

where k is the cotangent of the common angle of the spirals, and a is a constant varying from one spiral to another.

Assuming $\psi = f(a)$, and using the polar differential equation for ψ, we obtain
$$a\frac{d^2\psi}{da^2} + \frac{d\psi}{da} = 0, \tag{2}$$

$$\therefore \psi = A\log a, \tag{3}$$

or expressing ψ in terms of the co-ordinates of a point,
$$\psi = A(\log r - k\theta). \tag{4}$$

We have for the components of velocity at any point, along and perpendicular to the radius vector respectively,

$$a = \frac{d\psi}{rd\theta} = -\frac{Ak}{r},\qquad(5)$$

$$\beta = -\frac{d\psi}{dr} = -\frac{A}{r};\qquad(6)$$

so that if ϵ is the constant angle of the spirals ($\cot^{-1} k$), and V the resultant velocity,

$$V = \frac{A}{r\sin\epsilon}.\qquad(7)$$

The potential function, ϕ, is found from (5) and (6); and since $\frac{d\phi}{dr} = \frac{d\psi}{rd\theta}$, $\frac{d\phi}{rd\theta} = -\frac{d\psi}{dr}$, we have

$$\phi = -A(k\log r + \theta).\qquad(8)$$

The curve $\phi = $ const. is also an equiangular spiral,

$$r = ce^{-\frac{\theta}{k}},\quad [c = e^{-\frac{\phi}{kA}}]$$

whose angle is the complement of the constant angle ϵ.

The velocity at the pole is infinite. To find the circulation along any curve surrounding the pole, describe a circle of radius r round it, and take the circulation round this circle. It is $r\int\beta d\theta$, where β is given by (6). Hence the circulation is

$$2\pi A.$$

It may be well to call the attention of the student to a possible error.

Suppose it proposed *to find the nature of the motion so that the stream-lines shall be a series of equiangular spirals* of different constant angles *all having the same initial radius a*; that is, the number k in (1) is now the changing constant which determines each spiral. If we seek to form a differential equation for ψ in terms of k, analogous to (2), we find that such is possible, and the equation is

$$(1+k^2)\frac{d^2\psi}{dk^2} + 2k\frac{d\psi}{dk} = 0; \text{ therefore } \psi = A\tan^{-1}k = A\tan^{-1}\left(\frac{1}{\theta}\log\frac{r}{a}\right).$$

Now $\tan^{-1} k$ has an infinite number of values, so that an infinite number of these spirals will pass through any one point, P, in the plane of motion. This signifies that at *every* point the resultant velocity takes place in an infinite number of directions. But such motion is possible only at a source or a sink, and we should be compelled to conceive what Clifford calls a 'squirt' as existing at every point in the plane. Such is clearly impossible. We must take care, then, that the form which we assume for the flow-function does not at *every* point give us an infinite number of stream-lines.

The spirals assumed in the example which we have just solved—those in which a is different and k the same for all—do not labour under this objection, since, except at the pole, no two of them can ever intersect.

10. Determine the circumstances of motion so that the lines of flow shall be a series of parabolas having a common focus and coincident axes.

Ans. If r, θ are the polar co-ordinates of any point in the plane, referred to the focus and axis,

$$\psi = Ar^{\frac{1}{2}} \sin \frac{\theta}{2}; \quad \phi = Ar^{\frac{1}{2}} \cos \frac{\theta}{2}.$$

The equipotential lines are another series of confocal parabolas, having the same focus and axis as the first set. Let $F\infty$ be the axis through the focus F from which θ is measured, and $F\infty'$ this axis produced in the opposite sense through the focus. Then the velocity at every point on $F\infty$ is along this line and towards F; the velocity at every point on $F\infty'$ is perpendicular to this line; the velocity at F is infinite, so that there is a stream along the line ∞F, the liquid thus coming escaping down through F, but there is no escape through F of fluid coming in any other direction, so that F is not a *sink*; and if we calculate the total quantity of liquid passing per unit of time into any circle described round F, i.e., $\int v \, ds$ (Art 92), we find it to be zero, since only an infinitesimal amount escapes from the line ∞F at F.

11. If ρ_1 and ρ_2 are two variables defined by the equations

$$\rho_1 = f_1(x, y), \quad \rho_2 = f_2(x, y),$$

such that the curves $\rho_1 = $ const., $\rho_2 = $ const. are orthogonal, transform the equation $\nabla^2 \phi = 0$ into one involving differential coefficients with respect to ρ_1 and ρ_2 only.

At the point P (fig. 60, p. 165) draw an element, Pp, of the curve for which ρ_1 is const., and an element, Pq, of the curve for which ρ_2 is const.; draw also the close curves, qQ and pQ, for which the values are $\rho_1 + d\rho_1$ and $\rho_2 + d\rho_2$, respectively. Regarding ϕ as a liquid velocity potential, the equation which is equivalent to $\nabla^2 \phi = 0$ will simply express that the total quantity of liquid passing, per unit of time, out of the contour of the little area $PqQp$ is zero; and we shall obtain the transformed equation by expressing this fact. If (x, y) are the co-ordinates of P, and $(x + dx, y + dy)$ those of q, we have $\rho_1 + d\rho_1 = f_1(x + dx, y + dy)$ and $\rho_2 = f_2(x + dx, y + dy)$,

or $\quad \dfrac{d\rho_1}{dx} dx + \dfrac{d\rho_1}{dy} dy = d\rho_1; \quad \dfrac{d\rho_2}{dx} dx + \dfrac{d\rho_2}{dy} dy = 0.$

Solving these for dx and dy, and denoting $\begin{vmatrix} \dfrac{d\rho_1}{dx}, & \dfrac{d\rho_1}{dy} \\ \dfrac{d\rho_2}{dx}, & \dfrac{d\rho_2}{dy} \end{vmatrix}$ by J, and

putting $\overline{\dfrac{d\rho_1}{dx}}^2 + \overline{\dfrac{d\rho_1}{dy}}^2 = h_1^2$, $\overline{\dfrac{d\rho_2}{dx}}^2 + \overline{\dfrac{d\rho_2}{dy}}^2 = h_2^2$, we have $J \times Pq = h_2 d\rho_1$;

and similarly by determining the co-ordinates of p, we have $J \times Pp = h_1 dp_2$. Denote Pq by ds_2 and Pp by ds_1; then

$$Jds_1 = h_1 dp_2; \quad Jds_2 = h_2 dp_1.$$

But it is easy to prove that $J = h_1 h_2$; hence

$$ds_1 = \frac{1}{h_2} dp_2; \quad ds_2 = \frac{1}{h_1} dp_1.$$

Hence $qQ = \frac{1}{h_2} dp_2 - \frac{1}{h_2^2} \frac{dh_2}{dp_1} dp_1 dp_2$, and $pQ = \frac{1}{h_1} dp_1 - \frac{1}{h_1^2} \frac{dh_1}{dp_2} dp_1 dp_2$.

Now the flow inwards through the side Pp is $\frac{d\phi}{ds_2} \cdot Pp$, and the flow outwards through the side qQ is $\left(\frac{d\phi}{ds_2} + \frac{d^2\phi}{ds_2^2} ds_2\right) \cdot qQ$; therefore the flow outwards resulting from these together is

$$\frac{1}{h_2} \frac{d^2\phi}{ds_2^2} dp_2 ds_2 - \frac{1}{h_2^2} \frac{dh_2}{dp_1} \frac{d\phi}{ds_2} dp_1 dp_2, \text{ or } \frac{h_1}{h_2} \left(\frac{d^2\phi}{dp_1^2} - \frac{1}{h_2} \frac{d\phi}{dp_1} \frac{dh_2}{dp_1}\right) dp_1 dp_2.$$

Expressing in the same way the total flow through the sides Pq and pQ, and adding, we have

$$h_1{}^2 \frac{d^2\phi}{dp_1{}^2} + h_2{}^2 \frac{d^2\phi}{dp_2{}^2} - \frac{h_1{}^2}{h_2} \frac{d\phi}{dp_1} \frac{dh_2}{dp_1} - \frac{h_2{}^2}{h_1} \frac{d\phi}{dp_2} \frac{dh_1}{dp_2} = 0,$$

which is the required transformed equation.

12. Determine the circumstances of motion (irrotational) so that the stream-lines shall be the series of curves obtained by varying a in the equation $r^n \sin n\theta = a^n$.

13. Determine the motion so that the stream-lines shall be a series of rectangular hyperbolas having the same centre and asymptotes, and varying only the magnitudes of their axes.

Ans. $\psi = A(x^2 - y^2); \quad \phi = -2Axy$.

14. Is it possible to determine a velocity-potential function (or a flow-function) of the form $x^2 f\left(\frac{y}{x}\right)$?

Ans. Yes; such a function will be of the form $A(y^2 - x^2) + Bxy$, where A and B are arbitrary constants.

[Putting $\nabla^2 \cdot x^2 f\left(\frac{y}{x}\right) = 0$, we obtain the equation

$$(1 + \lambda^2) f'' - 2\lambda f' + 2f = 0,$$

where λ is used for $\frac{y}{x}$, and f', f'' stand for $\frac{df(\lambda)}{d\lambda}$, $\frac{d^2 f(\lambda)}{d\lambda^2}$, respectively. To integrate this, observe that $f'' = \frac{1}{\lambda} \frac{d}{d\lambda}(\lambda f' - f)$; so that the equation becomes $\frac{1+\lambda^2}{\lambda} \frac{d}{d\lambda}(\lambda f' - f) - 2(\lambda f' - f) = 0$, which gives $\frac{\lambda f' - f}{1 + \lambda^2} = A$. Again, $\lambda f' - f = \lambda^2 \frac{d}{d\lambda}\left(\frac{f}{\lambda}\right)$, $\therefore \frac{d}{d\lambda}\left(\frac{f}{\lambda}\right) = A\left(1 + \frac{1}{\lambda^2}\right)$, whence f is at once found.]

Examples.

15. Show that the equations $u = \dfrac{ax-by}{x^2+y^2}$, $v = \dfrac{ay+bx}{x^2+y^2}$ express an irrotational liquid motion, and that the velocity-potential function is obtained by combining the well-known functions $\log r$ and θ in the form

$$\phi = a \log r + b\theta,$$

a and b being constants.

16. A liquid is whirling round a fixed centre so that each particle describes a circle round the centre with an angular velocity varying as the nth power of the radius; find the ratio of the vortical spin of the particle to its angular velocity about the centre.

Ans. $\dfrac{n+2}{2}$. [Use the equation $\nabla^2 \psi = -2\omega$ in polar co-ordinates, and observe that $\dfrac{d\psi}{d\theta} = 0$, $\dfrac{d\psi}{dr} = r\dot{\theta} = kr^{n+1}$, &c. Otherwise, put

$$u = -\dot{\theta}y, \quad v = \dot{\theta}x, \quad \&c.]$$

17. Round any point, P, in the plane of motion of a gas moving with uniplanar motion, a small closed curve is described; prove that the mean tangential is to the mean normal velocity on the curve as 2ω is to θ, where ω and θ are the vortical spin and expansion at P.

18. Investigate the two-dimensional motion of a gas when it consists of a squirt from a fixed origin with condensation varying inversely as the nth power of the distance from the origin.

Ans. If $-\theta$ (the condensation) $= \dfrac{\mu}{r^n}$, the velocity potential (which exists) is $k \log r - \dfrac{\mu}{(n-2)^2 r^{n-2}}$, where k is a constant.

19. When the two-dimensional steady motion of a gas consists of a squirt from a fixed origin, and the law of condensation (as a function of the distance from the origin) is assigned, show how to determine the velocity and density at each point.

Ans. Let $u = xf(r)$, $v = yf(r)$, $-\theta =$ condensation; then

$$f(r) = \dfrac{k}{r^4} - \dfrac{1}{r^2}\int \theta r \, dr.$$

Also since $\dfrac{d\rho}{dt} = 0$ (Art. 94), the equation for ρ is

$$\dfrac{1}{\rho}\left(u \dfrac{d\rho}{dx} + v \dfrac{d\rho}{dy}\right) = \theta.$$

But $\dfrac{d}{dx} = \dfrac{x}{r}\dfrac{d}{dr}$; $\therefore \dfrac{d\rho}{\rho} = \dfrac{\theta \, dr}{rf(r)}$, which determines ρ.

20. Show that in the steady motion of a gas a flow-function may be found by means of an integrating factor for $u\,dy - v\,dx$.

[The density, ρ, at each point is the required integrating factor. See Art. 93.]

178   Kinematics of Fluids.

21. In a steady squirt motion of a gas the compression at any distance from the origin is $\frac{\mu}{r}\sin\frac{r}{a}$, where μ and a are constants; determine the density, velocity, etc. at each point.

22. The steady motion of a fluid is a whirl round a fixed axis, the vortical spin at each point being assigned; find the velocity and condensation at each point.

Ans. The condensation $= 0$. To get the velocity, let
$$u = -yf(r), \quad v = xf(r).$$

112. Superposition of Stream Lines and Equipotential Curves. Suppose that any cause of motion of a fluid gives a

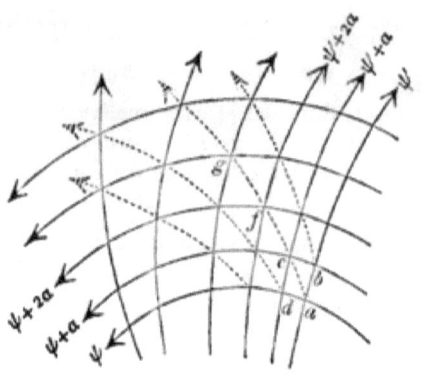

Fig. 62.

series of stream lines denoted in fig. 62 by ψ, $\psi+a$, $\psi+2a$, ... *constructed at intervals so that the time-rate of flow across the sections of the channels of flow* (Art. 107) *contained between successive pairs of them is the same and equal to* a (the values of the flow function for these lines being ψ, $\psi+a$, $\psi+2a$, ...); suppose also that another cause, acting independently of the first, would give a series of stream lines denoted by ψ', $\psi'+a$, $\psi'+2a$, ... *so constructed at intervals that the time-rate of flow across their channels is also* a (the values of their flow functions being ψ', $\psi'+a$, $\psi'+2a$, ...); it is required to construct from this dual system the series of stream lines which will result if both causes of motion simultaneously agitate the fluid.

[The stream lines of each system are supposed to be constructed at very small intervals.]

Superposition of Stream Lines.

Let a be the point at which the two lines ψ and ψ' intersect, and b, c, d the remaining intersections of the lines ψ, $\psi + a$, ψ', $\psi' + a$. Then $abcd$ may be considered as a small parallelogram. Also let the velocity at a due to the first cause be from a to d (of course along ψ), as indicated by the arrows, and the velocity at a due to the second from a to b. Denote these velocities by v_1 and v_2, respectively. Then if n denotes the length of the perpendicular from a on the curve $\psi + a$, we have (Art. 104)

$$v_1 = \frac{a}{n}.$$

Also if n' denotes the perpendicular from a on the curve $\psi' + a$,

$$v_2 = \frac{a}{n'}.$$

Hence $\frac{v_1}{v_2} = \frac{n'}{n}$. But the area of the little triangle $abc = \frac{1}{2} n' \times ab$, and also it $= \frac{1}{2} n \times ad$; $\therefore \frac{n'}{n} = \frac{ad}{ab}$; hence

$$\frac{v_1}{v_2} = \frac{ad}{ab},$$

and therefore the resultant of v_1 and v_2 is directed in the diagonal ac.

Similarly at c the resultant velocity is directed in the diagonal cf; and that at f in the diagonal fg. Hence $acfg\ldots$ is the resultant stream line at a.

All other resultant stream lines are similarly constructed, by drawing the diagonal of any other little parallelogram in the map and continuing it as above.

We shall now prove the following theorem—

In the stream line map the successive differences of the flow function from each resultant stream line to the next consecutive is equal to the constant difference of flow function which was assumed in drawing the map.

Let V = resultant velocity at a = velocity along ac; N = normal distance between the stream line $acfg$ and the next consecutive = perpendicular from b on ac; ω = angle bad. Then

$$V^2 = \frac{n^2 + n'^2 + 2nn' \cos \omega}{n^2 n'^2} \cdot a^2.$$

Also $ab = \dfrac{n}{\sin \omega}$, $ad = \dfrac{n'}{\sin \omega}$, $\therefore ac^2 = \dfrac{n^2 + n'^2 + 2nn' \cos \omega}{\sin^2 \omega}$,

$\therefore V = \dfrac{ac \cdot \sin \omega}{nn'} \cdot a$; and $ac \times N = 2 \cdot \text{area } abc$

$$= ab \cdot ad \sin \omega = \dfrac{nn'}{\sin \omega};\quad \text{therefore}$$

$$V = \dfrac{a}{N}.$$

But $V = \dfrac{\text{difference of flow function}}{N}$, $\therefore$ the difference of two consecutive resultant flow functions is equal to a, the difference assumed for the two superposed systems in drawing the map.

Exactly the same theorem holds with reference to the superposition of two separate systems of equipotential curves; but the diagonal of each little parallelogram must in this case be taken in a way different from the preceding.

For, let ψ, $\psi + a$, ... ψ', $\psi' + a$, ... be now two sets of equipotential curves.

Then the velocity at a due to the first cause is *at right angles* to ad and equal to $\dfrac{a}{n}$ (Art. 100); similarly the second velocity at a is at right angles to ab and equal to $\dfrac{a}{n'}$. These components are, as before, *directly* proportional to ad and ab, respectively, therefore, by the triangle of velocities, the resultant velocity at a is at *right angles* to bd; and therefore, if the intervals are sufficiently small, the resultant velocity at every point on bd is at right angles to bd, which is therefore an element of an equipotential curve. This element continued from point to point by the same rule becomes an equipotential curve.

Moreover, exactly as before, the successive potential differences along the resultant curves are all equal to a, the difference assumed in drawing the map of the component systems.

This graphic method of superposition is found in Clerk-Maxwell's *Electricity and Magnetism* (Vol. i. p. 265, 2nd ed.); it has also been employed by Professors G. Carey Foster and O. J. Lodge for the study of the effects of a multiple system of sources and sinks in the flow of electricity over a uniform plane

Energy in non-vortical motion. 181

conducting surface. (See the *Proceedings of the Physical Society*, Feb., 1875.)

113. Energy of non-vortically moving Liquid. Let $ADBC$ (fig. 63) be any closed curve containing non-vortically moving liquid. [Of course we do not assume this curve to be a boundary of the liquid. See Art. 111.] We propose to find an expression for the energy of the liquid enclosed by it. Divide its area into an indefinitely large number of narrow stream channels, and consider the liquid in one of them, AB. At any point P draw the normal section of the channel, and in the time Δt suppose that the mass of liquid crossing the section at P is that contained between the normal section at P and the normal section at a close point, Q. Let this mass be Δm; let $PQ = \Delta s$; ϕ = velocity potential at Q. Then the energy of the mass Δm is

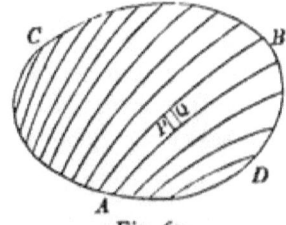

Fig. 63.

$$\tfrac{1}{2}\left(\frac{d\phi}{ds}\right)^2 \Delta m, \quad \text{or} \quad \tfrac{1}{2}\frac{d\phi}{ds}\cdot\frac{d\phi}{ds}\Delta m.$$

But since the particles in the section at Q have gone over Δs in the time Δt, we may put $\dfrac{\Delta s}{\Delta t}$ for $\dfrac{d\phi}{ds}$; and since Δm is the quantity which crosses *every* section of the channel in time Δt, the energy, ΔE, in the channel is $\tfrac{1}{2}\dfrac{\Delta m}{\Delta t}\Sigma\dfrac{d\phi}{ds}\Delta s$, the summation extending from B to A; i.e., $\Delta E = \tfrac{1}{2}(\phi_A - \phi_B)\dfrac{dm}{dt}$. But $\dfrac{dm}{dt}$ is the time rate of flow across the section of the limiting curve at A, which (Art. 92) = density multiplied by normal velocity multiplied by element of arc at $A = \rho\left(\dfrac{d\phi}{d\nu}\right)_A \cdot ds_A$. Similarly, measuring the element of normal *outwards* at B, $\dfrac{dm}{dt} = -\rho\left(\dfrac{d\phi}{d\nu}\right)_B \cdot ds_B$.

Hence $\Delta E = \tfrac{1}{2}\rho\phi_A\left(\dfrac{d\phi}{d\nu}\right)_A ds_A + \tfrac{1}{2}\rho\phi_B\left(\dfrac{d\phi}{d\nu}\right)_B ds_B;$

so that if we take all the channels of flow, and measure dv, the element of normal, consistently *outwards* all over the curve $ADBC$, we have for E, the total energy contained,

$$E = \tfrac{1}{2}\rho \int \phi \frac{d\phi}{dv} ds, \qquad (a)$$

the integration extending over the whole contour of the curve.

From (a) we derive another proof of the result given in Art. 105, viz., that irrotational motion is impossible inside a boundary which has at no point a velocity along the normal to it; for $E = \tfrac{1}{2}\rho \int (u^2+v^2)dS$, and if at every point on the boundary $\dfrac{d\phi}{dv} = 0$, we must have $u=0$, $v=0$ at every point inside. Hence if we take a liquid at rest; then set its boundary in motion in any way whatever, altering its shape to any extent; and then suddenly stop all motion of the boundary, the *whole* of the fluid inside must *instantly* come to rest. (Thomson, *Vortex Motion*, Edin. Trans., 1869.)

114. Green's Equation. Let U and V be any two functions of x, y, the co-ordinates of a point P inside any closed curve in the plane xy; let $\nabla^2 \equiv \dfrac{d^2}{dx^2} + \dfrac{d^2}{dy^2}$; let dS be the element of area at P; let ds denote the element of arc at any point of the curve, and dv the corresponding element of normal drawn outwards; then will

$$\int U\nabla^2 V\,dS = \int U\frac{dV}{dv}ds - \int\left(\frac{dU}{dx}\frac{dV}{dx} + \frac{dU}{dy}\frac{dV}{dy}\right)dS. \qquad (a)$$

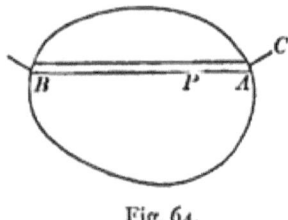

Fig. 64.

First let us see how the line-rate of variation of V (any function) along the normal at any point A of a curve is deduced from the line-rates $\dfrac{dV}{dx}$ and $\dfrac{dV}{dy}$ at A. Let l and m be the cosine and sine of the angle which the normal AC makes with the axis of x. Then if $AC = dv$ and (α, β) are the co-ordinates of A, those of C are $(\alpha + l\,dv, \beta + m\,dv)$; so that if $V = f(x, y)$, we have
$$V_A = f(\alpha, \beta),$$

Energy in vortical motion.

$$V_C = f(a+ldv, \beta+mdv) = V_A + (l\frac{dV}{da} + m\frac{dV}{d\beta})dv;$$

hence $\quad \dfrac{V_C - V_A}{dv},\quad$ or $\quad \dfrac{dV}{dv} = l\dfrac{dV}{da} + m\dfrac{dV}{d\beta}.$

Now $\int U\nabla^2 V\,dS = \iint U(\dfrac{d^2V}{dx^2} + \dfrac{d^2V}{dy^2})\,dx\,dy.$ First take $\int U\dfrac{d^2V}{dx^2}\,dx$, and perform the integration between B and A, the points in which a line through P parallel to the axis of x meets the curve. Thus we get $(U\dfrac{dV}{dx})_A - (U\dfrac{dV}{dx})_B - \int\dfrac{dU}{dx}\dfrac{dV}{dx}\,dx.$

Multiply this by dy, and observe that at A we have $dy = lds$, and at B, $dy = -lds$; taking therefore the terms relative to A and B in one, we get

$$\iint U\frac{d^2V}{dx^2}\,dx\,dy = \int lU\frac{dV}{dx}\,ds - \iint\frac{dU}{dx}\frac{dV}{dx}\,dx\,dy.$$

Similarly, $\iint U\dfrac{d^2V}{dy^2}\,dx\,dy = \int mU\dfrac{dV}{dy}\,ds - \iint\dfrac{dU}{dy}\dfrac{dV}{dy}\,dx\,dy.$

Adding,

$$\iint U\nabla^2 V\,dx\,dy = \int U(l\frac{dV}{dx} + m\frac{dV}{dy})\,ds - \iint(\frac{dU}{dx}\frac{dV}{dx} + \frac{dU}{dy}\frac{dV}{dy})\,dx\,dy,$$

which is identical with (a) above.

Cor. Let U and V be the same; then

$$\int U\nabla^2 U\,dS = \int U\frac{dU}{dv}\,ds - \int\{\overline{\frac{dU}{dx}}\Big|^2 + \overline{\frac{dU}{dy}}\Big|^2\}\,dS. \qquad (\beta)$$

115. Energy of Vortically moving Liquid. Let the liquid contained within the contour $ADBC$ (fig. 63) be in vortical motion, and suppose ψ to be the flow function at any point P. Divide the area into elements of the type $dx\,dy$, i.e., small rectangles with their sides parallel to two fixed rectangular axes. Then the square of the resultant velocity at P is $(\dfrac{d\psi}{dx})^2 + (\dfrac{d\psi}{dy})^2$, $\therefore$ the energy of the element of mass at P is

$$\tfrac{1}{2}\rho\{\overline{\frac{d\psi}{dx}}\Big|^2 + \overline{\frac{d\psi}{dy}}\Big|^2\}\,dx\,dy.$$

Hence $\quad E = \tfrac{1}{2}\rho\iint\{\overline{\dfrac{d\psi}{dx}}\Big|^2 + \overline{\dfrac{d\psi}{dy}}\Big|^2\}\,dx\,dy;$

and this formula may be put into another shape by means of Green's equation. From (β) of last Article we have

$$\int \psi \nabla^2 \psi \, dS = \int \psi \frac{d\psi}{d\nu} ds - \frac{2}{\rho} E.$$

But $\nabla^2 \psi = -2\omega$ (Art. 102), therefore

$$E = \tfrac{1}{2}\rho \int \psi \frac{d\psi}{d\nu} ds + \rho \int \omega \psi \, dS. \tag{1}$$

The expression for E given in Art. 113 for the case in which ϕ exists follows at once in the same way, since $\overline{\dfrac{d\phi}{dx}}^2 + \overline{\dfrac{d\phi}{dy}}^2$ is the square of the velocity at P, so that we get the result from (1) by writing ϕ for ψ and putting $\omega = 0$. In non-vortical motion, therefore, the term in the energy consisting of a *surface*-integral[1] over the area disappears, but not so in vortical motion.

If the liquid is at rest over the boundary $\dfrac{d\psi}{d\nu} = 0$, and we have simply $E = \rho \int \omega \psi \, dS$; and if there is not vortical motion throughout the whole area, but only local vortices, this integral will reduce to a simple sum of terms equal in number to the number of vortices.

116. Resistance of a Flow Channel. Suppose AB, fig. 63, to be a channel of flow bounded by two very close streamlines. Then, in analogy with Ohm's Law in the theory of electric current flow, the *resistance* of this channel to the flow is defined to be *the difference of potential of its extremities divided by the time-rate of flow across any of its normal sections*. Hence if $\Delta \psi$ is the difference of the flow function for its bounding stream-lines, its resistance is

$$\frac{\phi_A - \phi_B}{\Delta \psi}. \tag{1}$$

Consider now any area, $ABCD$, fig. 65, bounded by two equipotential curves, AB and CD, on which the velocity

---

[1] This is in reality a *volume*-integral; dS stands for a volume with unit height standing on the area dS; and similarly ds is really an *area* with unit height.

Lagrangian Method.

potentials are ϕ_1 and ϕ_2, respectively, and by two lines of flow, AD and BC, on which the stream functions are ψ_1 and ψ_2, respectively. Then to find the resistance of this area, we may consider it as broken up into an indefinitely great number of flow channels, or an indefinitely great number of equipotential strips. Adopting the latter method, let $pqrs$ be one of the strips, the potential along rs being ϕ, and that along pq being $\phi + \Delta\phi$. The flow takes place *normally* to this strip, and its amount (Art. 103) is $\psi_1 - \psi_2$; hence the resistance of the strip is

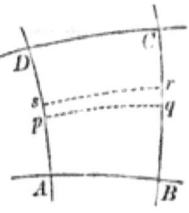

Fig. 65.

$$\frac{\Delta\phi}{\psi_1 - \psi_2}. \qquad (2)$$

Also the liquid encounters the strips *in succession*, so that their combined resistance, R, is the *sum* of their resistances; therefore

$$R = \frac{\phi_1 - \phi_2}{\psi_1 - \psi_2}. \qquad (3)$$

The same result would have been obtained if we had chosen to break the area up into flow channels. In this case the liquid does not encounter the strips in succession but *side by side;* so that, other things being the same, a multiplication of such channels diminishes the resistance to flow, since a *wider* channel is thus provided for the flow. Their combined resistance is found from the fact that its *reciprocal* is equal to the sum of the *reciprocals* of their separate resistances. Integrating the reciprocal of (1), and taking the reciprocal of the result, we get (3).

The energy of the liquid contained in the area $ABCD$ is at once obtained from (1) of last Article. The portions of the integral on the right-hand side contributed by the lines AD and BC are zero; that contributed by AB is $\frac{1}{2}\rho\phi_1(\psi_2 - \psi_1)$, and that contributed by CD is $\frac{1}{2}\rho\phi_2(\psi_1 - \psi_2)$; so that

$$E = -\tfrac{1}{2}\rho(\phi_1 - \phi_2)(\psi_1 - \psi_2). \qquad (4)$$

117. Lagrangian, or Historical Method. In this method, as already explained (Art. 89), we follow the course of an individual fluid particle, P. At any time, t, let (x, y) be the co-ordinates of this particle, at this time let ρ be the fluid density at P, and (a, b) be the co-ordinates of the point, A, occupied by P at the origin of time. Then x and y are supposed to be expressed as functions of a, b, and t. Suppose

$$x = f(a, b, t), \quad y = g(a, b, t), \qquad (1)$$

where f and g are symbols of functionality.

We shall now find the form which the equation of continuity takes.

Let B be a particle very near A when $t = 0$, the co-ordinates of B being $(a+\alpha, b+\beta)$; let ρ_0 be the density at A at this time; also at this time let C be another particle very close to A, the co-ordinates of C being $a+\alpha'$, $b+\beta'$; at the time t let the particles B and C occupy the positions Q and R, respectively, which are both very close to P; and let the co-ordinates of Q and R be $(x+\xi, y+\eta)$ and $(x+\xi', y+\eta')$, respectively.

Then we shall express the fact that the *mass* covering the triangular area QPR is equal to the *mass* covering BAC. Now the area QPR is $\frac{1}{2}\begin{vmatrix} \xi, \eta \\ \xi', \eta' \end{vmatrix}$, and that of BCA is $\frac{1}{2}\begin{vmatrix} \alpha, \beta \\ \alpha', \beta' \end{vmatrix}$; hence

$$\rho \begin{vmatrix} \xi, \eta \\ \xi', \eta' \end{vmatrix} = \rho_0 \begin{vmatrix} \alpha, \beta \\ \alpha', \beta' \end{vmatrix} = \text{constant.} \qquad (2)$$

Again, $x+\xi = f(a+\alpha, b+\beta, t) = f + \alpha\dfrac{df}{da} + \beta\dfrac{df}{db}$, where f stands for $f(a, b, t)$; so that we have

$$\xi = \alpha\frac{df}{da} + \beta\frac{df}{db}, \qquad (3)$$

$$\eta = \alpha\frac{dg}{da} + \beta\frac{dg}{db}, \qquad (4)$$

and similar values of ξ' and η'.

Substituting these in (2), we get

$$\rho \begin{vmatrix} \dfrac{df}{da}, & \dfrac{df}{db} \\ \dfrac{dg}{da}, & \dfrac{dg}{db} \end{vmatrix} = \rho_0, \qquad (5)$$

or
$$J\rho = \rho_0, \qquad (6)$$

where $\mathcal{J}$ is used for the determinant, which is the Jacobian of the functions f and g with respect to the co-ordinates a, b.

Equation (6) is, therefore, the Lagrangian form of the equation of continuity.

For a liquid ρ (the density of an invariable particle) is constant and $=\rho_0$, so that

$$\mathcal{J} = 1, \text{ for a liquid.} \tag{7}$$

118. Lagrangian Equation for Vortical Spin. We shall now calculate the vortical spin, or rotation, ω, at the particle P. It may be calculated by transformation of variables; but the following method relies on elementary principles.

The method consists in calculating the components of *relative* velocity of any very close particle, Q, with respect to P. The component velocities of P are, of course, $\frac{dx}{dt}$ and $\frac{dy}{dt}$, or $\dot{f}$ and $\dot{g}$, respectively; and since t, a, and b are completely independent variables, the order of differentiations with respect to them is interchangeable, so that the component velocities of Q are $\dot{f}(a+a, b+\beta, t)$, $\dot{g}(a+a, b+\beta, t)$, (Q being originally at B, whose co-ordinates are $a+a$, $b+\beta$). Hence the components for Q are

$$\dot{f} + a\frac{d\dot{f}}{da} + \beta\frac{d\dot{f}}{db} \text{ and } \dot{g} + a\frac{d\dot{g}}{da} + \beta\frac{d\dot{g}}{db};$$

so that the components of *relative* velocity are

$$a\frac{d\dot{f}}{da} + \beta\frac{d\dot{f}}{db}, \tag{1}$$

$$a\frac{d\dot{g}}{da} + \beta\frac{d\dot{g}}{db}. \tag{2}$$

The components of strain at P are obtained, of course, by expressing these in terms of the co-ordinates, ξ, η, of Q relatively to P.

Now from (3) and (4) of last Article, we have

$$\mathcal{J}a = \xi\frac{dg}{db} - \eta\frac{df}{db}, \tag{3}$$

$$\mathcal{J}\beta = -\xi\frac{dg}{da} + \eta\frac{df}{da}. \tag{4}$$

Substituting in (1) and (2), we get for the components

$$\frac{1}{\mathcal{J}}[(\frac{d\dot{f}}{da}\frac{dg}{db}-\frac{d\dot{f}}{db}\frac{dg}{da})\xi-(\frac{d\dot{f}}{da}\frac{df}{db}-\frac{d\dot{f}}{db}\frac{df}{da})\eta], \qquad (5)$$

$$\frac{1}{\mathcal{J}}[(\frac{d\dot{g}}{da}\frac{dg}{db}-\frac{d\dot{g}}{db}\frac{dg}{da})\xi-(\frac{d\dot{g}}{da}\frac{df}{db}-\frac{d\dot{g}}{db}\frac{df}{da})\eta]. \qquad (6)$$

We must now put these into forms showing a pure strain and a rotation (Art. 96), i.e., we must put them into the forms

$$\lambda\xi+s\eta-\omega\eta, \qquad (7)$$

$$s\xi+\mu\eta+\omega\xi, \qquad (8)$$

where $2s$ and ω will be the shear and vortical spin, respectively.

The values of ω and s are at once given by the equations

$$\omega = -\frac{1}{2\mathcal{J}}\begin{vmatrix}\frac{df}{da}, & \frac{df}{db} \\ \frac{d\dot{f}}{da}, & \frac{d\dot{f}}{db}\end{vmatrix} - \frac{1}{2\mathcal{J}}\begin{vmatrix}\frac{dg}{da}, & \frac{dg}{db} \\ \frac{d\dot{g}}{da}, & \frac{d\dot{g}}{db}\end{vmatrix}, \qquad (9)$$

$$s = \frac{1}{2\mathcal{J}}\begin{vmatrix}\frac{df}{da}, & \frac{df}{db} \\ \frac{d\dot{f}}{da}, & \frac{d\dot{f}}{db}\end{vmatrix} - \frac{1}{2\mathcal{J}}\begin{vmatrix}\frac{dg}{da}, & \frac{dg}{db} \\ \frac{d\dot{g}}{da}, & \frac{d\dot{g}}{db}\end{vmatrix}, \qquad (10)$$

while for the dilatation, or expansion, $\lambda+\mu$, or θ, we get

$$\theta = \frac{1}{\mathcal{J}}\frac{d\mathcal{J}}{dt}. \qquad (11)$$

This last we might at once have deduced from (6) of last Article; for (Art. 95) $\theta = -\frac{1}{\rho}D_t\rho$.

119. Generalised Co-ordinates. We have assumed that x, y, the co-ordinates of an invariable fluid particle, are expressed as functions of t and the initial *rectangular co-ordinates* of the particle, a and b. This latter restriction is not necessary. Instead of a and b, the initial rectangular co-ordinates, we may express x and y in terms of t and *any* two constant quantities whatever, K and L, which serve to identify the particular particle; so that we may take

$$x = F(K, L, t); \quad y = G(K, L, t). \qquad (1)$$

For K and L will each be some function of a and b, so that we shall, if we choose, be able to express x and y in the very same forms, (1), Art. 117, as before; and the results (6), Art. 117, and (9), (10), (11), Art. 118, will still hold; but we must express them in terms of K and L.

Now it follows at once that

$$\begin{vmatrix} \dfrac{df}{da}, & \dfrac{df}{db} \\ \dfrac{dg}{da}, & \dfrac{dg}{db} \end{vmatrix} = \begin{vmatrix} \dfrac{dF}{dK}, & \dfrac{dF}{dL} \\ \dfrac{dG}{dK}, & \dfrac{dG}{dL} \end{vmatrix} \times \begin{vmatrix} \dfrac{dK}{da}, & \dfrac{dK}{db} \\ \dfrac{dL}{da}, & \dfrac{dL}{db} \end{vmatrix}, \qquad (2)$$

with similar values of the determinants in (9) and (10) of last Article, the multiplier $\begin{vmatrix} \dfrac{dK}{da}, & \dfrac{dK}{db} \\ \dfrac{dL}{da}, & \dfrac{dL}{db} \end{vmatrix}$, or $\mathcal{J}_0$ suppose, coming out in all; so that equation (6) of Art. 117 will be simply

$$\mathcal{J}\rho = \mathcal{J}_0 \rho_0,$$

where $\mathcal{J}$ now stands for the first of the two determinants on the right side of (2); and the values of ω, s, and θ of precisely the same forms in K and L as in a and b. Hence *we may regard a and b as any two constants whatever identifying a fluid particle—* e.g., its initial polar co-ordinates.

120. Graphic representation of Expansion, Shear, and Vortical Spin. Expressing the rectangular co-ordinates, x, y, of any particle, P, in terms of any two co-ordinates (simple or generalised), K and L, in the forms (1) of last Article, take a point, P_1, whose rectangular co-ordinates are

$$\frac{dF}{dK} \quad \text{and} \quad \frac{dF}{dL};$$

and also take a point, P_2, whose co-ordinates are

$$\frac{dG}{dK} \quad \text{and} \quad \frac{dG}{dL}.$$

Then if O is the origin, $\mathcal{J}$ is obviously equal to double the area of the triangle $P_1 O P_2$; so that equation (11), Art. 118, informs us that *the expansion at P at any instant is equal to the areal expansion of the triangle $P_1 O P_2$*. Also if h_1 and h_2 denote double the areas swept out per unit of time by P_1 and P_2 round O, we have by (9) and (10) of that Article

$$\omega = -\frac{1}{2j}(h_1 + h_2). \tag{1}$$

$$s = \frac{1}{2j}(h_1 - h_2). \tag{2}$$

If the fluid be a liquid, it follows that the two derived points determine with the origin a triangle of constant area.

121. Invariability of Vortices. On account of its importance in the general theory of fluid motion, it is thought advisable to introduce here a proposition which is not kinematical but kinetical. The proposition is this—

If any fluid in which the density at any particle is either constant or a function of the intensity of pressure at the particle, moves under the action of external forces which have a potential, then, if at any time whatever there was vortical spin in any particle, this particle will always continue to have vortical spin; and if at any time the particle had no vortical spin, it can never acquire it.

Suppose that V is the potential, per unit mass, of the external forces, at the point x, y, where there is a particle of density ρ, and where p is the intensity of fluid pressure, the equations of motion of the particle are

$$\frac{d^2 x}{dt^2} = \frac{dV}{dx} - \frac{1}{\rho}\frac{dp}{dx};$$

$$\frac{d^2 y}{dt^2} = \frac{dV}{dy} - \frac{1}{\rho}\frac{dp}{dy}.$$

Hence if $\rho = f(p)$, these equations can be written

$$\frac{d^2 x}{dt^2} = \frac{d\Omega}{dx}, \quad \frac{d^2 y}{dt^2} = \frac{d\Omega}{dy}, \tag{1}$$

where Ω is a function of x, y. Multiplying these last by $\frac{dx}{dK}$ and $\frac{dy}{dK}$, respectively, and adding, and also regarding x and y as given in terms of the constants K and L by equations (1) of Art. 119, we have

Invariability of Vortices.

$$\ddot{F}\frac{dF}{dK} + \ddot{G}\frac{dG}{dK} = \frac{d\Omega}{dK}, \qquad (2)$$

$\ddot{F}$ denoting $\frac{d^2F}{dt^2}$, &c. Similarly multiplying the expressions (1) by $\frac{dx}{dL}$ and $\frac{dy}{dL}$, and adding, we get

$$\ddot{F}\frac{dF}{dL} + \ddot{G}\frac{dG}{dL} = \frac{d\Omega}{dL}. \qquad (3)$$

Since K and L are independent quantities, we get identical results by differentiating (2) with respect to L and (3) with respect to K. Doing so, we get by subtraction

$$\frac{dF}{dK}\frac{d\ddot{F}}{dL} - \frac{dF}{dL}\frac{d\ddot{F}}{dK} + \frac{dG}{dK}\frac{d\ddot{G}}{dL} - \frac{dG}{dL}\frac{d\ddot{G}}{dK} = 0. \qquad (4)$$

But the first two terms of this equation are obviously

$$\frac{d}{dt}\left(\frac{dF}{dK}\frac{d\dot{F}}{dL} - \frac{dF}{dL}\frac{d\dot{F}}{dK}\right), \quad \text{or} \quad \frac{dh_1}{dt} \text{ (Art. 120)};$$

and the last two are $\frac{dh_2}{dt}$; so that (4) is the same as

$$\frac{d}{dt}(h_1 + h_2) = 0,$$

$$\therefore \ \frac{d}{dt}(\mathcal{J}\omega) = 0, \qquad (5)$$

$$\therefore \ \mathcal{J}\omega = \mathcal{J}_0 \omega_0, \qquad (6)$$

which proves the proposition. For, if ω_0 (which may be taken as any previous value of the vortical spin of the particle) is not zero, ω can never become zero unless $\mathcal{J}$ becomes ∞; and if $\omega_0 = 0$, ω is always zero.

Thus, supposing the fluid to be compressible, so that $\mathcal{J}$ is not constant, the vortical spin of an element will get quicker as the element becomes more compressed, and slower as the element becomes less dense, since

$$\frac{\omega}{\rho} = \frac{\omega_0}{\rho_0}, \qquad (7)$$

in virtue of Art. 119; and if the fluid is a liquid the vortical spin of each element remains constant throughout the motion.

122. Acceleration at a point. Supposing, as in Art. 96, that the components, u and v, of velocity at a point, P, are expressed by the equations

$$u = f_1(x, y, t), \quad v = f_2(x, y, t), \tag{1}$$

then at the end of the time-interval Δt, the fluid *particle* which at the time t was at P will be at the point $(x + u\Delta t, y + v\Delta t)$; and by (1) its components of velocity will then be

$$f_1(x + u\Delta t, y + v\Delta t, t + \Delta t) \text{ and } f_2(x + u\Delta t, y + v\Delta t, t + \Delta t);$$

so that the gain of x-velocity by the particle is

$$\frac{df_1}{dx} u\Delta t + \frac{df_1}{dy} v\Delta t + \frac{df_1}{dt} \Delta t;$$

or

$$\left(u \frac{du}{dx} + v \frac{du}{dy} + \frac{du}{dt}\right) \Delta t;$$

and the gains, $\ddot{a}$, $\ddot{\beta}$, of velocity parallel to the axes, per unit of time, i.e., the components of acceleration of the particle, are therefore

$$\ddot{a} = \frac{du}{dt} + u \frac{du}{dx} + v \frac{du}{dy}, \tag{2}$$

$$\ddot{\beta} = \frac{dv}{dt} + u \frac{dv}{dx} + v \frac{dv}{dy}. \tag{3}$$

If the motion is referred to polar co-ordinates, let λ and μ be the velocities along and perpendicular to the radius vector OP (p. 165); at the end of the time-interval Δt, suppose the particle which was at P to have reached Q; let also

$$\lambda = f_1(r, \theta, t), \quad \mu = f_2(r, \theta, t).$$

Then the velocities, λ' and μ', of the particle along and perpendicular to OQ are

$$\lambda' = f_1\left(r + \lambda\Delta t, \; \theta + \frac{\mu}{r}\Delta t, \; t + \Delta t\right);$$

$$\mu' = f_2\left(r + \lambda\Delta t, \; \theta + \frac{\mu}{r}\Delta t, \; t + \Delta t\right);$$

or

$$\lambda' = \lambda + \left(\lambda \frac{d\lambda}{dr} + \frac{\mu}{r}\frac{d\lambda}{d\theta} + \frac{d\lambda}{dt}\right) \Delta t;$$

$$\mu' = \mu + \left(\lambda \frac{d\mu}{dr} + \frac{\mu}{r}\frac{d\mu}{d\theta} + \frac{d\mu}{dt}\right) \Delta t.$$

Now the gain of velocity along OP is

$$\lambda' - \lambda - \mu'\Delta\theta, \text{ or } \lambda' - \lambda - \mu'\frac{\mu}{r}\Delta t;$$

so that the acceleration along the radius vector is

$$\frac{d\lambda}{dt} + \lambda\frac{d\lambda}{dr} + \frac{\mu}{r}\frac{d\lambda}{d\theta} - \frac{\mu^2}{r}. \qquad (4)$$

Similarly the gain of velocity along Pp is $\mu' - \mu + \lambda\Delta\theta$, and the acceleration is
$$\frac{d\mu}{dt} + \lambda\frac{d\mu}{dr} + \frac{\mu}{r}\frac{d\mu}{d\theta} + \frac{\lambda\mu}{r}. \qquad (5)$$

SECTION II.—**Multiply-Connected Spaces.**

123. Single-Valued and Many-Valued Function. A function of one, two, or any number of variables is said to be a *single-valued function* if it can have only one definite value when definite values are assigned to the variables. Thus, $\sin\frac{y}{x}$ is a single-valued function of x and y; but $\sin^{-1}\frac{y}{x}$ is a many-valued function, because there are several angles each of which has its sine equal to $\frac{y}{x}$.

The velocity potential at any point in the case of a whirl (p. 167) is a many-valued function; for it is of the form $A\theta$, or $A\tan^{-1}\frac{y}{x}$; but in this case the stream function, $A\log r$, is single-valued. The reverse takes place in the case of a squirt.

We have seen (Art. 105) that the flow from a point A to another, B, is a definite quantity independent of the path pursued if the velocity potential function is single-valued; but if it is many-valued, the flow is not definite, but depends on the particular path, drawn from A to B, along which it is estimated. Generally, whenever there are vortical regions anywhere in a fluid, the velocity-potential function (which exists, of course, only in the non-vortical regions) is a many-valued function; but its differential coefficients—which express components of velocity at any point—are necessarily single-valued.

Hence theorems, such as Green's (Art. 114), require modification when they are concerned with a function whose value at one and the same point is ambiguous.

In this case there is an artificial method of removing the ambiguity from the function; and this method, after a few preliminary definitions, we proceed to explain.

124. Simply and Multiply-Connected Regions. Let *DEF* (fig. 66) be a contour enclosing any portion of a moving fluid. We may speak of the whole of this space as a *region*. Within this region there may be several smaller regions, such as *A*, *B*, *C*, within each of which the nature of the motion differs in some essential particular from that of the motion in the space outside it.

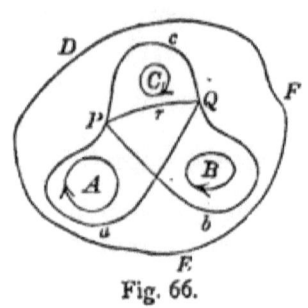

Fig. 66.

For example, the motion may be vortical within *A*, *B*, and *C*, and non-vortical in the space outside them. By the region *DEF* we shall now understand *only that portion of the space bounded by the contour DEF which is not included in any of the sub-regions A, B, C*. Now in the region take any two points, *P* and *Q*, and connect them by any path, *PrQ*, every point of which lies in the region. This path *PrQ* is said to be reconcileable with any other path, *PqQ*, connecting *P* and *Q* if we can imagine the closed curve formed by the two paths to be capable of shrinking up into a point without requiring any of its points to leave the region—or (which comes to the same) if the second path, *PqQ*, can be imagined to change into *PrQ* by a gradual motion of all its points without any of its points ever leaving the given region; and two paths connecting *P* and *Q* are said to be *irreconcileable* if this cannot be done. Thus the paths *PrQ* and *PcQ* are irreconcileable, because in order to bring the second into coincidence with the first by gradual changes, it would be necessary to bring a portion of it through the separate region *C*.

Similarly, the paths PaQ and PbQ are irreconcileable with PrQ, and with each other.

Hence our figure represents the case in which four irreconcileable paths can be drawn between any two points in the region. This region is therefore said to be *quadruply connected*. Similarly, if in a region n irreconcileable paths can be drawn between two points, the region is said to be 'n-ply connected.' A *simply* connected region is one in which *all* paths drawn between the same two points are mutually reconcileable. Thus, if the sub-regions A, B, and C all vanish, the region DEF will be simply connected.

The term 'circuit' will be used to signify any closed path.

Now it has been proved (Art. 105) that the circulation round any circuit is equal to twice the surface-integral of the vortical spin over its area. Hence if this surface-integral over the region A is $\tfrac{1}{2}k_1$, measured in the sense of the arrow, the circulation along the path $PrQa$ is k_1. Denote the flow from P to Q along the path PrQ by $\overline{PrQ}$; then obviously, since $\overline{QaP} = -\overline{PaQ}$, we have

$$\overline{PrQ} = \overline{PaQ} + k_1. \tag{1}$$

But the flow (being the line-integral of the velocity) is

$$\int (u\,dx + v\,dy),$$

and this integral is what we have denoted by ϕ, the velocity-potential. Suppose that we assign an arbitrary value, h, to the velocity potential at P, and let ϕ_Q be its value at Q. Then $\overline{PrQ} = \phi_Q - h$, and equation (1) gives

$$\overline{PaQ} = \phi_Q - h - k_1;$$

but $\overline{PaQ}$ is in the same way to be put equal to potential at Q-potential at P; hence denoting this new potential at Q by $\phi_{Q'}$, we should have $\quad \phi_{Q'} = \phi_Q - k_1,$

so that the velocity-potential at Q has not a single definite value.

In the same way, if k_2 and k_3 are the circulations round the regions B and C, we shall find different values for the velocity-potential at Q by taking the integration round the paths PbQ

and PcQ, and a different value again by going from P to Q along a path which, taken with the path PrQ, forms a circuit enclosing two of the sub-regions.

The velocity-potential at any point in the given region is therefore multiple-valued, or indeterminate to the extent of a quantity of the form $n_1 k_1 + n_2 k_2 + n_3 k_3$, where n_1, n_2, n_3 are integer numbers.

In a simply-connected region no such ambiguity exists, since $\int (u\,dx + v\,dy)$ is the same along all paths drawn from P to Q—i.e., ϕ has a definite value at every point.

125. Method of Barriers. The potential at any point can be made single-valued by making the region simply connected; and this is effected by drawing barriers of arbitrary shapes across from the regions A, B, C to the region DEF. Suppose these barriers to be drawn across to the points a, β, γ (fig. 67), and assume that *both* sides of a barrier are portions of the boundary of the region DEF, the other portions of the boundary being the contours of the regions A, B, C, and the contour DEF itself. Then starting from any point (suppose D) on the boundary it is possible to travel *continuously* over the whole of the boundary, thus completed by the barriers, and to come back to the starting-point; one portion of this motion will consist, for example, of a motion from a along the left-hand side of the barrier up to the surface of A, then a motion round A, and a return motion from A to a along the right-hand side of the barrier.

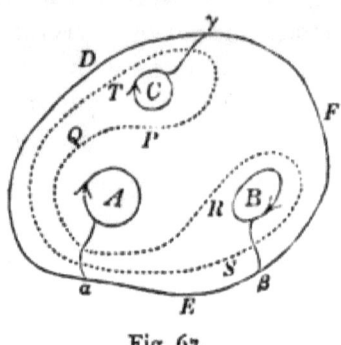

Fig. 67.

Supposing, then, that we take any two points, P and Q, within this modified region, and that we estimate the flow from P to Q along any curve (not represented in the figure) belonging *strictly* to the region, i.e., any curve, however complicated, *which crosses none of the barriers;* this flow will be the same for all such paths; for if we draw two such paths, they will form a circuit which does not include any vortical region, and therefore the flows along them from P to Q are the same. Hence we do not get a multiple-valued potential at Q.

Moreover, the circulation along any circuit, $PQRSTP$, drawn in the unmodified region, i.e., a circuit which cuts the barriers any number of times, is easily found. Thus, the circulations k_1, k_2, k_3, being estimated in the senses of the arrows, the circulation

$$\overline{PQRSTP} = k_2 + k_3; \text{ for } \overline{PQR} = \overline{PQRP} + \overline{PR} = -k_1 + \overline{PR};$$

and the flow from R to S along the path $= k_2 + \overline{RS}$; then the flow along the path from S to $T = k_1 + \overline{ST}$, so that $\overline{PQRST} = k_2 + \overline{PT}$ (since $\overline{PR} + \overline{RS} + \overline{ST} = \overline{PT}$); finally, the flow from T to P along the path $= k_3 - \overline{PT}$; therefore the whole circulation in the circuit $= k_2 + k_3$.

The same method applied to any other path shows that the circulation in it is

$$n_1 k_1 + n_2 k_2 + n_3 k_3 + \ldots \quad (\epsilon)$$

there being any number of vortical regions, *and n_1 will be the number of times that the path crosses the corresponding barrier, a, in the direction of the circulation round the corresponding region, A.* Thus, in the path drawn in the figure, a point travelling along it from P in the sense assumed above crosses the barrier a twice, first in the sense opposed to k_1 and afterwards in the sense of k_1, so that $n_1 = 0$.

Supposing, then, that we artificially modify the region by drawing barriers, the velocity-potential has one definite value at every point; and *the values of ϕ at two points indefinitely close to each other but on opposite faces of a barrier, a, differ by the corresponding circulation, k_1.* For, take two such points, N, N', on opposite sides of the barrier a. Then $\overline{PN} - \overline{PN'} = k_1$; but $\overline{PN} = \phi_N - \phi_P$, and $\overline{PN'} = \phi_{N'} - \phi_P$, $\therefore \phi_N - \phi_{N'} = k_1$.

Hence when we draw the barriers, we may select any one point, O, in the region and assign the value zero to the velocity-potential at it; and then *the value of the potential at any other point, P, in the region is the flow from O to P along any path which does not cut any barrier.* In this way all ambiguity is removed from the value of ϕ.

For example, take the case of the whirl, example 6, p. 167. Here $\phi = A\theta$; but θ, as determining the direction of a line OP, is multiple-valued, the same direction being given by $\theta + 2n\pi$. However, we may draw any line Ox, which we can regard as a barrier, assign a zero value to ϕ at any point on the upper side (or aspect) of this line, and then at any point P, the value of ϕ will be $A \times$ circular measure of $\angle POx$, this angle being always less than 2π.

The quantities $k_1, k_2, k_3, \ldots$ are called the *cyclic constants* of ϕ.

126. Thomson's Modification of Green's Equation. Green's equation (a), p. 182, fails in definiteness if the function U is multiple-valued. Consequently the ambiguity in U must be removed by the method of barriers.

Confining our attention to the equation

$$\frac{2}{\rho} T = \int \phi \frac{d\phi}{dn} ds, \qquad (1)$$

which is derived from Green's general equation, and which expresses the energy, at any instant, of the liquid contained within a given closed curve, DEF (fig. 67), and assuming ϕ to be multiple-valued—or, in other words, that there are vortical regions within the contour—the equation will be applicable without ambiguity only when the region is modified by barriers, and its boundary consists of the contour DEF, the contours of the vortical regions, and both sides of every barrier. Selecting any arbitrary point in the modified region, and assigning a zero potential to it, the value of ϕ at every point is known. Now the value of the right-hand side of (1) given by integration over both aspects of the barrier a taken together is $-k_1 \int_a \frac{d\phi}{dn} ds$, where the suffix merely indicates the barrier over which the integration is performed. For, taking two points, N, N', indefinitely close but at opposite sides of the barrier, the elements of length, ds, at them would contribute the term

$(\phi_N \frac{d\phi_N}{dn} - \phi_{N'} \frac{d\phi_{N'}}{dn}) ds$, or (Art. 125) the term $-k_1 \frac{d\phi_N}{dn} ds$,

Examples.

for there is no ambiguity about the differential coefficients of ϕ, so that $\frac{d\phi_{N'}}{dn} = \frac{d\phi_N}{dn}$. Hence, the suffixes $A, B, \ldots$ denoting integration over the contours of the regions $A, B, \ldots$

$$\frac{2}{\rho} T = \int \phi \frac{d\phi}{dn} ds + \int_A \phi \frac{d\phi}{dn} ds + \int_B \phi \frac{d\phi}{dn} ds + \ldots$$
$$- k_1 \int_\alpha \frac{d\phi}{dn} ds - k_2 \int_\beta \frac{d\phi}{dn} ds - \ldots \quad (2)$$

the first term on the right side referring to the given contour DEF.

This modification of Green's equation was given by Sir W. Thomson (*Vortex Motion*, Edin. Trans., 1869).

EXAMPLES.

1. Of all the motions of a liquid contained within a given contour, which give at each point of the contour the same velocity component along the normal, non-vortical motion produces the least energy (Sir W. Thomson).

[We give the proof for uniplanar motion, but the student has only to introduce a third co-ordinate into the equations to prove it for motion in three dimensions.] Let T be the energy with the velocity-system $\left[\frac{d\phi}{dx}, \frac{d\phi}{dy}\right]$, i.e., with non-vortical motion; and let U be the energy with the velocity system $[u, v]$ such that u and v are not the differential coefficients of a potential function.

Then

$$\frac{2}{\rho}(U - T) = \iint \left(u^2 - \frac{d\phi^2}{dx^2} + v^2 - \frac{d\phi^2}{dy^2}\right) dx\, dy$$

$$= \iint \left[\left(u - \frac{d\phi}{dx}\right)^2 + \left(v - \frac{d\phi}{dy}\right)^2 + 2\frac{d\phi}{dx}\left(u - \frac{d\phi}{dx}\right) + 2\frac{d\phi}{dy}\left(v - \frac{d\phi}{dy}\right)\right] dx\, dy$$

$$= \iint \left[\left(u - \frac{d\phi}{dx}\right)^2 + \left(v - \frac{d\phi}{dy}\right)^2\right] dx\, dy$$

$$+ 2\iint \left[\frac{d\phi}{dx}\left(u - \frac{d\phi}{dx}\right) + \frac{d\phi}{dy}\left(v - \frac{d\phi}{dy}\right)\right] dx\, dy.$$

The first term on the right-hand side is essentially positive, and the second term is zero, as is thus seen. Taking the term

$$\iint \frac{d\phi}{dx}\left(u - \frac{d\phi}{dx}\right) dx\, dy,$$

and integrating it first with respect to x, we have

$$\int \frac{d\phi}{dx}\left(u - \frac{d\phi}{dx}\right) dx = \left[\phi\left(u - \frac{d\phi}{dx}\right)\right]'' - \int \phi\left(\frac{du}{dx} - \frac{d^2\phi}{dx^2}\right) dx,$$

where the first term refers to the points on the bounding contour determined by the limits of x. Hence, if (l, m) are the direction-cosines of the outward drawn normal to the contour,

$$\iint \frac{d\phi}{dx}\left(u - \frac{d\phi}{dx}\right) dx\,dy = \int l\phi\left(u - \frac{d\phi}{dx}\right) ds - \iint \phi\left(\frac{du}{dx} - \frac{d^2\phi}{dx^2}\right) dx\,dy.$$

Similarly,

$$\iint \frac{d\phi}{dy}\left(v - \frac{d\phi}{dy}\right) dx\,dy = \int m\phi\left(v - \frac{d\phi}{dy}\right) ds - \iint \phi\left(\frac{dv}{dy} - \frac{d^2\phi}{dy^2}\right) dx\,dy.$$

The sum of the right-hand sides of these equations is zero; for $lu + mv$ is the normal component of velocity at the contour in the one system, and $l\frac{d\phi}{dx} + m\frac{d\phi}{dy}$ is the same thing in the other. Also $\frac{du}{dx} + \frac{dv}{dy}$, being the expansion at any point inside, is zero, as is also $\nabla^2\phi$. Therefore, &c.

2. The motion of a liquid being irrotational, except at a given system of vortical centres, prove that all values of the velocity-potential can be found in the immediate vicinity of each vortex, and that therefore every equipotential curve passes through every vortex.

3. Prove that Thomson's correction of Green's equation is independent of the particular system of barriers drawn.

4. Find the energy of the liquid contained between two concentric circles, each with centre at the vortex, in the case of a whirl.

5. Prove the following expression for the energy, at any instant, of a gas moving irrotationally within a given contour,

$$T = \tfrac{1}{2}\int \rho\phi \frac{d\phi}{dn} ds + \int \phi \frac{d\rho}{dt} dS.$$

Section III.—Motion due to Sources and Vortices. Electrical Flow.

127. Recapitulation. When the moving fluid is incompressible, there is at every point a flow function, ψ, which satisfies the equation

$$\nabla^2\psi = -2\omega; \qquad (1)$$

and in those regions (if any such exist) in which there is no vortical spin, there is, in addition, a velocity-potential function, ϕ, which satisfies the equation

$$\nabla^2\phi = 0. \qquad (2)$$

We have shown (Art. 110) how in the case of a compressible fluid the components, u, v, of velocity are found at every point— by the determination of two quantities, P and N—when the law of expansion and that of vortical spin are assigned for each point.

In the present section we shall confine ourselves to the case in which the fluid is incompressible and without inertia; and under this head we shall include the case of the flow of electricity—to which this section is specially devoted.

Pursuing the same analogy with the case of a gravitation potential, we write the equation (1) in the form

$$\nabla^2 \psi = -4\pi \cdot \frac{\omega}{2\pi},$$

which shows that ψ may be regarded as the gravitation potential at any point produced by a distribution of matter along right lines perpendicular to the plane (x, y) of motion, and of practically infinite length above and below this plane, in such a manner that *the density at any point in space is* $\frac{1}{2\pi}$ *times the value of the vortical spin at the orthogonal projection of the point on the plane of motion.*

Now we are not always concerned with the value of ψ itself at any point, but we are concerned with the velocity at the point, or with its two components (u, v); and this analogical method of regarding ψ as an attraction potential, enables us to find the velocity components without calculating ψ itself; for there are methods in Statics for calculating the resultant attraction of matter, or its components, without finding the potential.

Now the components of the attraction at any point, parallel to the axes of x and y, of matter arranged as above are $\frac{d\psi}{dx}$ and $\frac{d\psi}{dy}$, which we know to be, respectively,

$$-v \text{ and } u,$$

the velocity components at the point. Hence these latter are at once given from the attraction.

128. Motion due to a single Vortex. Suppose that at a point A there is a very small circular area, dS, in which the mean value of the rotation is ω; then $2\omega dS$ is defined to be the *strength* of the vortex, so that the strength of the vortex is the circulation round any closed curve enclosing the vortex once. Denote it by m. To calculate the components, u, v, of velocity at all points in the plane we are to imagine an infinitely long solid cylinder described on dS as base, whose mass per unit length is represented by $\dfrac{m}{4\pi}$. Now if we calculate the attraction potential of this cylinder at any point, P, in the plane, we shall find its value to be infinite (see *Statics*, p. 403). The attraction on a unit particle at P is however easily found. It is (see *Statics*, p. 417) directed from P along the perpendicular to the cylinder, i.e., along PA, and is equal to

$$\frac{m}{2\pi \cdot PA}. \tag{1}$$

If the co-ordinates of P are (x, y) and those of A are (a, β), this attraction is

$$\frac{m}{2\pi \sqrt{(x-a)^2 + (y-\beta)^2}}. \tag{2}$$

And we have just seen that the x and y components of this attraction are, respectively,

$$-v \text{ and } u.$$

Hence, denoting PA by r,

$$u = -\frac{m}{2\pi} \frac{y-\beta}{r^2}, \tag{3}$$

$$v = \frac{m}{2\pi} \frac{x-a}{r^2}. \tag{4}$$

Now $d\phi = u\,dx + v\,dy = \dfrac{m}{2\pi} \dfrac{(x-a)\,dy - (y-\beta)\,dx}{r^2}$

$$= \frac{m}{2\pi} d\tan^{-1}\frac{y-\beta}{x-a}. \tag{5}$$

Hence if θ denotes the angle made by the line PA with the axis of x, the velocity-potential at P due to the vortex m at A is given by the equation

$$\phi = \frac{m}{2\pi} \cdot \theta. \tag{6}$$

The resultant velocity at P is $\sqrt{u^2+v^2}$, or $\dfrac{m}{2\pi r}$, which is infinite at A.

This resultant is at right angles to the line PA, and in the same sense as that of the rotation in the vortex at A.

Again,
$$d\psi = udy - vdx = -\frac{m}{2\pi}\frac{(x-a)dx+(y-\beta)dy}{r^2} = -\frac{m}{2\pi}\frac{dr}{r};$$
therefore
$$\psi = -\frac{m}{2\pi}\log r,$$
omitting a constant[1].

Hence the stream lines are a series of concentric circles having the vortex for centre, and the equipotential lines are right lines diverging from A, so that we are led to the case already discussed (p. 167), which we now see to be a case of motion due to a vortex of small area placed at the origin.

The plane of motion may either extend to infinity in all directions, or be bounded by any circle having A for centre. (See Art. 111. No other boundary would leave the motion unchanged.) If the stream lines are drawn so that the values of the stream function for them proceed with a common difference, they will be a series of circles with radii in G. P.

129. Electrical Equivalent of a Vortex. The magnitude and direction of the resultant velocity at P due to the vortex at A answer exactly to the magnitude and direction of the action, on a magnetic pole of unit strength, of an electric current transversing the infinitely long wire, or cylinder, which we have imagined to extend from A perpendicularly to the plane, above and below it; the strength of the current being proportional to the strength, m, of the vortex.

Hence all our results for vortices in a liquid are directly applicable to electro-magnetic phenomena, if we imagine indefinitely long straight currents to replace our vortices, the strength of each current being proportional to the strength of the corresponding vortex; and the plane of motion must be

---

[1] In reality the value of ψ, calculated as the attraction potential of an infinite slender cylinder, contains an *infinite constant term*. See Lamb's *Treatise on the Motion of Fluids*, p. 162.

considered as a field at any point of which, instead of a liquid particle, we imagine a magnetic pole.

Or again, for the vortex at A we may substitute a magnetic pole of strength proportional to m; and at each point, P, in the plane imagine a little element of electric current running perpendicularly to the plane.

130. Motion due to two Vortices. Let there be two vortices of strengths m_1 and m_2 at two points, A_1 and A_2, in the plane of motion; then if ϕ is the velocity-potential resulting from both together, and ϕ_1, ϕ_2 those due to each separately, we know that $\dfrac{d\phi}{dx} = \dfrac{d\phi_1}{dx} + \dfrac{d\phi_2}{dx}$, and $\dfrac{d\phi}{dy} = \dfrac{d\phi_1}{dy} + \dfrac{d\phi_2}{dy}$, since the resultant x-velocity is the sum of the separate x-velocities, etc. Hence $\phi = \phi_1 + \phi_2$; and similarly $\psi = \psi_1 + \psi_2$, where ψ is the stream function. We have, therefore, by last Art.,

$$\phi = \frac{m_1}{2\pi}\theta_1 + \frac{m_2}{2\pi}\theta_2, \tag{1}$$

$$\psi = -\frac{m_1}{2\pi}\log r_1 - \frac{m_2}{2\pi}\log r_2, \tag{2}$$

where θ_1 and θ_2 are the angles made with any fixed right line by the lines, PA_1, PA_2, joining the point P, to which ϕ and ψ refer, to the vortices A_1, A_2; and where also $PA_1 = r_1$, $PA_2 = r_2$.

The stream lines are curves given by the equation

$$r_1^{m_1} \cdot r_2^{m_2} = C,$$

where C is a constant. If $m_1 = m_2$, or the two vortices are of equal strength, this becomes $r_1 r_2 = k^2$, which gives the ovals of Cassini already discussed.

Also in this case $\theta_1 + \theta_2 = \dfrac{2\pi}{m}\phi$, $\therefore$ the curves $\phi = $ const. are rectangular hyperbolas, as found in p. 173. If we wish the plane of motion to be of limited extent, we must make its boundary one of the Cassinian ellipses.

The case in which $m_2 = -m_1$ is interesting. In this case the curves $\psi = $ const. are given by the equation

$$\frac{r_1}{r_2} = k,$$

i.e., they are a series of circles, loci of vertices of triangles standing on $A_1 A_2$ as base and having their sides in a constant ratio. A *limited* field must be bounded by one of them. The equipotential curves become

$$\theta_1 - \theta_2 = \text{const.} = a,$$

i.e., a series of circles, loci of vertices of triangles having $A_1 A_2$ as base and a constant vertical angle, a, which has different values for the different loci.

Every one of these equipotential circles passes through *both* vortices; and observe that also in the last Art. every one of the equipotential lines passes through the vortex. This is a general property of all equipotential lines. (See ex. 8, p. 173).

The case in which $m_1 = m_2$, i.e., in which there are in the field two vortices of equal strengths and the same sense, is realised by making two holes, A_1, A_2, in a flat plate and passing two very long straight wires through these holes, the wires conveying the same electric current *in the same sense*, i.e., both up through the plate or both down through it. The *magnetic* action of this current system, i.e., its action on magnetic poles lying in the plane of the plate, is represented by the system of ellipses of Cassini (as lines of electro-magnetic force) and the system of rectangular hyperbolas (as lines of equal potential).

The case in which $m_1 = -m_2$ requires the wires passing through the holes A_1 and A_2 to convey the same current in opposite senses—i.e., one up and the other down through the plate.

It will be good exercise for the student to map out the resultant electro-magnetic fields, which correspond to the two cases discussed in this Article, by means of the graphic method of superposition explained in Art. 112.

Observe that in the second case ($m_1 = -m_2$), if we draw the velocities along the lines of flow (circles) corresponding to the vortex A_1 from right to left, those along the lines of flow (circles) corresponding to A_2 must be drawn from left to right. If from a vortex A_1 we draw any number of rays dividing the whole angle 2π into equal parts, and then from any other vortex, A_2, we draw a number of rays on which the velocity-potential has successively the values which it has on the rays of the first pencil, the strengths of the vortices are proportional to the numbers of rays in their respective pencils.

131. Any number of Vortices. Let there be any number of vortices, $A_1, A_2, \ldots A_n$, of strengths $m_1, m_2, \ldots m_n$; then, as shown in last Article, their functions of potential must be added to produce the resultant potential at any point, and similarly for their flow-functions. Hence

$$2\pi \cdot \phi = m_1 \theta_1 + m_2 \theta_2 + \ldots + m_n \theta_n, \tag{1}$$

$$2\pi \cdot \psi = -m_1 \log r_1 - m_2 \log r_2 - \ldots - m_n \log r_n. \tag{2}$$

132. Comparison of Vortices with Sources and Sinks. We have already proved (p. 166) that if at the point A_1 there is a source of strength m_1, the stream-lines being of course right lines passing through A_1, we have

$$\phi_1 = \frac{m_1}{2\pi} \log r_1,$$

$$\psi_1 = \frac{m_1}{2\pi} \theta_1;$$

and if, in addition, there are sources at points $A_2, A_3, \ldots$ of strengths $m_2, m_3, \ldots$ (some of which may be sinks, for which m will be negative), the complete values of ϕ and ψ at any point in the plane will be given by the equations

$$2\pi \cdot \phi = m_1 \log r_1 + m_2 \log r_2 + \ldots + m_n \log r_n, \tag{a}$$

$$2\pi \cdot \psi = m_1 \theta_1 + m_2 \theta_2 \ldots + m_n \theta_n. \tag{β}$$

So that *the values of ϕ and ψ for sources and sinks are exactly the values of ψ and ϕ, respectively, for vortices.*

The electrical equivalent of sources and sinks is obtained by taking a plane along which electricity can flow, i.e., a thin metallic sheet, and inserting electrodes, some positive and some negative, i.e., some sources and some sinks, at any arbitrary points $A_1, A_2, A_3, \ldots A_n$ in the sheet. The flow will take place along the curves given by equating ψ in (β) to a constant; and observe that *every flow-line passes through every source and every sink*, a fact which at once follows from the property proved for vortices, since *vortices and sinks interchange* their equipotential and flow functions. But it may be seen directly that a locus defined by the equation

$$m_1 \theta_1 + m_2 \theta_2 + m_3 \theta_3 + \ldots = \text{const.}$$

passes through every one of the points $A_1, A_2, A_3, \ldots$. For if (x, y) are the co-ordinates of the variable point P, and $(x_1, y_1) \ldots$ those of $A_1 \ldots$, the equation is the same as

$$m_1 \tan^{-1} \frac{y - y_1}{x - x_1} + m_2 \tan^{-1} \frac{y - y_2}{x - x_2} + \ldots = \text{const.},$$

which is manifestly satisfied by $x = x_1, y = y_1$, since we may take all the terms except the first together, and write the equation $\tan^{-1} \frac{y - y_1}{x - x_1} = U$, or $y - y_1 = (x - x_1) \tan U$. Similarly for all the other points.

In the particular case in which the strengths $m_1, m_2, \ldots$ are all equal (whether with or without the same sign), every complete curve of flow is a curve of the nth degree (there being n centres); and in general the degree of a complete flow-curve is Σm, taken *arithmetically*. If the poles are unequal, each stream-line passes more than once through the stronger ones. It passes through each as often as its strength contains the G. C. M. of the set.

133. Infinite Velocity. We shall find in all cases (see example 1, following) that the velocity at each point-source and also at each point-sink is infinite. It is not at all surprising that our investigation of the motion should lead to such a result; for one explanation, at least, appears to be that we have all through regarded a source as a mere *point*, in the strictest geometrical sense; and we have assumed that through this point (which is a mere extentless entity of the imagination) a *finite quantity* of something measurable (fluid, electricity, etc.) is discharged per unit of time. The two conceptions are manifestly contradictory of each other; and we are compelled therefore to admit some one of three things—viz., (1) that our sources and sinks are not geometrical points (and that therefore none of the quantities $r_1, r_2, \ldots$ can ever possibly be zero); or (2) that their strengths, $m_1, m_2, \ldots$ which are measures of the quantities they discharge, are all infinitely small, so that, for example, when we meet such an expression as $\dfrac{m}{r}$, we must not assume it to be infinite when $r = 0$, because m is also zero; or, lastly, (3) that if a finite quantity is discharged through a *point*, an infinite velocity is required at the point. The first of these choices is, of course, the one to be adopted—every source and every sink is in every actual and physically possible case a little equipotential *area*—not a *point*.

134. Flow in a given Bounded Plane. We have already mentioned (Art. 111) the case in which the plane or field of motion is limited in extent and bounded by a given invariable contour, while inside the field there are given sources and sinks. The values of ϕ and ψ given in the previous Articles are not at once applicable here, because they suppose the field to be either *unlimited* or bounded in a very particular manner—viz., by some one of the lines of flow due to the *given* causes of motion; whereas in the present case the boundary is to be taken at random. Undoubtedly the boundary, whether taken at random or not, is in all cases a stream-line of the fluid; but the values of ϕ and ψ just given and calculated from the *given* causes alone will not make it so. Hence the problem to find the velocity, etc., at each point in this bounded field, due to the given causes (sources, sinks, etc.), is much more difficult than it would be if the field were unlimited.

Still, the two problems are solved on the same lines. What we do is this. Imagine the field to be unlimited. With the given causes $(A_1, A_2, ...)$ of motion we then combine certain others $(B_1, B_2, ...)$, which we must completely determine in such a way that, taking them all together, they would, in the infinite field, make the given boundary curve a stream-line.

The velocity-potential and stream-function, at any point either inside the given bounded field or outside it, will then be the sums $(\phi_A + \phi_B$ and $\psi_A + \psi_B$, respectively) of those due to the given causes (A) and the determined causes (B).

Of course the whole difficulty consists in finding out the system (B)—which is called the *image* of the given system (A)

in the given boundary, in analogy with the language of Optics, because the image-system (B), if it existed, would produce the same effect *within the given limited field* as the boundary does—'just as,' to quote Professor Lodge[1], 'the illumination inside a mirror-walled room containing candles would be imitated in unlimited space' [i.e., would still be produced inside the room if walls and mirrors were removed] 'by placing extra candles at all the points occupied by the images of the original candles in the mirrors.'

For example, let the flow considered be that of electricity, the plane of flow being a sheet of tinfoil.

Now for a source, A_1, and an equal sink, A_2, we have

$$\phi = \frac{m}{2\pi} \log \frac{r_1}{r_2}; \quad \psi = \frac{m}{2\pi}(\theta_1 - \theta_2),$$

in an *unlimited* plane; so that the stream-lines are circles, each of which passes through A_1 and A_2.

If, then, we wish to make an experiment with a *limited* sheet, using the electrodes from the poles of a battery for the supply and removal of electricity, we see that we must cut the sheet into a circular shape, and place the ends of the electrodes at any two points on the circular boundary. This done, the velocity and all other circumstances at any point in the sheet are found from the above values of ϕ and ψ. The equipotential curves are circles ($\frac{r_1}{r_2}$ = constant).

For the method of practically verifying this, and tracing out the equipotential curves, see an account of very elegant experiments by Professors G. Carey Foster and O. J. Lodge in the *Proceedings of the Physical Society* (Feb., 1875), or *Phil. Mag.* (Dec., 1875).

Again, suppose the plane of flow to be bounded by a given right line—which must therefore be made a stream-line by properly arranging sources and sinks. The expression for ψ in Art. 130 shows that this can be done by placing a source or sink A_1 at any point in the sheet and another source, or sink

[1] *Phil. Mag.* vol. i. 1876.

A_2, of equal strength at the optical image of A_1 in the line (considered as a reflecting surface); or we may put a source A_1 (a positive electrode) at any point, and an equal sink, B_1 (negative electrode), at any other point, and an equal source, A_2, and sink, B_2, at the optical images of A_1 and B_1 in the line. When we use the two sources and two sinks, the line will be necessarily only a *part* of a stream-line (Art. 133). Thus the image of a pole (source or sink) in a straight stream-line is a pole of equal strength and same kind (source or sink) at the optical image of the first.

If, on the other hand, we wish to make the given right line an equipotential line, the value of ϕ shows that this can be done by placing a source, A_1, at any point and an equal sink, B_1, at the optical image of the first; and similarly we may use any number of sources, each accompanied by an equal sink. Another method of making the line an equipotential line is to fix along the line a metallic band of practically infinite conductivity. Unless this is done, the sheet may not be cut away along an equipotential line, as it may along a stream-line.

Examples.

1. To draw the direction of flow at any point in a plane containing any number of sources and sinks.

Let P be the point at which we wish to find the direction of flow, and let the sources and sinks be A_1, A_2, A_3, ... of strengths m_1, m_2, m_3, ... (the strength corresponding to a sink being, of course, negative). Then

$$2\pi\phi = m_1 \log r_1 + m_2 \log r_2 + m_3 \log r_3 + \ldots \quad (1)$$

and the direction of flow is that of the normal to the surface $\phi = \text{const.}$ Now the normal to this surface is found by measuring from P along the lines PA_1, PA_2, PA_3, ... lengths, Pa_1, Pa_2, Pa_3, ... proportional to $\dfrac{m_1}{r_1}$, $\dfrac{m_2}{r_2}$, $\dfrac{m_3}{r_3}$, ... and finding the *resultant*, PG, of these lines. If A_1 is a source, the length Pa_1 will be measured from P in the sense $\overline{PA_1}$, and if it is a sink, the length will be measured from P in the sense $\overline{A_1P}$. For the proof of this theorem see *Statics*, p. 76, 2nd ed.

The points a_1, a_2, a_3, ... may be practically determined in various ways. One way is this: round P as centre and with any convenient radius, k, describe a circle, C; through A_1 draw a circle, C_1, cutting C orthogonally; then if PA_1 meets C_1 in b_1, we have $Pb_1 = \dfrac{k^2}{r_1}$; hence by measuring a line Pa_1 equal to $m_1 . Pb_1$ from P either towards or from A_1, according as A_1 is a source or a sink, the point a_1 is found. Similarly for all the other points.

The velocity at P may be regarded as the resultant of the velocities which would be produced by $A_1, A_2, A_3, \ldots$ separately, i.e., the resultant of $\dfrac{m_1}{r_1}, \dfrac{m_2}{r_2}, \ldots$; and if there are n centres the resultant of $Pa_1, Pa_2, \ldots$ is $n . PG$ (*Statics*, p. 16).

2. To investigate resistance to electrical flow in an unlimited sheet of tinfoil containing a source and an equal sink.

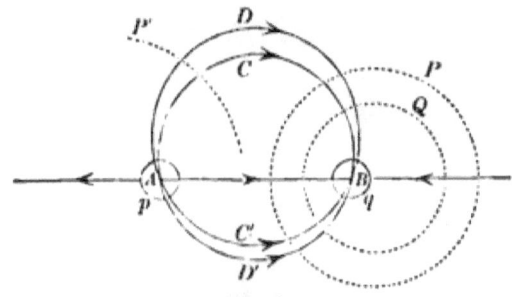

Fig. 68.

Let m be the strength of the source, A, and $-m$ that of the sink, B (fig. 68).

Then the distances of any point from A and B being r and r',

$$\phi = \frac{m}{2\pi} \log \frac{r}{r'},$$

$$\psi = \frac{m}{2\pi} (\theta - \theta').$$

Let C and D be any two stream-lines (circles), and P, Q, P' any equipotential lines (also circles); then (Art. 116) the resistance of the strip contained between the circles C, D, and terminated by the circles P, P' is $\dfrac{\phi_P - \phi_{P'}}{\psi_D - \psi_C}$. Now $\dfrac{r}{r'}$ for points near A is very small, while for points near B it is very large; hence *at* A and B the value of ϕ is infinite. Also at each of these points the value of ψ is indeterminate; hence the above expression for the resistance cannot be applied to the whole of the crescent-shaped strip included between the circles C and D. The difficulty is avoided by drawing round A an equipotential circle, p, of very small radius, and another, q, round B and considering them as electrodes. The expression for the resistance of the strip terminated by these circles will not be indeterminate. Let ϵ be the value of the ratio $\dfrac{r}{r'}$ on the circle p, and $\dfrac{1}{\epsilon}$ its value on q; also let the value of $\theta - \theta'$ on C be α, and on D be β. Then the resistance of the strip is

$$\frac{2 \log \epsilon}{\alpha - \beta}.$$

More generally, the resistance will be $\dfrac{\log p + \log p'}{\alpha - \beta}$, if we take as electrodes any two equipotential circles (one round each pole), where p and p' are the values of $\dfrac{r}{r'}$ for the circles.

One stream-line is AB, and for it $\theta - \theta' = \pi$, so that the resistance of the area included between the circle D, the line AB, and the circles p, q is

$$\frac{2 \log \epsilon}{\pi - \beta}.$$

For the remainder, D', of the circle D, the value of $\theta - \theta'$ is $\pi - \beta$, and if R is the resistance of the whole area of the circle D, excluding the portions within the small circles p, q,

$$\frac{1}{R} = \frac{\pi - \beta}{2 \log \epsilon} + \frac{\pi - (\pi - \beta)}{2 \log \epsilon} = \frac{\pi}{2 \log \epsilon},$$

$$\therefore R = \frac{2 \log \epsilon}{\pi}, \tag{1}$$

which is independent of the circle, D, selected.

In the same way the resistance of the area included between any two equipotential circles, P, Q, may be calculated. The resistance of as much of this strip as is bounded by the circles C and D is

$$\frac{\phi_Q - \phi_P}{\psi_D - \psi_C} = \frac{\phi_Q - \phi_P}{\beta - \alpha},$$

and if β and α are nearly equal, so that $\beta - \alpha = d\alpha$, the reciprocal of the resistance of the interval between P and Q bounded by the upper side of the line AB is

$$\int_0^\pi \frac{d\alpha}{\phi_Q - \phi_P} = \frac{\pi}{\phi_Q - \phi_P},$$

since for points on AB between A and B the value of α is π, and for points on AB beyond B the value is 0. Hence the resistance of the upper half of the strip between P and Q is $\dfrac{1}{\pi}(\phi_Q - \phi_P)$; and the resistance of the lower half being the same, the total resistance of the strip is $\dfrac{1}{2\pi}(\phi_Q - \phi_P)$ —i.e., *one half* the resistance of the half strip, since the strips are *abreast* and not *in series.*

Examples.

Hence the resistance of the area between P and q is $\frac{1}{2\pi}(\log\frac{1}{\epsilon} - \log\lambda)$, if λ is the value of $\frac{r}{r'}$ on P; or $\frac{1}{2\pi}\log\frac{1}{\lambda\epsilon}$. If we put $\lambda = 1$, P will be the right line bisecting AB perpendicularly, and the resistance of half the infinite sheet, leaving out the area of q, is $\frac{1}{2\pi}\log\frac{1}{\epsilon}$. The other half is, as regards flow, arranged *in series* with this half, so that their resultant resistance is their sum, or

$$\frac{\log \epsilon}{\pi},$$

(neglecting sign); and this is half the resistance in (1), which denotes the resistance of the area of any circle of flow.

To find the velocity at any point, P, we may regard the actual motion as a superposition of two separate motions, one due to the source and the other due to the sink. If $PA = r$, $PB = r'$, the first motion would give a velocity $\frac{m}{2\pi}\cdot\frac{1}{r}$ along $\overline{AP}$, and the other a velocity $\frac{m}{2\pi}\cdot\frac{1}{r'}$ along $\overline{PB}$; hence the resultant velocity at P is

$$\frac{m}{2\pi}\cdot\frac{AB}{rr'}.$$

The curves of constant velocity are therefore ellipses of Cassini.

The arrows in the figure indicate the senses of flow.

If A and B were both sources, or both sinks, the velocity at any point would be $\frac{m}{\pi}\cdot\frac{\rho}{rr'}$, where ρ is the distance of the point from the middle point of AB.

The system of equipotential circles, P, Q, ... may be drawn by taking any point on the line AB (produced) as centre and describing a circle with radius equal to the length of the tangent from the point to any one, C, of the circles of flow.

3. Given two sources, A, B, each of strength m, and a sink C in the right line AB, also of strength m, show that one stream-line consists partially of a circle.

Let A and B be the two sources (fig. 69) and C the sink; let P be any point in the plane, $\angle PAx = \theta$, $\angle PBx = \theta'$, $\angle PCx = \theta''$. Then if the flow-function is ψ, we have

$$\psi = \frac{m}{2\pi}(\theta + \theta' - \theta'').$$

Consider the value $\psi = 0$. Then $\angle PAx = \angle BPC$; therefore if a circle is described through A, B, P, the line CP is a tangent to it at P. Hence if $\psi = 0$, P lies on the locus of points of contact

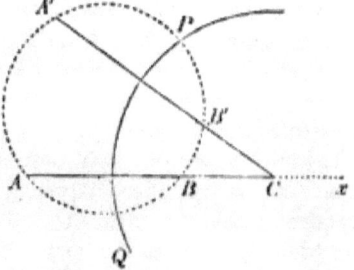

Fig. 69.

of tangents from C to circles passing through A and B, i.e., the locus of P is a circle, PQ, with C as centre and radius $= \sqrt{CA \cdot CB}$, which is the required result. But the right line AB is also part of this stream-line ($\psi = 0$), since at all points on it $\theta = 0$, $\theta' = \theta'' = \pi$. Also the line BC is part of a stream-line for which $\psi = -\tfrac{1}{2}m$; the production of BA beyond A is part of a stream-line for which $\psi = \tfrac{1}{2}m$; and the production of BC through C is part of a stream-line for which $\psi = 0$. Each *complete* stream-line is a curve of the third degree (Art. 132).

Now, in addition to this arrangement, let there be at two points, A', B', which are collinear with C, and which lie on any circle passing through A and B, two sinks, each of strength m, and at C a source of strength m. Of this new system the circle PQ will also be a stream-line; hence it will be a stream-line if we superpose the systems; and, as C is thus obliterated, the superposition gives two sources, A, B, and two sinks, A', B'. We can regard this as produced by bifurcating the wire from the $+$ pole of a battery and connecting it with A and B, and similarly bifurcating the wire from the $-$ pole and connecting it with A' and B'; and adjusting the resistances in the circuit so that the current splits equally between the branches.

Of course the circle PQ is only a part of the stream-line, because if we measure θ from the line Ax as an initial line and let the angles made with it by the lines PA, PB, PA', and PB' be, respectively, θ_1, θ_2, θ_1', θ_2', we have

$$\psi = \frac{m}{2\pi}(\theta_1 + \theta_2 - \theta_1' - \theta_2'),$$

and the curve $\psi = $ const. is (Art. 132) of the fourth degree.

To find the value of ψ on the circle PQ, let $\angle PCx$ be χ, and $\angle ACA' = a$; then $\psi = \dfrac{m}{2\pi}(\theta_1 + \theta_2 - \chi + \chi - \theta_1' - \theta_2')$; but $\theta_1 + \theta_2 - \chi = 0$, and $\chi - \theta_1' - \theta_2' = a$,

$$\therefore \psi = m\frac{a}{2\pi}$$

all over the circle PQ. Moreover this is also the value of ψ at all points on the circle passing through $ABB'A'$, as is easily seen from the property that angles in the same segment of a circle are equal.

Hence the complete stream-line considered is the system of *two* circles represented in the figure.

The other stream-lines will each consist of two separate portions—one (more or less oval-shaped) passing through B and B' and completely contained within the circle PQ, and the other passing through A and A' and wholly external to the circle PQ, since, except at the poles, no two different stream-lines can intersect.

A drawing of these curves is given in the paper by Professors Foster and Lodge (*Proceedings of the Physical Society*, Feb., 1875).

It is evident that the velocity at P, and also at the other point of intersection of the circle PQ with the circle of poles, is zero. For the velocity at P along the arc PA' is the line-rate of variation of ψ along the arc PQ (p. 154), and this is zero. Also the velocity at P along the arc PQ is zero for a similar reason. The velocity at P may be graphically represented (example 1) as the resultant of two forces from P *towards* A and B, inversely proportional to PA and PB, and two forces *from* A' and B', inversely as PA' and PB'. Such forces are therefore in equilibrium.

Again, we see (example 2) that the quantities $\dfrac{BB'}{PB.PB'}$ and $\dfrac{AA'}{PA.PA'}$ are equal, since we may regard the state of affairs at P as due to a superposition of the effects of the source and sink (B, B') and the source and sink (A, A'). And again, $\dfrac{PM}{PA.PB} = \dfrac{PM'}{PA'.PB'}$, where M, M' are the middle points of AB and $A'B'$, respectively.

But if r, s, r', s' denote the distances of any point from A, B, A', B', respectively,
$$\phi = \frac{m}{2\pi} \log \frac{rs}{r's'}.$$

Hence the potential at $P = \dfrac{m}{2\pi} \log \dfrac{PM}{PM'}$. The equipotential curves are given by the equation $\dfrac{rs}{r's'} = k$, k being a constant changing from one curve to another. They are therefore of the fourth degree, except the curve for which $k = 1$.

The equipotential curve through P is one having P for a double point.

If we produce AA' and BB' to meet in a point, D, the circle with D as centre cutting the circle of poles orthogonally is part of an equipotential curve; for (last example) the source and sink (A, A') and the source and sink (B, B') would separately give this as an equipotential curve. This is easily seen otherwise; for since A, A' are inverse points with regard to this circle, if J is any point on the circle, $\dfrac{JA}{JA'} = \dfrac{\sqrt{DA}}{\sqrt{DA'}}$; similarly
$$\frac{JB}{JB'} = \frac{\sqrt{DB}}{\sqrt{DB'}}; \quad \therefore \quad \frac{JA.JB}{JA'.JB'} = \sqrt{\frac{DA.DB}{DA'.DB'}} = \text{const.,}$$
so that ϕ is constant all over the circle with D as centre cutting the circle of poles orthogonally.

The remaining portion of this equipotential locus is easily seen to be the (imaginary) circle which cuts the circle of poles orthogonally and has for centre the point, D', of intersection of the lines AB' and BA'. For, by expressing the ratios of the lines involved in terms of the sines of angles, we at once obtain the equality $\dfrac{DA.DB}{DA'.DB'} = \dfrac{D'A.D'B}{D'A'.D'B'}$; and a circle with D' as centre and radius equal to $\sqrt{-D'A.D'B'}$ will be a locus (imaginary) of the vertex of a triangle having AB' for base, with the constant ratio $\sqrt{-\dfrac{DA}{DB'}}$ between its sides; similarly for the pair of points B, A'. The imaginary circle with D' as centre cutting the circle of poles orthogonally is then, by last example, an equipotential circle for the source and sink (A, B') and for the source and sink (B, A').

When $k = 1$, the equipotential curve becomes a circular cubic, and it obviously passes through C and through the centre of the circle of poles.

Now in the case of the four poles consider the nature of the equipotential curve on which the value of the potential is a very great positive quantity. If the distances of any point P from A, B, A', B' are s, r, s', r', respectively, we have $\phi = \dfrac{m}{2\pi} \log \dfrac{sr}{s'r'}$; and if this is a very great positive quantity, we must have $\dfrac{sr}{s'r'} = \dfrac{1}{k}$, where k is very small. Now $\dfrac{s'r'}{sr}$ will be very small if s' is very small, i.e., if P is very near A'. By taking A' as origin of polar co-ordinates (s', θ), expressing s, r, r' in terms of s', θ, and constants, and neglecting such small quantities as s'^2 and ks', the equation $s'r' = ksr$ gives
$$A'B' \cdot s' = k \cdot A'B \cdot A'A,$$
which shows that P must lie on a small circle with A' as centre, its radius being $k \dfrac{A'A \cdot A'B}{A'B'}$. Similarly, ϕ will have the same large positive value if r' is very small, i.e., if P lies on a small circle described round B' with radius $= k \dfrac{B'B \cdot B'A}{B'A'}$. Hence the complete equipotential curve for which ϕ has the constant large value $\dfrac{m}{2\pi} \log \dfrac{1}{k}$ consists, approximately, of *two* small circles surrounding the poles A and B'; and the curve on which ϕ has the constant large negative value $\dfrac{m}{2\pi} \log k$ consists of two circles surrounding the poles A and B.

Let us now make a boundary along the whole circumference of the circle PQ; i.e., suppose that we have a circular sheet of tinfoil with the positive pole of a battery connected with it at B, and the negative pole at B'. The circumstances of flow are to be calculated from the above case of four poles. The resistance of the sheet may be thus calculated. Since the circumference of the sheet is a stream-line, no electricity crosses it, and the whole of the electricity discharged from B flows into B'. The quantity discharged by B in all directions and absorbed by B', per unit of time, is m. Consider then the whole area of the sheet, with the exception of the two small circles, one of large and the other of small potential, surrounding B and B'. If the radii of these circles be each λ, we shall have on the circle surrounding B'
$$\phi_{B'} = \dfrac{m}{2\pi} \log \dfrac{AB' \cdot BB'}{\lambda \cdot A'B'},$$
and for that round B
$$\phi_B = \dfrac{m}{2\pi} \log \dfrac{\lambda \cdot AB}{A'B \cdot BB'},$$

Examples. 217

and the resistance of the whole sheet, leaving out the two small circles, is
$\frac{\phi_{B'} - \phi_B}{m}$, or $\frac{1}{2\pi} \log \frac{AB' \cdot A'B \cdot BB'^2}{\lambda^2 \cdot AB \cdot A'B'}$.

4. In example 3 show that the velocity at any point, I, on the circle of poles and very near P is equal to $\frac{m}{4\pi a}(aa' - \beta\beta') \cdot \theta$, where $\theta = \angle$ subtended by PI at the centre of the circle of poles, a = radius of this circle, and a, a', β, β' are the cotangents of the halves of the angles made with the diameter through P by the radii drawn to A, A', B, B', respectively, these angles being all measured round in the same sense (that of watch-hand rotation).

[The consideration of this case gives the following theorem: If from any point C outside a circle a line be drawn cutting it in A, A', and a tangent touching at P, the sum of the cotangents of the halves of the angles made with the diameter through P by the radii drawn to A, A' is constant.]

5. In an unlimited sheet of tinfoil containing any number, n, of equal sources, $A_1, A_2, A_3, \ldots$, and the same number of equal sinks, $B_1, B_2, B_3, \ldots$, prove that the resistance of the whole portion included between the equipotential curve passing through any point P and the equipotential curve passing through any point Q is

$$\frac{1}{2n\pi} \log_e \frac{PA_1 \cdot PA_2 \ldots PA_n \cdot QB_1 \cdot QB_2 \ldots QB_n}{PB_1 \cdot PB_2 \ldots PB_n \cdot QA_1 \cdot QA_2 \ldots QA_n}.$$

$\left[\text{It is } \frac{\phi_P - \phi_Q}{nm}.\right]$

6. Find the expression if the strengths are not all equal.

[Consider unequal poles as superpositions of equal ones. See Prof. Lodge's paper on the *Flow of Electricity in a Plane*, Phil. Mag. vol. i. 1876.]

7. An indefinitely long rectangular strip of tinfoil of breadth b is taken, and the electrodes of a battery are connected with it at two points A, B, one on one edge and the other on the opposite edge of the strip, the line AB being at right angles to the edges; prove that the resistance of the strip is of the form

$$k \log \frac{2b}{\pi \lambda},$$

where k is a constant depending on the thickness and specific conductivity of the foil and λ the radius of each (circular) electrode. (Prof. Lodge, *ibid.*)

[Use the (infinite number of) images both of A and B in the edges of the strip, and apply the result in example 5.]

8. Find the resistance of an infinite rectangular strip when the two electrodes from a battery are connected with two points on its middle line, in terms of their distance apart, the breadth of the strip, and the radii of the (small) electrodes.

9. The velocity at P (fig. 69) being zero, explain how there can be flow along the paths PQ, PB', and QB.

(See Art. 133; *physical* stream-lines must be distinguished from *mathematical*.)

10. Show that a similar difficulty occurs when a given right line is made a stream-line by means of a source and its image, the velocity at a particular point on the line being zero.

11. Instead of the arrangement of sources and sinks in example 3, consider an arrangement in which the sources are at A and B' and the sinks at B and A'; and show that the circle of poles will still be a stream-line, while the orthogonal circles with centres C and D will be both equipotential curves, the imaginary circle with centre D' being a stream-line.

12. Discuss the case of an isosceles right-angled triangle with the poles at the extremities of the hypothenuse; and also that of a sector of a circle with the poles at the extremities of the arc. (Prof. Lodge.)

13. A river flows with uniform velocity; a source of given strength is placed at a given point on its surface; draw the resultant stream-lines.

14. What is the physical meaning of a *source* placed at a given point of a magnetic field?

15. In a magnetic field are placed a source (m) at A and a vortex (m) at B; draw the resultant lines of force.

[If P is any point in the field, $BP = r$, $\angle PAB = \theta$, the lines of force have for equation $r = e^{k\theta}$.]

16. Find the resultant force at any point in the last example.

$\left[\text{If } AP = \rho, \ \angle APB = a, \text{ the force} = \dfrac{m}{2\pi} \dfrac{\sqrt{r^2 - 2\rho r \sin a + \rho^2}}{\rho r}. \right]$

17. Show from elementary considerations that the resistances are the same for all strips opening out at the same angle from a pole and terminated by the same equipotential curve.

[If successive stream-lines are drawn proceeding by a constant difference between their stream-functions, then infinitely near any pole they will form an equiangular pencil, since at such points the distant poles produce no effect.]

135. Theory of Linear Flow. The *resistance* of a portion of any channel of flow bounded by two equipotential curves is (Art. 116) $\dfrac{\phi_2 - \phi_1}{\psi_2 - \psi_1}$, or in other words—

the potential difference of its extremities divided by the quantity which flows, per unit of time, across any section of the channel. (a)

The boundary walls of the channel, i.e., the flow curves ψ_1, ψ_2, may be very close together throughout the whole length of the channel; and, all other things being the same, the effect of narrowing the channel is to increase its resistance to the flow.

The passage of electricity through conductors follows very accurately the laws which we have investigated for the flow of a theoretically perfect liquid—the *channels* along which electricity passes being the substances of metallic bodies, copper and silver being those best suited for its conveyance.

Now a *wire* of copper or other metal supplies us with the narrow channel of which we have just spoken; and Ohm proved experimentally that the resistance of a given length of a given wire at a constant temperature, as above quantitatively defined, is a definite physical property of the wire; i.e., if we take a length AB of the wire, and pass a current through it, measuring the potential difference, $\phi_B - \phi_A$, between its extremities (by an electrometer), and at the same time the time-rate of flow, C, across a section of it (by a voltameter and watch, or a galvanometer), the ratio $\dfrac{\phi_B - \phi_A}{C}$ is always constant. The time-rate of flow, C, is called *the strength of the current*, and Faraday proved experimentally that this is the same at all sections of the wire (Jamin, *Cours de Physique*, vol. 3). If, then, the resistance of the wire between A and B is denoted by R, we have

$$C = \frac{\phi_B - \phi_A}{R}, \qquad (\beta)$$

expressing the current strength in the wire in terms of the potential difference of its ends and its resistance.

This equation assumes that there is no *sudden* alteration of potential at any point of the wire between A and B.

Suppose, however, that at any one point, P, there occurs a *sudden* change in the potential; or that sudden changes occur at two or more points. Fig. 70 takes the case of two points P and Q at which sudden changes occur, and, for greater clearness, the places of sudden change are spread out; the points P and P' are in reality coincident, as also Q and Q'.

Fig. 70.

Let the resistance of BP be ρ, that of $P'Q'$ being ρ', and that of QA being ρ''.

Then, by Faraday's Law the current strength will be the same all through.

Hence we have $C = \dfrac{\phi_B - \phi_P}{\rho} = \dfrac{\phi_{P'} - \phi_{Q'}}{\rho'} = \dfrac{\phi_Q - \phi_A}{\rho''}$; therefore

$$C = \frac{\phi_B - \phi_A + (\phi_{P'} - \phi_P) + (\phi_Q - \phi_{Q'})}{\rho + \rho' + \rho''}. \qquad (\gamma)$$

The resistance, R, of the whole is, of course, $\rho + \rho' + \rho''$, and if we denote the sudden changes of potential by Δ_1 and Δ_2, these being, as we see, measured in the *same sense*, we have

$$C = \frac{\phi_B - \phi_A + \Delta_1 + \Delta_2}{R}. \qquad (\delta)$$

The sum of the sudden changes of potential, $\Delta_1 + \Delta_2$, measured in the *same sense* (that of the flow) is called the *Electromotive Force* between A and B. Denote it by E; then

$$C = \frac{\phi_B - \phi_A + E}{R}, \qquad (\epsilon)$$

which is the modification of (β) when sudden changes of potential occur.

If the extremities, A and B, of the wire are brought together, there being still sudden changes at P and Q, the wire forms a closed circuit, and $\phi_B - \phi_A = 0$; then (ϵ) becomes

$$C = \frac{E}{R}, \qquad (\zeta)$$

which is *Ohm's Law*. The same expression holds if any number of sudden changes of potential occur. Actually such changes are produced by inserting a voltaic cell[1] in the interval PQ, with its zinc pole (suppose) at P and its copper pole at Q.

In any circuit a sudden alteration of potential occurs wherever contact occurs between two different metals, two different liquids, or a metal and a liquid. Thus in the case of a zinc-dilute acid-copper cell inserted between P and Q, assuming the

---

[1] Or an Electromotor doing work by means of the current.

conducting wire, $QAPB$, to be of copper, *two* sudden changes occur at P—a copper-zinc difference and a zinc-liquid difference—and *one* sudden change occurs at Q, viz., a liquid-copper difference; and, as has been said, the sum of these three differences, each taken in the sense of the flow, is the Electromotive Force of the arrangement; and it obviously depends on the nature of the substances employed, and not on the length of the liquid column, the areas of the plates, or the length of the conducting wire.

136. Kirchhoff's Corollaries to Ohm's Law. A very important case occurs when several wires conveying different currents meet in a point.

Thus, let fig. 71 represent an arrangement of three batteries whose electromotive forces are E_1, E_2, E_3, their interpolar wires being connected at A and B.

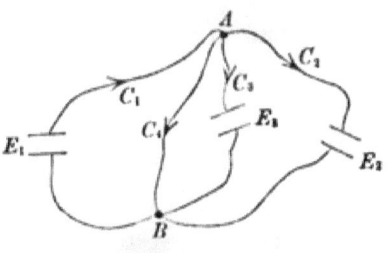

Fig. 71.

Kirchhoff's first Corollary is that—*the sum of the strengths of all the currents running into a point is equal to the sum of the strengths of all those running out of it;* or, in other words, the *algebraic* sum of the currents starting *from* any point is zero.

Thus, assuming the currents in the figure as running in the senses indicated by the arrows, we should have $C_1 = C_2 + C_3 + C_4$.

This principle is obvious, since it is a re-statement of the equation of continuity (Art. 93), there being no creation and no annihilation of electricity at the point.

Kirchhoff's second Corollary is that—*if we take any closed circuit, and estimate all the currents traversing it in the* SAME SENSE, *and also the various electromotive forces within it in this same sense, the sum obtained by multiplying the current strength in each branch of the circuit chosen by the resistance of this branch and adding these products together, is equal to the sum of the electromotive forces in the circuit.*

The general proof of this will be easily understood by taking the closed circuit AE_2BE_3A above. Throughout the whole branch AE_2B the current strength is C_2, and suppose the whole resistance of this branch (battery E_2 included) to be R_2. Then, if ϕ_A and ϕ_B are the potentials at A and B, we have by last Article

$$C_2 R_2 = \phi_A - \phi_B + E_2.$$

Similarly for the branch AE_3B,

$$C_3 R_3 = \phi_A - \phi_B + E_3.$$

Hence $C_2 R_2 - C_3 R_3 = E_2 - E_3$. If we estimate the currents and electromotive forces all in the sense of watch-hand rotation, we can write this equation $R_2 C_2 + R_3(-C_3) = E_2 + (-E_3)$, in which form it is included in the general analytical expression of the second Corollary, viz.,

$$\Sigma CR = \Sigma E,$$

for any closed circuit whatever.

137. Current Power. The expression already given for the power of a flow (Art. 116) becomes in the case of linear flow in an interpolar between two points, A and B, $C(\phi_B - \phi_A)$.

The power of the battery $P'Q'$ (fig. 70) is $C(\phi_{P'} - \phi_{Q'})$; this added to the power in the interpolar gives $C(\Delta_1 + \Delta_2)$, or

$$CE,$$

as the total power. If B is the resistance of the battery, and R that of the interpolar, the whole power is $(B+R)C^2$.

Examples.

1. Show that if between any two points, A, B (fig. 71), there are inserted n wires of resistances $r_1, r_2, \ldots r_n$, their resultant resistance, ρ, is given by the equation $\dfrac{1}{\rho} = \dfrac{1}{r_1} + \dfrac{1}{r_2} + \ldots + \dfrac{1}{r_n}$.

Let the current, C, of any battery break up at A into currents $c_1, c_2, \ldots c_n$ along the wires, respectively. Then $C = c_1 + c_2 + \ldots + c_n$. Also $\dfrac{\phi_A - \phi_B}{r_1} = c_1$; writing down n such equations and adding, we have $(\phi_A - \phi_B)\Sigma\dfrac{1}{r} = C$. But as C is the total current passing between A and B, we must have $\dfrac{\phi_A - \phi_B}{\rho} = C$; therefore $\dfrac{1}{\rho} = \Sigma\dfrac{1}{r}$.

Examples. 223

2. Calculate the separate magnitudes of the branch currents, $c_1, c_2 \ldots c_n$.

Take the closed circuit formed by the wires r_1 and r_2; then since there are no electromotive forces in the wires, Kirchhoff's second corollary gives $c_1 r_1 - c_2 r_2 = 0$, $\therefore c_1 : c_2 = \dfrac{1}{r_1} : \dfrac{1}{r_2}$; and similarly

$$c_1 : c_2 : c_3 : \ldots : c_n = \frac{1}{r_1} : \frac{1}{r_2} : \frac{1}{r_3} : \ldots : \frac{1}{r_n}.$$

Hence $c_1 = \dfrac{\dfrac{1}{r_1}}{\Sigma \dfrac{1}{r_1}}$. C, etc. [Branch wires inserted in this way between two points are said to be arranged in *Multiple Arc*.]

3. Show that the battery current at A breaks up in such a way that the sum of the energies in the branch wires is less than it would be if the same total current were broken up in any other manner.

The energies in the branches are $c_1^2 r_1$, $c_2^2 r_2$, ... no matter what the values of $c_1, c_2, \ldots$ are; therefore if X is the sum of the energies,

$$X = c_1^2 r_1 + c_2^2 r_2 + \ldots + c_n^2 r_n;$$

and if X is made a minimum by the variables $c_1, c_2, \ldots$, we must have

$$r_1 c_1 dc_1 + r_2 c_2 dc_2 + \ldots + r_n c_n dc_n = 0. \tag{1}$$

Also we have
$$dc_1 + dc_2 + \ldots + dc_n = 0, \tag{2}$$

since $c_1 + c_2 + \ldots + c_n$ is given. Multiplying the second by λ, subtracting from the first, and equating to zero the coefficients of dc_1, dc_2, we have

$$r_1 c_1 - \lambda = 0, \; r_2 c_2 - \lambda = 0 \ldots r_n c_n - \lambda = 0; \;\; \therefore \; c_1 : c_2 : \ldots = \frac{1}{r_1} : \frac{1}{r_2} : \ldots$$

which shows that the current *actually* breaks up in the required manner.

4. To represent graphically the fall of potential along a wire, being given the potentials at its extremities.

Let A and B be the extremities, and ϕ_A, ϕ_B the potentials at them; let P be any point on the wire between A and B; let ϕ_P be the potential at P. Then the current strength, C, being the same all through the wire, and the resistance of any portion of the wire being proportional to its length, we must have

$$\frac{\phi_A - \phi_P}{AP} = \frac{\phi_P - \phi_B}{BP}, \;\; \therefore \; \phi_P = \frac{BP \times \phi_A + AP \times \phi_B}{AB},$$

which shows that if we draw any right line, $A_1 B_1$, to represent the resistance of the given wire, at A_1 and B_1 erect two perpendiculars, $A_1 A'$ and $B_1 B'$, to represent ϕ_A and ϕ_B, respectively, take between A_1 and B_1 a point P_1 dividing $A_1 B_1$ so that $\dfrac{A_1 P_1}{B_1 P_1} = \dfrac{AP}{BP}$, and at P_1 erect the perpendicular $P_1 P'$ meeting the line $A'B'$ in P', then the ordinate $P_1 P'$ represents the potential ϕ_P.

5. Two wires, APB and AQB, connect, in multiple arc, two points, A and B; the current from any battery enters at A and emerges at B; the ends of a wire are attached at P and Q to the two first mentioned; find the condition that there shall be no current in the wire PQ.

It follows at once from last example that we must have

$$\frac{\text{resistance } AP}{\text{resistance } BP} = \frac{\text{resistance } AQ}{\text{resistance } BQ},$$

since if no current flows through PQ, the current strength is constant through APB, and also constant through AQB, and moreover ϕ_P must be equal to ϕ_Q. [This is the principle of *Wheatstone's Bridge*.]

138. Motion of Plane Vortices. We close the present section with a proposition relating to the motion produced in each other by indefinitely long straight vortex filaments whose axes are parallel, generated originally by any causes and then left to act upon each other [1].

It has been shown (Art. 128) that a vortex of strength m_1 at the point (a_1, β_1) produces at the point (x, y) a resultant velocity $\frac{m_1}{2\pi r}$. Suppose now that we have any number of such vortices of strengths $m_1, m_2, \ldots$ at points $(a_1 \beta_1), (a_2 \beta_2), \ldots$; then the motion of m_1 is due to the action of the remaining vortices, and if (u_1, v_1) are its velocity components, we have

$$2\pi u_1 = -m_2 \frac{\beta_1 - \beta_2}{r_{12}^2} - m_3 \frac{\beta_1 - \beta_3}{r_{13}^2} - \ldots \quad (1)$$

$$2\pi v_1 = m_2 \frac{a_1 - a_2}{r_{12}^2} + m_3 \frac{a_1 - a_3}{r_{13}^2} + \ldots \quad (2)$$

Writing down, in the same way, the velocity components $(u_2, v_2), (u_3, v_3), \ldots$ of the remaining vortices, we have at once

$$m_1 u_1 + m_2 u_2 + m_3 u_3 + \ldots = 0, \text{ or } \Sigma mu = 0, \quad (3)$$

$$m_1 v_1 + m_2 v_2 + m_3 v_3 + \ldots = 0, \text{ or } \Sigma mv = 0. \quad (4)$$

Now the strengths $m_1, m_2, \ldots$ all remain constant throughout the motion (p. 191), and these equations hold at every instant. Take at any time t the point, G, which is the centre of mean position of the points $(a_1, \beta_1), (a_2, \beta_2), \ldots$ for the system of

---

[1] A fuller account of this subject is given in Mr. Greenhill's papers (*Quarterly Journal*) and in his article on Hydrodynamics (*Encyclop. Brit.*); and also in Lamb's *Treatise on the Motion of Fluids*, chap. vi.

multiples $m_1, m_2, \ldots$. If its co-ordinates (referred to the fixed axes of co-ordinates) are $(\bar{x}, \bar{y})$, we have

$$\bar{x} \Sigma m = m_1 a_1 + m_2 a_2 + \ldots ,$$
$$\bar{y} \Sigma m = m_1 \beta_1 + m_2 \beta_2 + \ldots$$

Take the position of this point, G', again at the time $t + \Delta t$; let its co-ordinates be $(\bar{x}', \bar{y}')$.

Now at this time the co-ordinates of m_1 will be

$$(a_1 + u_1 \Delta t, \; \beta_1 + v_1 \Delta t),$$

and similarly for the other vortices. Then

$$\bar{x}' \Sigma m = m_1(a_1 + u_1 \Delta t) + m_2(a_2 + u_2 \Delta t) + \ldots ,$$
$$= \bar{x} \Sigma m + \Delta t \Sigma mu,$$
$$= x \Sigma m;$$

hence $\bar{x}' = \bar{x}$, and similarly $\bar{y}' = \bar{y}$, so that G' is the same as G. Hence—*the 'vortical centre' of any system of freely moveable and mutually influencing vortices is fixed in space throughout the whole motion*—a result identical with that which holds for the centre of mass of any material system which is left to its own mutual actions.

[The student must be careful to observe that the fixity of the vortical centre (or of the centre of mass of a mutually attracting system) does not imply that the velocity of the fluid which exists at this point is always zero (or, in the second case, that the resultant attractive force at the point is always zero).]

The signs of the quantities $m_1, m_2, \ldots$ depend on the directions of their spins.

Examples.

1. Find the motion of *two* mutually influencing vortices.

Ans. Each describes a circle about their vortical centre, with constant velocity. If $m_1 = -m_2$, G is at infinity, each describes a right line, and the line joining them retains a constant direction.

2. Find the motion of a single vortex, m, in front of a fixed smooth wall perpendicular to the plane of the fluid motion.

[Take the image of the vortex, which will be another vortex, $-m$, at the optical image of the first. Thus we have the last case; and the vortex moves parallel to the wall with constant velocity.]

3. Prove that a single vortex moving inside the space bounded by two plane walls enclosing an angle $\dfrac{\pi}{2n}$ describes the Cotes spiral $r \cos n\theta = a$. (Mr. Greenhill.)

4. Prove that a single vortex of strength m inside a circular cylinder of radius a, at a distance c from the centre will move with velocity due to an image of strength $-m$ at a distance $\dfrac{a^2}{c}$ from the centre, and that it will describe a circle of radius c with velocity $\dfrac{m}{2\pi} \dfrac{c}{a^2 - c^2}$. (Mr. Greenhill.)

5. Prove that the energy of any system of moveable vortices is

$$\tfrac{1}{2} \sum \frac{m_1 m_2 (m_1 + m_2)}{r_{12}^2} + 8 \sum \frac{m_1 m_2 m_3 \Delta_{123}^2}{r_{12}^2 r_{23}^2 r_{31}^2},$$

where r_{12} means the distance between the vortices m_1 and m_2 at any instant, and Δ_{123} means the area of the triangle formed by m_1, m_2, and m_3.

Section IV.—Conjugate Functions.

139. Definition of two Conjugate Functions. If ϕ and ψ are two functions, each of them a function of x and y, such that $\phi + \psi \sqrt{-1}$ is a function of $x + y\sqrt{-1}$, then ϕ and ψ are said to be two *conjugate functions* of x and y[1].

In the case of two such functions, then, we have

$$\phi + \psi\sqrt{-1} = f(x + y\sqrt{-1}). \qquad (1)$$

We may, of course, convert the proposition and say that if we take any function of $x + y\sqrt{-1}$, suppose $f(x + y\sqrt{-1})$, and express this function as the sum of a real and an imaginary term in the form $\phi + \psi\sqrt{-1}$, the quantities ϕ and ψ will be two conjugate functions of x and y.

Before proceeding to deduce some forms of conjugate functions, it is desirable to prove general properties characteristic of them.

140. General properties of Conjugate Functions. (1) *Each of two conjugate functions satisfies Laplace's equation.* For if we differentiate equation (1) of last Article twice with respect

---

[1] It is thought advisable not to use the symbol i for $\sqrt{-1}$, since the former has been so firmly established as the designation for a *unit vector*.

Properties of Conjugate Functions.

to x and twice with respect to y, and add the results, we have $\nabla^2\phi + \sqrt{-1}\,\nabla^2\psi = 0$, which requires

$$\nabla^2\phi = 0 \text{ and } \nabla^2\psi = 0. \tag{1}$$

(2) The following relations hold between two conjugate functions as defined by equation (1) of last Article—

$$\frac{d\phi}{dx} = \frac{d\psi}{dy}; \quad \frac{d\phi}{dy} + \frac{d\psi}{dx} = 0. \tag{2}$$

For, using z, for a moment, instead of $x + y\sqrt{-1}$, and differentiating (1) of last Article with respect to x, we have

$$\frac{d\phi}{dx} + \sqrt{-1}\,\frac{d\psi}{dx} = f'(z);$$

differentiating it with respect to y, we have

$$\frac{d\phi}{dy} + \sqrt{-1}\,\frac{d\psi}{dy} = \sqrt{-1}\,f'(z);$$

hence from these two equations

$$\sqrt{-1}\,\frac{d\phi}{dx} - \frac{d\psi}{dx} = \frac{d\phi}{dy} + \sqrt{-1}\,\frac{d\psi}{dy},$$

which, by equating the real and the imaginary parts, gives the equations (2); *and these two equations may be taken as giving a new form of definition of conjugate functions*, since equations (1) both of this and of last Article obviously follow from them.

(3) *If ϕ and ψ are any two conjugate functions of x and y, the curves*

$$\phi = C, \quad \psi = C',$$

where C and C' are any two constants, intersect at right angles.

For, if R^2 denotes $\overline{\dfrac{d\phi}{dx}}\Big|^2 + \overline{\dfrac{d\phi}{dy}}\Big|^2$, which is the same as

$$\overline{\frac{d\psi}{dx}}\Big|^2 + \overline{\frac{d\psi}{dy}}\Big|^2,$$

the cosine of the angle between the curves at any point of intersection is $\dfrac{\dfrac{d\phi}{dx}\dfrac{d\psi}{dx} + \dfrac{d\phi}{dy}\dfrac{d\psi}{dy}}{R^2}$, and by (2) the numerator of this is zero; therefore, &c.

The foregoing properties prove that we can regard ϕ and ψ

as a velocity-potential function and a function of flow, respectively, or *vice versa*, since if $\frac{d\phi}{dx} = u$, $\frac{d\phi}{dy} = v$, we have

$$\frac{d\psi}{dx} = -v, \quad \frac{d\psi}{dy} = u.$$

(4) *If ϕ and ψ are any two conjugate functions of x and y, and if at any point we describe the curves $\phi = const.$ and $\psi = const.$, the line-rate of increase of the function ϕ along the curve $\psi = const.$ is equal to the line-rate of increase of ψ along the curve $\phi = const.$*

Regard ϕ and ψ as the velocity function and function of flow in the irrotational motion of a liquid, and the equality enunciated is exactly the equation of Art. 104. Of course all other properties which have been proved for equipotential and flow functions hold for any two conjugate functions.

(5) *If ϕ and ψ are any two conjugate functions of x and y, and if ξ and η are any two other conjugate functions of x and y; then by putting ξ and η instead of x and y in the values of ϕ and ψ, we get two new conjugate functions of x and y.*

For, let $\phi \equiv f_1(x, y)$, $\psi \equiv f_2(x, y)$. In these put ξ for x and η for y, and let $\phi' = f_1(\xi, \eta)$, $\psi' = f_2(\xi, \eta)$; then

$$\frac{d\phi'}{dx} = \frac{df_1}{d\xi}\frac{d\xi}{dx} + \frac{df_1}{d\eta}\frac{d\eta}{dx}; \quad \frac{d\psi'}{dy} = \frac{df_2}{d\xi}\frac{d\xi}{dy} + \frac{df_2}{d\eta}\frac{d\eta}{dy}.$$

But since $\frac{df_1}{dx} = \frac{df_2}{dy}$, and $\frac{df_1}{dy} + \frac{df_2}{dx} = 0$, it is evident that

$$\frac{df_1}{d\xi} = \frac{df_2}{d\eta}, \quad \text{and} \quad \frac{df_1}{d\eta} + \frac{df_2}{d\xi} = 0.$$

Also by supposition $\frac{d\xi}{dx} = \frac{d\eta}{dy}$, and $\frac{d\xi}{dy} + \frac{d\eta}{dx} = 0$. Hence

$$\frac{d\phi'}{dx} = \frac{d\psi'}{dy}, \quad \text{and similarly} \quad \frac{d\phi'}{dy} + \frac{d\psi'}{dx} = 0,$$

which are necessary and sufficient to define two conjugate functions of x and y.

This is a very important property, which we may enunciate thus in general terms, considering it with reference to fluid motion—

Properties of Conjugate Functions.

If the circumstances of any irrotational fluid motion, which takes place over the plane xy, are at any point (x, y) expressed by the velocity and flow functions ϕ and ψ, the circumstances of another possible state of irrotational fluid motion over the same plane and at the same point (x, y) will be expressed by velocity and flow functions derived from ϕ and ψ by replacing, in their expressions, the co-ordinates x, y by any two conjugate functions of x and y whatever.

Instead of regarding the matter from the point of view of fluid motion or electrical current flow, we may regard it from that of *electrostatic distribution*. The functions ϕ and ψ will then be the attraction potential and force-direction functions; and we may say that *if $[\phi, \psi]$ expresses a possible electrostatic distribution at any point x, y in a given plane, we shall get another possible electrostatic distribution at the same point x, y, by writing for x and y, in the expressions for ϕ and ψ, any two conjugate functions of x and y.*

Applications of this very fruitful principle will presently be given.

(6) *If ξ and η are two conjugate functions of x and y, then, conversely, x and y are two conjugate functions of ξ and η.*

For, let $f(x+y\sqrt{-1}) = \xi + \eta\sqrt{-1}$. Differentiate this equation first with respect to ξ and then with respect to η, and divide the one result by the other. Thus we get

$$\left(\frac{dx}{d\xi} + \sqrt{-1}\frac{dy}{d\xi}\right)\sqrt{-1} = \frac{dx}{d\eta} + \sqrt{-1}\frac{dy}{d\eta},$$

which involves $\frac{dx}{d\xi} = \frac{dy}{d\eta}$ and $\frac{dx}{d\eta} + \frac{dy}{d\xi} = 0$,

the necessary and sufficient conditions that x and y should be conjugate functions of ξ and η.

(7) Let ξ and η be any two conjugate functions of x and y, the latter being the co-ordinates of any point, P, referred to two rectangular axes Ox and Oy; draw any two new rectangular axes, $O'\xi$ and $O'\eta$, and in this new system lay down the point, P', whose co-ordinates (ξ, η) have reference to the values of x and y at P; and let this be done for all the points P in the first figure. Then, corresponding to any curve or continuous space

in the first figure we shall have a curve or continuous space in the second figure. [Were it not for the confusion of figures, we might lay off the values of ξ and η along the given axes Ox and Oy; but it will be better to imagine the axes Ox, Oy and the points P laid down on one piece of paper, and the axes $O\xi$, $O\eta$ and the corresponding points P' laid down on a separate piece of paper.] Now any two curves passing through P in the first figure will transform into two curves passing through P' in the second figure, and *the angle between the curves at P' will be the same as that between the two corresponding curves at P.*

For, let A be a point very close to P on one curve, and B a point very close to P on the other; let A' and B' be the corresponding points on the corresponding curves through P' in the second figure. Let (x, y) be the co-ordinates of P; $(x+a, y+\beta)$ those of A; $(x+a', y+\beta')$ those of B; and (ξ, η) those of P'. Then the co-ordinates of A' are

$$(\xi + a\frac{d\xi}{dx} + \beta\frac{d\xi}{dy};\ \eta + a\frac{d\eta}{dx} + \beta\frac{d\eta}{dy}),$$

similar expressions holding for the co-ordinates of B'.

Also, if the lengths PA and PB are ds_1 and ds_2, and the lengths $P'A'$ and $P'B'$ are ds_1' and ds_2', we have at once

$$\left.\begin{array}{l} ds_1' = K ds_1 \\ ds_2' = K ds_2 \end{array}\right\} \qquad (3)$$

where $\quad K^2 \equiv \overline{\dfrac{d\xi}{dx}}\Big|^2 + \overline{\dfrac{d\xi}{dy}}\Big|^2 = \overline{\dfrac{d\eta}{dx}}\Big|^2 + \overline{\dfrac{d\eta}{dy}}\Big|^2.$

Again, the direction-cosines of $P'A'$ (with reference to $O'\xi$, $O'\eta$) are

$$\frac{1}{K ds_1}\left(a\frac{d\xi}{dx} + \beta\frac{d\xi}{dy}\right) \text{ and } \frac{1}{K ds_1}\left(a\frac{d\eta}{dx} + \beta\frac{d\eta}{dy}\right),$$

similar expressions holding for the direction-cosines of $P'B'$.

Hence, in virtue of equations similar to (2), p. 227, holding between ξ and η, we have

$$\cos \angle B'P'A' = \frac{aa' + \beta\beta'}{ds_1 ds_2} = \cos \angle BPA, \qquad (4)$$

which proves the proposition.

Again, if dS is any small element of area in the first figure, the

Properties of Conjugate Functions.

area of the corresponding small element in the second figure is $K'^2 dS$. For, if P is any point inside dS, the radius vector, r, from P to any point on the contour of dS transforms into $K' \cdot r$ drawn from P' to the contour of the corresponding area, dS'; and the angle between two radii vectores from P remains unaltered; hence

$$dS' = K'^2 \cdot dS. \tag{5}$$

It follows, in particular, that any two curves orthogonal at P transform into two curves orthogonal at P'; and that any small closed curve at P transforms into a *similar* small curve at P'.

[It is scarcely necessary to observe that the axes of reference $O'\xi$, $O'\eta$ are not the transformed equivalents of the axes Ox, Oy. These latter transform, in general, into *curved* lines in the second figure.]

(8) Let $f(x, y)$ denote any quantity having reference to a point P, whose co-ordinates are x, y. Transform this quantity by substituting $f_1(x, y)$ for x and $f_2(x, y)$ for y, where f_1 and f_2 are any two conjugate functions of x and y; so that the new function connected with P is

$$f[f_1(x,y), f_2(x,y)],$$

or, briefly, $F(x, y)$.

It is required to find a value for

$$\left(\frac{d^2}{dx^2} + \frac{d^2}{dy^2}\right) F(x, y),$$

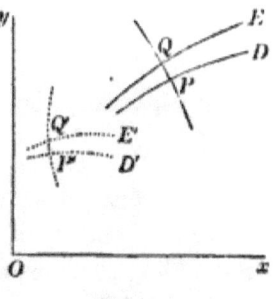

Fig. 72.

in terms of the original form, f, of the function.

We have

$$\frac{dF}{dx} = \frac{df[f_1(x,y), f_2(x,y)]}{dx} = \frac{df(f_1, f_2)}{df_1}\frac{df_1}{dx} + \frac{df(f_1, f_2)}{df_2}\frac{df_2}{dx},$$

where we have used f_1 and f_2, for abbreviation, instead of $f_1(x, y)$ and $f_2(x, y)$. Hence

$$\frac{d^2F}{dx^2} = \frac{d^2f(f_1, f_2)}{df_1^2}\overline{\frac{df_1}{dx}}\bigg|^2 + \frac{d^2f(f_1, f_2)}{df_2^2}\overline{\frac{df_2}{dx}}\bigg|^2 + 2\frac{d^2f(f_1, f_2)}{df_1 df_2}\frac{df_1}{dx}\frac{df_2}{dx}$$
$$+ \frac{df(f_1, f_2)}{df_1}\frac{d^2f_1}{dx^2} + \frac{df(f_1, f_2)}{df_2}\frac{d^2f_2}{dx^2}.$$

Writing down, in the same way, the value of $\dfrac{d^2 F}{dy^2}$, and adding the two results, we have

$$\frac{d^2 F}{dx^2} + \frac{d^2 F}{dy^2} = K^2 \left\{ \frac{d^2 f(f_1, f_2)}{df_1^2} + \frac{d^2 f(f_1, f_2)}{df_2^2} \right\}, \qquad (6)$$

in virtue of the relations just proved between the conjugate functions f_1 and f_2, and using K^2 to denote the common value of

$$\overline{\left|\frac{df_1}{dx}\right|}^2 + \overline{\left|\frac{df_1}{dy}\right|}^2 \quad \text{and} \quad \overline{\left|\frac{df_2}{dx}\right|}^2 + \overline{\left|\frac{df_2}{dy}\right|}^2.$$

Now let P' be the point whose co-ordinates (deduced from those of P) are $f_1(x, y)$ and $f_2(x, y)$; then the quantity in brackets in (6) is the value of $\left(\dfrac{d^2}{dx^2} + \dfrac{d^2}{dy^2}\right) f(x, y)$ at P'. Hence—

transformed value of $\nabla^2 f(x, y)$ at $P = K^2 \times$ untransformed value of $\nabla^2 f(x, y)$ at P', (7)

where P' is the point 'corresponding' to P in the sense explained in No. 7.

Again, if dS denotes any small area surrounding P, and dS' the 'corresponding' small area surrounding P', we proved in No. 7 that $dS = \dfrac{1}{K^2} dS'$. Hence the product of the transformed value of $\nabla^2 f(x, y)$ at P and a small area surrounding P is equal to the product of the untransformed value at P' and the corresponding small area.

Again, *let it be required to find the line-rate of variation of the transformed quantity, $F(x, y)$, at P in any direction, PQ, in terms of the original form, f, of the function.*

Let the elementary length PQ be ds; then $\dfrac{dF}{ds}$ is the line-rate of variation of F along PQ. But

$$\frac{dF}{ds} = \frac{df(f_1, f_2)}{df_1} \frac{df_1}{ds} + \frac{df(f_1, f_2)}{df_2} \frac{df_2}{ds}.$$

Let Q' be the point corresponding to Q, as P' corresponds to P. Then if $P'Q' = ds'$, we have from equation (3) $ds' = K ds$, so that

$$\frac{dF}{ds} = K \frac{df}{ds'}, \qquad (8)$$

or *transformed value of line-rate at* $P = K$ (*untransformed value of line-rate at* P').

The application of these theorems to electrostatic distribution is obvious.

Suppose that at all points on the same right line parallel to the axis of z the electrical density, force, &c. are the same; then clearly if V is the attraction-potential function, the general equation $\frac{d^2 V}{dx^2} + \frac{d^2 V}{dy^2} + \frac{d^2 V}{dz^2} = 0$ becomes $\frac{d^2 V}{dx^2} + \frac{d^2 V}{dy^2} = 0$; and the volume-density, ρ, at any point, P, is $-\frac{1}{4\pi}\nabla^2 V$, i.e., $\rho\, dS$ is the quantity in a cylinder of unit height standing on the small area dS surrounding the point P, the base dS being parallel to the plane x, y.

Now transform the whole distribution by putting for x and y, in the expression for V, any two conjugate functions of x and y, and let P' be the point 'corresponding' to P. The result is a law of a new possible distribution. Then the quantity of electricity in a cylinder with unit height standing on any small area at P in the new distribution is the same as the quantity in a cylinder standing on the corresponding small area at P' in the old distribution; and the new volume-density, ρ, at P is connected with the old volume-density, ρ', at P' by the equation
$$\rho = K^2 \rho'. \tag{9}$$

It may happen that the electrical density, force, &c. are the same not only at all points on the same line parallel to the axis of z, but also at all points on the same line parallel to the axis of x; and in this case we have simply

$$\frac{d^2 V}{dy^2} = 0.$$

This latter is the case when we have two infinite (or practically infinite) electrified plane surfaces held parallel to each other at a small distance apart, the origin of co-ordinates being at the middle of one plane, and the axis of y perpendicular to the planes. Our supposition will then hold good for all points in the region between the plates which are not very near the edges.

In like manner, we may suppose $f(x, y)$ to be the flow function, ψ, at a point, in a given motion of a liquid, so that the spin at P is $-\frac{1}{2}\nabla^2\psi$; and we have at once the strengths of vortices at all points in a new motion given as equal to the strengths of vortices at corresponding points in the old motion.

The second theorem, expressed by equation (8), gives a result with regard to the transformation of the *surface*-density of a given electrostatic distribution—just as the first theorem relates to the transformation of *volume*-density. For, supposing the function $f(x, y)$ to express electrostatic potential at (x, y), which is constant along a surface whose section by the plane of xy is the curve PD (fig. 72), and that PQ is an element of the normal to this surface at P; then if σ is the surface-density at P when the potential is transformed into $f(f_1, f_2)$, or $F(x, y)$, we know (see *Statics*, p. 440) that

$$\sigma = -\frac{1}{4\pi}\frac{dF}{ds}.$$

And as $f(f_1, f_2)$ is the value of the potential at the point (f_1, f_2), i.e., at P', in the old distribution, in which let σ' be the surface-density at P', equation (8) gives

$$\sigma = K\sigma'. \qquad (10)$$

It will be convenient to speak of this quantity K as a *modulus of transformation*.

Of course the potential at P in the new distribution has the same value as the potential at P' in the old; for, the potential at any point being $f(x, y)$ in the old distribution, the potential at P' in this distribution is $f(f_1, f_2)$, and this is by hypothesis what the potential at P becomes in the new; in terms of the co-ordinates (x, y) of P this is $F(x, y)$.

Instead of calling $-\frac{1}{4\pi}\frac{dF}{ds}$ the 'surface-density' at a point, in the application of this theory to electrostatics, it would be better to call it the *electric displacement*, at the point, per unit area of the surface to which ds is normal—the surface in question being, of course, *perpendicular* to the plane xy. This designation for σ is preferable, because the point to which it refers may be a point in the dielectric and not one on a *metallic* surface. The quantity σ is the quantity of positive electricity driven, or separated from negative, across unit area of the surface (real or imagined in the dielectric) to which ds is normal.

Cor. Equipotential curves and curves of flow (or curves of induction in electrostatics and electromagnetics) 'correspond' to, or transform into, curves of the same nature.

(9) With exactly the same suppositions and notation as in No. 8, we have

$$\overline{\frac{dF}{dx}}\Big|^2 + \overline{\frac{dF}{dy}}\Big|^2 = K^2 \left\{ \overline{\frac{df}{df_1}}\Big|^2 + \overline{\frac{df}{df_2}}\Big|^2 \right\}, \qquad (8)$$

so that if $f(x, y)$ denote a flow function of a moving fluid, we have the result that—

The energy of transformed motion inside any small area at $P =$ the energy of the original motion inside the small corresponding area at P'.

(10) If (ξ, η) are any two conjugate functions of x and y, and (ξ', η') any other two conjugate functions of x and y; then, n being any positive integer, the functions [1]

$$\left(\xi'\frac{d}{dx} + \eta'\frac{d}{dy}\right)^n \xi \quad \text{and} \quad \left(\xi'\frac{d}{dx} + \eta'\frac{d}{dy}\right)^n \eta$$

are two conjugate functions of x and y.

For, let $U = \left(\xi'\frac{d}{dx} + \eta'\frac{d}{dy}\right)\xi$ and $V = \left(\xi'\frac{d}{dx} + \eta'\frac{d}{dy}\right)\eta$. Then, in virtue of the fundamental equations (2) of No. 2, we have at once

$$\frac{dU}{dx} - \frac{dV}{dy} = 0 \quad \text{and} \quad \frac{dU}{dy} + \frac{dV}{dx} = 0,$$

which give $\nabla^2 U = 0$, $\nabla^2 V = 0$ at once, and prove U and V to be conjugate functions.

Applying the same process to U and V, we have

$$\left(\xi'\frac{d}{dx} + \eta'\frac{d}{dy}\right)U \quad \text{and} \quad \left(\xi'\frac{d}{dx} + \eta'\frac{d}{dy}\right)V$$

---

[1] These derived conjugate functions are, in the language of modern Higher Algebra, *Emanants* of the given functions ξ and η.

conjugate functions of x and y, i.e.,

$$\left(\xi'\frac{d}{dx} + \eta'\frac{d}{dy}\right)^2 \xi \quad \text{and} \quad \left(\xi'\frac{d}{dx} + \eta'\frac{d}{dy}\right)^2 \eta$$

are conjugate functions. Repeating the operation any number of times, the theorem enunciated above is proved.

For example, $x^2 - y^2$ and $2xy$ are obviously conjugate functions, and if (ξ, η) are any two conjugate functions of x and y, so will $\xi^2 - \eta^2$ and $2\xi\eta$ be (No. 5). Hence if (ξ', η') are any other two conjugate functions of x and y, so will be the Emanants $\xi\xi' - \eta\eta'$ and $\xi\eta' + \xi'\eta$
(proved otherwise in Clerk-Maxwell, vol. i., Art. 187).

(11) If ϕ is any function of (x, y) satisfying Laplace's equation $\nabla^2 \phi = 0$, then will $\dfrac{d\phi}{dx}$ and $-\dfrac{d\phi}{dy}$ be two conjugate functions of x and y.

For they satisfy the (necessary and sufficient) equations of No. 2. Hence the components of velocity of a liquid at any point of a non-vortical region—the sign of one component being altered—are two conjugate functions of the co-ordinates of the point.

141. General Formulae of Transformation. In the old distribution let $(V', \overline{F}', \sigma', \rho', ds', dS', E')$ be respectively the potential, force, electrical displacement, volume-density, element of length, element of area, and quantity of electricity, at the point P'; and let $(V, \overline{F}, \sigma, \rho, ds, dS, E)$ be the same quantities at P in the new distribution, K being the modulus of transformation. Then, collecting the results just proved, we have the following relations—

$$V = V',$$
$$\overline{F} = K\overline{F'},$$
$$\sigma = K\sigma',$$
$$\rho = K^2 \rho',$$
$$ds = \frac{1}{K} ds',$$
$$dS = \frac{1}{K^2} dS',$$
$$E = E'.$$

We have above interpreted the quantities involved in these equations with reference to electrostatic distribution, but it is manifest that they might have been used with reference to liquid motion, electrical flow, or electromagnetic action.

142. Clerk-Maxwell's Graphic Method of Transformation. Granted that a certain function, ϕ, of (x, y) is constant all along a curve whose equation in (x, y) is given; being given also two equations connecting x and y with two other quantities (ξ, η); it is required to trace out graphically the values of ξ and η corresponding to the given constant value of ϕ.

[Analytically, of course, the method of doing this is plain. We have only to solve the given relations for x and y, expressing these explicitly in terms of ξ and η, to put these values for (x, y) into the equation of the given curve, and to trace out the new curve with reference to two assumed axes of ξ, η.]

Mark a large number of (x, y) points,

$$(a, b), (a', b'), (a'', b''), \ldots$$

on the given curve, for which ϕ is constant; then draw two axes of (ξ, η); eliminate y from the two relations between (x, y) and (ξ, η); suppose that this gives $f(\xi, \eta, x) = 0$; now trace out a series of curves

$$f(\xi, \eta, a) = 0, f(\xi, \eta, a') = 0, f(\xi, \eta, a'') = 0, \qquad (\alpha)$$

obtained by making x constant and successively equal to $a, a', a'', \ldots$; again, eliminate x between the given relations and suppose the result to be $g(\xi, \eta, y) = 0$; then trace out a new series of curves

$$g(\xi, \eta, b) = 0, g(\xi, \eta, b') = 0, g(\xi, \eta, b'') = 0, \qquad (\beta)$$

obtained by making y successively equal to $b, b', b'', \ldots$.

Now take the point (or points) of intersection of the first of the series (α) and the first of (β), and we obtain a (ξ, η) point (or points) corresponding to the given value of ϕ. Similarly the intersection of any curve of the series (α) with the corresponding curve of the series (β) will give another required (ξ, η) point; &c.

This method applies, of course, whatever be the natures of

ϕ, ξ, η; but it is especially useful in the case in which ϕ is a potential (or flow) function, and (ξ, η) are two conjugate functions of (x, y). In this case ϕ will also be a potential (or flow) function of the new variables (ξ, η), and the curve thus graphically traced out will be a transformed equipotential (or flow) curve.

143. Determination of Conjugate Functions. An infinite number of forms of two conjugate functions of (x, y) can be found from the fact (Art. 139) that the real part and the coefficient of $\sqrt{-1}$ in the expansion of $f(x+y\sqrt{-1})$ are conjugate functions, whatever be the form of f.

Thus
$$f(x) - \frac{y^2}{\lfloor 2} f''(x) + \frac{y^4}{\lfloor 4} f^{IV}(x) - \ldots \equiv \xi, \quad (1)$$

and
$$yf'(x) - \frac{y^3}{\lfloor 3} f'''(x) + \ldots \equiv \eta, \quad (2)$$

are two conjugate functions of (x, y).

The following are some examples of most frequent occurrence.

(α). Let $f(x) \equiv x^2$. Put $x + y\sqrt{-1}$ for x, and we get
$$\xi \equiv x^2 - y^2; \quad \eta \equiv 2xy.$$

(β). Let $f(x) \equiv \log \frac{x}{a}$. Put $x + y\sqrt{-1}$ for x; then put $x = r \cos \theta$, $y = r \sin \theta$, where $r = \sqrt{x^2 + y^2}$, $\theta = \tan^{-1} \frac{y}{x}$.

Then $\log \frac{x + y\sqrt{-1}}{a} = \log \frac{r}{a} + \log (\cos \theta + \sqrt{-1} \sin \theta)$.

But $\cos \theta + \sqrt{-1} \sin \theta = e^{\theta \sqrt{-1}}$; therefore
$$\log \frac{x + y\sqrt{-1}}{a} = \log \frac{r}{a} + \theta \sqrt{-1};$$

$$\therefore \xi \equiv \log \frac{r}{a} \equiv \log \frac{\sqrt{x^2 + y^2}}{a}; \quad \eta \equiv \theta \equiv \tan^{-1} \frac{y}{x},$$

which are the well-known conjugate functions which play such an important part in the cases of *whirls* and of *sources* and *sinks* (see Section I).

Determination of Conjugate Functions.

If we put $\frac{r}{a} = e^\rho$, we may write $x = ae^\rho \cos\theta$, $y = ae^\rho \sin\theta$; then ρ and θ are conjugate functions of x and y; and conversely, x and y are conjugate functions of ρ and θ—as may be seen at once, since $\frac{dx}{d\rho} = \frac{dy}{d\theta}$ and $\frac{dx}{d\theta} + \frac{dy}{d\rho} = 0$.

COR. 1. Since (No. 11, Art. 140) $\frac{d\phi}{dx}$ and $-\frac{d\phi}{dy}$ are conjugate functions of (x, y), if ϕ is any function satisfying $\nabla^2\phi = 0$, we have $\log \bar{v}$ and θ, conjugate functions, where $\bar{v}$ *is the resultant velocity at any point and θ its direction, in the case of a non-vortically moving liquid.*

If V is an electrostatic potential for a two-dimensional distribution (i.e., one which is the same all along the same line parallel to the axis of z, p. 233), we have exactly the same result for the *electric force* and its *direction*, since $\frac{dV}{dx}$ and $-\frac{dV}{dy}$ are conjugates.

COR. 2. If n is any constant, it is obvious that $n\rho$ and $n\theta$ are conjugate functions of x and y, since ρ and θ are so.

Hence we derive a very useful transformation. If $F(x, y)$, or $F(ae^\rho \cos\theta, ae^\rho \sin\theta)$ is any liquid velocity-potential function, flow function, or electrostatic potential, at any point P, the function
$$F(ae^{n\rho} \cos n\theta,\ ae^{n\rho} \sin n\theta)$$
will also be a function of the same nature at the same point in a possible new distribution. Thus we may put $\frac{r^n}{a^{n-1}}$ for r, and $n\theta$ for θ.

(γ). Let $f(x) \equiv \cos x$. Put $x + y\sqrt{-1}$ for x, and we shall find, by using the well-known exponential values of the sine and cosine,
$$\cos(x + y\sqrt{-1}) = \tfrac{1}{2}(e^y + e^{-y})\cos x - \tfrac{1}{2}\sqrt{-1}(e^y - e^{-y})\sin x,$$
$$= \cosh y \cos x - \sqrt{-1} \sinh y \sin x,$$
so that when
$$\xi = \cosh y \cos x \quad \text{and} \quad \eta = -\sinh y \sin x,$$

ξ and η are conjugate functions. This could have been inferred from the results in (β), since if (ξ, η) and (ξ', η') are pairs of conjugate functions, the pairs $(\xi \pm \xi', \eta \pm \eta')$ are also conjugates.

144. Electric Inversion. We propose to apply the results in Cor. 2, Art. 143, and in No. 8, Art. 140, to a particular case of transformation.

Let $F(x, y)$ be the electrostatic potential at any point P; then, choosing $n = -1$ in Cor. 2, Art. 143, we see that

$$F(ae^{-\rho}\cos\theta, -ae^{-\rho}\sin\theta)$$

is the potential at P in the new distribution. This is the same as

$$F\left(\frac{a^2}{r}\cos\theta, -\frac{a^2}{r}\sin\theta\right),$$

r being the radius-vector from the origin, O, to P. Now

$$\left(\frac{a^2}{r}\cos\theta, \frac{a^2}{r}\sin\theta\right)$$

are the co-ordinates of a point inverse to P on the line OP; hence P' is the reflection of the inverse of P in the axis of x.

In this case $\xi = \dfrac{a^2 x}{x^2 + y^2}$, $\eta = -\dfrac{a^2 y}{x^2 + y^2}$, and we have

$$K = \frac{a^2}{r^2};$$

so that if we use this value of K in the equations of Art. 140, we get the complete connection between the two distributions.

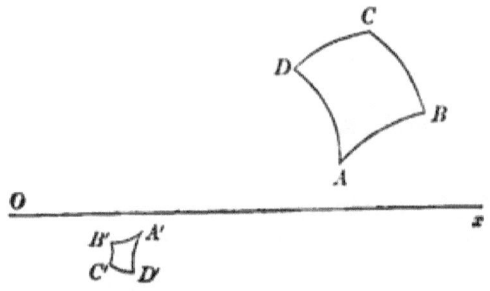

Fig. 73.

Thus, let $ABCD$ (fig. 73) be any figure at any point on the contour of which the potential has a value $f(x, y)$; and let the potential be transformed by the inverse substitution $\dfrac{a^2}{r}$ and

$-\theta$ for r and θ. Then over the same contour $ABCD$ in the new distribution the potential at any point will have the same value as, in the old distribution, it had at the corresponding point of the contour $A'B'C'D'$, which latter is obtained by taking the reflection, in Ox, of an inverse of the contour $ABCD$.

If in the previous distribution there is, for example, a quantity e of electricity at the point D', there will in the new distribution be a charge e at the point D, and no charge at D'. This charge is not an electric 'displacement,' or surface-density, but the product of a volume-density and an element of volume. Again, if in the first distribution there is a charge e at D, this in the new distribution will be transferred to D', and there will be no charge at D.

In both distributions there will be equal *electric displacements*, σdS and $\sigma' dS'$, through corresponding areas of surfaces (imagined or metallic) at D, D', and all pairs of corresponding points—the surfaces being perpendicular to the plane of the figure.

We shall take as an example a case which is discussed by Clerk Maxwell (*Electricity and Magnetism*, vol. i. p. 269).

Let AB (fig. 74) be a section of an infinitely long hollow circular metallic cylinder of radius b by the plane of xy, which is perpendicular to its axis; let its axis (represented by C) be a slender metallic rod to which a charge of electricity has been given;

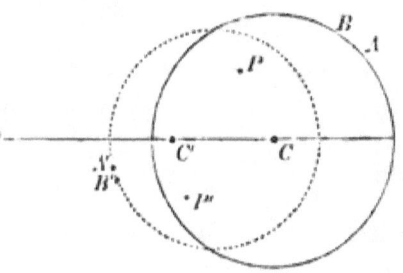

Fig. 74.

let e be the quantity of the charge at C, and let the potential of the metallic cylinder be V_0.

We may speak of the *circle AB* instead of the cylinder, and the *point C* instead of the axis.

We propose to investigate the new distribution of electrification, &c., which will be produced by inversion about the arbitrary origin O.

The actual distribution of potential, &c., in the given case is easily found. For, if V is the potential at any point, P, whose distance from C is k, V is simply a function of k, so that the differential equation of example 3, p. 165, becomes

$$\frac{d^2V}{dk^2} + \frac{1}{k}\frac{dV}{dk} = 0,$$

whose integral is
$$V = V_0 + A \log \frac{k}{b}.$$

To find A, we use the differential equation

$$\frac{dV}{dv} + \frac{dV}{dv'} = -4\pi\sigma,$$

which is satisfied at any point of any curve whatever, σ being the electric displacement across unit area, and dv, dv' being elements of normal drawn outwards and inwards. Applying this equation to the inner side of the circle AB, and observing that $\frac{dV}{dv} = 0$, while $\frac{dV}{dv'} = -\frac{dV}{dk}$, we have $\frac{A}{b} = 4\pi\sigma$; and since the charge at C plus the total charge on the inner side of the circle is zero, $2\pi\sigma b = -e$;

$$\therefore V = V_0 + 2e \log \frac{b}{k}. \qquad (1)$$

We must express this in terms of $OP(=r)$ and the angle $POC(=\theta)$, for the purpose of inversion about O. Then

$$V = V_0 + 2e \log \frac{b}{\sqrt{r^2 - 2cr\cos\theta + c^2}}, \qquad (2)$$

where $c = CO$. Now consider the points, C' and P', corresponding by inversion about O to C and P.

In the new distribution there will be at C whatever charge was at C' in the old—i.e., none; and in the new distribution there will be at C' whatever charge there was at C in the old—i.e. there will be e at C'. Again, at every point P in the new distribution the potential will be what it was at P' in the old; and as this latter is obtained by writing $\frac{a^2}{r}$ and $-\theta$ for r and θ in (2), we have

$$V = V_0 + 2e \log \frac{br}{\sqrt{c^2r^2 - 2ca^2 r\cos\theta + a^4}}, \qquad (3)$$

expressing the potential at any point due to a charge e at C' and an electrification over the metallic circle AB.

This is evidently the same as

$$V = V_0 + 2e \log \frac{b \cdot OP}{c \cdot C'P}. \qquad (4)$$

In the second distribution the potential at any point of the circle AB is equal to the potential, in the first, at the corresponding point on the circle $A'B'$—which latter is the inverse of the former with respect to O. Hence, in general, the circle AB ceases to be a curve of constant potential in the new distribution. But if we arrange so that the inverse, $A'B'$, is the circle AB itself, then AB will remain a curve of constant potential. This of course is done by taking a equal to the tangent from O to the circle $= \sqrt{c^2 - b^2}$. The same result follows at once by making $\frac{OP}{C'P} = \frac{c}{b}$ when P is any point on the circle.

The points O and C' are in this case inverse points with regard to the circle.

If A', B' correspond to A, B, there is the same electric displacement through the arc AB in the new distribution as through $A'B'$ in the old. When AB is its own inverse, the change of distribution is effected by drawing any two lines OA, OB, which cut the circle again in A'', B'', and then transferring the surface charge over $A''B''$ to the surface AB, and vice versa.

The value of V in (4) may be expressed in terms of the polar co-ordinates (R, ψ) of P referred to the centre C. If we assume the condition $ch = b^2$ and take h to represent the length CC', we get

$$V = V_0 + 2e \log \frac{1}{b} \sqrt{\frac{h^2 R^2 - 2 b^2 h R \cos \psi + b^4}{R^2 - 2 h R \cos \psi + h^2}} \qquad (5)$$

for the potential at any point P.

The value (4) shows that the new equipotential curves are circles—as is otherwise evident from the fact that in the original distribution they are circles. Also in the original distribution the lines of induction (lines of flow in case of liquid motion) are right lines emanating from C. Hence in the new distribution the lines of induction are circles, viz., the inverses, with respect to O, of the system of right lines passing through C.

This case, therefore, in electrostatics is the same as the case in p. 211 for liquid motion or electrical flow over a conducting plane; the lines of *actual* flow in these cases becoming *lines of induction* across the dielectric inside the circle—no continuous *flow* being here possible.

And the new state of affairs just deduced for electrostatic induction by inversion may be applied, with a merely suitable change of terms, to the corresponding cases of liquid motion and electrical flow—i. e., all such cases may be transformed by inversion.

Comparing the value of V in (4), viz.,

$$V_0 + 2e \log \frac{b}{c} + 2e \log OP - 2e \log C'P,$$

with the value in Art. 132, we see that the new distribution is equivalent to the action of a charge e at C' and a charge $-e$ at O.

145. Screw Motion of a Liquid. We shall slightly generalise the uniplanar motion which we have hitherto discussed, and assume that, while the motions of all the particles take place parallel to the plane of xy, the motions at corresponding points in two planes parallel to that of xy are not the same. In the general case of displacement (see *Statics*, p. 473, 2nd ed.) the spin at a point has three components, $\omega_1, \omega_2, \omega_3$, round axes parallel to those of x, y, z, expressed by the equations

$$\omega_1 = \tfrac{1}{2}\left(\frac{dw}{dy} - \frac{dv}{dz}\right); \quad \omega_2 = \tfrac{1}{2}\left(\frac{du}{dz} - \frac{dw}{dx}\right); \quad \omega_3 = \tfrac{1}{2}\left(\frac{dv}{dx} - \frac{du}{dy}\right).$$

Now, supposing the motion such that the axis of resultant vortical spin coincides with the line of resultant velocity at the point, we must have

$$\frac{\frac{dw}{dy} - \frac{dv}{dz}}{u} = \frac{\frac{du}{dz} - \frac{dw}{dx}}{v} = \frac{\frac{dv}{dx} - \frac{du}{dy}}{w}. \tag{a}$$

Screw Motion of a Liquid.

But as we are supposing no velocity parallel to the axis of z, we put $w = 0$, so that we must have

$$u\frac{du}{dz} + v\frac{dv}{dz} = 0 \quad \text{and} \quad \frac{dv}{dx} - \frac{du}{dy} = 0. \tag{β}$$

Moreover, assuming the fluid incompressible,

$$\frac{du}{dx} + \frac{dv}{dy} = 0. \tag{γ}$$

The first equation (β) gives $u^2 + v^2 = $ a function of x and y only; while (γ) and the second equation (β) show that u and v are conjugate functions of x and y.

To satisfy these conditions, assume

$$u = A \cos f(z) - B \sin f(z), \tag{δ}$$

$$v = A \sin f(z) + B \cos f(z), \tag{ϵ}$$

where A and B are functions of x and y only. Using equation (γ), we have $\dfrac{dA}{dx} + \dfrac{dB}{dy} = 0$ and $\dfrac{dA}{dy} - \dfrac{dB}{dx} = 0$; which prove that A and B are conjugate functions of x, y in the same way as u and v. Equations (δ) and (ϵ) can be put into the forms

$$u = \sqrt{A^2 + B^2} \cos\{f(z) + \eta\}, \quad v = \sqrt{A^2 + B^2} \sin\{f(z) + \eta\},$$

where $\eta = \tan^{-1}\dfrac{B}{A}$. Now we know that $\log\sqrt{A^2 + B^2}$ and η are conjugate functions of x, y. Hence if we put ξ for

$$\log\sqrt{A^2 + B^2},$$

$$u = e^{\xi} \cos\{f(z) + \eta\}, \tag{1}$$

$$v = e^{\xi} \sin\{f(z) + \eta\}, \tag{2}$$

express a liquid motion in which the axis of vortical spin at any point coincides with the line of resultant velocity at the point—all such axes being parallel to the plane of xy. This is a screw motion at every point.

Examples.

1. Two large plane metallic surfaces are electrified, and kept parallel to each other at a small distance apart, each being at a given potential; to determine the distribution of potential in the space between them, and to deduce therefrom other possible distributions.

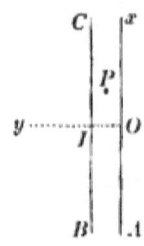

Fig. 75.

Let one plate, Ax, be at potential V_1 and the other, BC, at potential V_2.

Take the origin O at the middle of the first plate, and the axis of y perpendicular to the plates. Let y be the distance of any point P (not near the edges) between the plates from the first plate. Then if V is the potential at P, V is a function of y alone; hence $\frac{d^2V}{dy^2} = 0$, and if $OI = a =$ distance between the plates,

$$V = V_1 + (V_2 - V_1)\frac{y}{a}. \tag{1}$$

The electric displacement, σ, at any point is $-\frac{V_2 - V_1}{4\pi a}$.

Now if in (1) we substitute any function, ϕ, satisfying the equation $\nabla^2\phi = 0$ we get another possible distribution. Thus if we put $2xy$ for y, the surface Ax becomes two rectangular planes, and the surface BC a rectangular hyperbolic cylinder to which these planes are asymptotic. Again, if for y in (1) we put $r\sin\theta$, where θ is the angle POx, and then write $\frac{r^n}{a^{n-1}}$ for r and $n\theta$ for θ (Art. 143),

$$V = V_1 + (V_2 - V_1)\frac{r^n}{a^n}\sin n\theta \tag{2}$$

will express a new distribution at P.

The line BC in the first distribution will correspond to the curve

$$r^n \sin n\theta = a^n \tag{3}$$

in the second; and the line Ox in the first will correspond to Ox in the second, while OA in the first will correspond to a line, OD, making an angle, DOx, $\frac{\pi}{n}$ with Ox in the second. In the new distribution, therefore, the potential has the constant value V_2 over the curve (3), and the constant value V_1 over a system of two right lines intersecting at O. This system of two lines is, of course, in reality two *planes* perpendicular to the plane of the figure, and the curve (3) a cylindrical surface.

Since the equipotential lines in the space between the planes Ax and BC were lines parallel to Ax, the lines of force were lines parallel to OI, their equations being $r\cos\theta =$ const. Hence in the new distribution the lines of force are
$$r^n \cos n\theta = \text{const.} \tag{4}$$

and the equipotential lines $r^n \sin n\theta =$ const. $\tag{5}$

Examples.

It is well to observe that all this remains true if we have only *one* infinite metallic plane, Ax, electrified to the potential V_1. The value of V at any point would still be a function of y, so that $\frac{d^2V}{dy^2} = 0$, and we should have $V = Ay + V_1$ as the integral instead of (1).

The value of V at P cannot be completely determined unless we know its value, V_2, at some one point off the plane; and our assumption of *two* metallic planes the potential on each of which is assigned serves this purpose.

We have now the law of variation of potential corresponding to the electrification of two infinite metallic planes Ox, OD (fig. 76) intersecting in an edge at O; and we have found $\angle DOx = \frac{\pi}{n}$. Hence if a is the angle, we must have
$$n = \frac{\pi}{a},$$
and
$$V = V_1 + (V_2 - V_1) \left(\frac{r}{a}\right)^{\frac{\pi}{a}} \sin \frac{\pi}{a} \theta, \qquad (6)$$
where V_2 is the potential on the equipotential surface whose equation is

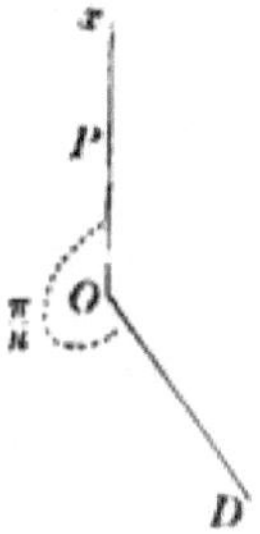

$$\left(\frac{r}{a}\right)^{\frac{\pi}{a}} \sin \frac{\pi}{a} \theta = 1. \qquad (7)$$

Fig. 76.

At any point P on the system of planes the surface-density σ is the value of $-\frac{1}{4\pi} \frac{dV}{r d\theta}$, with $r = OP$, $\theta = 0$; i.e.,

$$\sigma = -\frac{V_2 - V_1}{4a} \cdot \frac{r^{\frac{\pi}{a}-1}}{a^{\frac{\pi}{a}}}. \qquad (8)$$

The total charge on the portion of the plate Ox between O and P and having a breadth (perpendicular to the plane of the figure) equal to b is $b \int_0^r \sigma \, dr$.

2. To discuss the ellipses and hyperbolas of example 7, p. 167, from the point of view of conjugate functions.

It appears from (γ), p. 239, that the functions[1] $(e^\phi + e^{-\phi}) \cos \psi$ and $(e^\phi - e^{-\phi}) \sin \psi$ are conjugate functions of ϕ and ψ. Denote them by $\frac{2x}{c}$ and $\frac{2y}{c}$, so that

$$2x = c(e^\phi + e^{-\phi}) \cos \psi, \quad 2y = c(e^\phi - e^{-\phi}) \sin \psi. \qquad (1)$$

---

[1] It would, perhaps, be better to use $\frac{\phi}{A}$ and $\frac{\psi}{A}$ instead of ϕ and ψ, where A is a magnitude of the same kind as ϕ and ψ—for the sake of homogeneity.

Conversely, then [(6), p. 229], ϕ and ψ are conjugate functions of x, y.

Hence in fluid motion and electrical flow we may take ϕ and ψ as potential and flow-functions, and *vice versa;* and in electrostatics as potential and induction (or force) function, and *vice versa.* Let us then find the curves over which ϕ is constant. Eliminating ψ, we obtain

$$\frac{4x^2}{c^2(e^\phi + e^{-\phi})^2} + \frac{4y^2}{c^2(e^\phi - e^{-\phi})^2} = 1, \tag{2}$$

which denotes an ellipse when ϕ is constant.

Eliminating ϕ, we have

$$\frac{x^2}{c^2 \cos^2\psi} - \frac{y^2}{c^2 \sin^2\psi} = 1, \tag{3}$$

which denotes a hyperbola when ψ is constant.

One of the ellipses of the system is the line FF' joining the foci for which $\phi = 0$, the length FF' being $2c$, and one of the hyperbolas consists of the lines $F\infty$ and $F'\infty$, the value of ϕ, of course, ceasing to be constant beyond the points F and F' on the axis of x; while for points on $F\infty$ the value of ψ is constant and equal to 0, and for points on the line $F'\infty$, ψ is constant and equal to π (since for such points x is negative). This case has been already translated into the language of fluid and electrical flow. We shall now deal with it as expressing an electrostatical distribution.

Any equipotential surface may be taken as the surface of a conductor, statically charged; and the flow-lines will then be lines of electrostatic induction in the surrounding dielectric—or, in other words, lines such that the tangent to any one of them at any point is the direction in which electricity is separated or forced forward out of the particle of the dielectric at the point.

Taking ϕ to be the potential function and ψ that of induction, we may imagine FF' to represent a plane strip of metal of infinite length (of course extending perpendicularly to the plane of the paper both above and below it). It is at potential zero; and we can calculate the charge on it per unit length (measured perpendicularly to the paper). If σ is the surface-density at any point of FF', we have $4\pi\sigma = -\frac{d\phi}{dy}$, since dy is the element of normal to the equipotential surface FF'; $\therefore 4\pi\sigma = \frac{d\psi}{dx}$, $\therefore$ the charge on the length dx (and unit height) is $\frac{1}{4\pi} d\psi$, $\therefore$ the charge on the portion between any two points on FF' is $\frac{1}{4\pi}(\psi_2 - \psi_1)$, where ψ_1 and ψ_2 are induction functions at the points. Now at F' we have $\psi = \pi$, and at F, $\psi = 0$; $\therefore$ the charge on the strip FF' with unit height $= \frac{1}{4}$ of a unit.

The ellipses are the equipotential curves (sections of elliptic cylinders) in the dielectric due to the electrification supposed.

Again, we may take ψ as the potential function and ϕ as the electrostatic induction function; and then we may suppose any one of the hyperbolic cylinders to be the surface of a metallic conductor. In particular, the two separated planes $F\infty$, $F'\infty$ may be taken. One is at zero potential and the other at potential π, the space FF' between them being occupied only by the dielectric. The equipotential curves due to this system of two electrified planes are hyperbolas having F and F' for foci, and the potential, ψ, at any point (x, y) in the dielectric is given by equation (3).

We now proceed to deduce some important results by transforming this distribution by the method of Art. 140.

The functions ξ, η, such that

$$\xi = c e^{\frac{x}{b}} \cos \frac{y}{b}, \quad \eta = c e^{\frac{x}{b}} \sin \frac{y}{b}, \qquad (4)$$

are conjugate functions of x, y. If, then, in the values of ϕ and ψ we write ξ for x and η for y, we obtain a new distribution. Denoting the point (x, y) by P, and its *corresponding* point (p. 230) by P', we define the transformation for clearness thus—

$$P \equiv (x, y),$$
$$P' \equiv (c e^{\frac{x}{b}} \cos \frac{y}{b}, \ c e^{\frac{x}{b}} \sin \frac{y}{b}).$$

Let us find the points P at which, in the new distribution, ϕ has the value zero. They will be points at whose corresponding points, P', in the old distribution, ϕ was zero; i.e., P' must lie on FF' between F and F'. Hence η must be zero, i.e.,

$$y = 0, \ b\pi, \ 2b\pi, \ldots nb\pi, \qquad (5)$$

which shows that P is to be found on a series of lines parallel to the axis of x, at equal distances, $b\pi$. But *every* point on one of these lines will not do, because P' must lie between F and F', i.e., ξ must be numerically $< c$. Hence since for every point given by (5), $\xi = \pm c e^{\frac{x}{b}}$, it follows that x must lie between 0 and $-\infty$; consequently the points at which in the new distribution $\phi = 0$ lie on the lines $O\infty'$, $I_1\infty'$, $I_2\infty'$, ... at equal distances, $b\pi$, from each other (fig.[1] 77). [These lines represent, of course, *planes* perpendicular to the plane of the paper.]

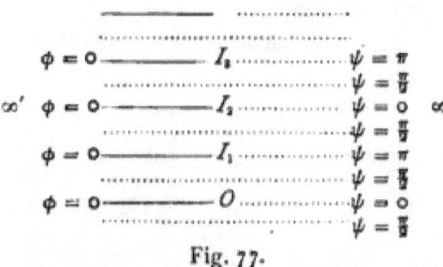

Fig. 77.

---

[1] In this figure the lines are supposed to be produced to infinity towards the right and left, and the symbols ∞, ∞' stand for the infinitely distant points on each line.

To find the points P at which in the new distribution ψ has the constant value zero, we must have P' on the line $F\infty$ (fig. 61, p. 171), i.e., $\eta = 0$ and $\xi > c$, $\therefore y$ is given by those values in (5) which make $\cos\frac{y}{b}$ positive, and x must lie between 0 and $+\infty$. Hence

$$y = 0,\ 2b\pi,\ 4b\pi, \tag{6}$$

and the required points lie on the lines $O\infty$, $I_2\infty$, ...

The points at which in the new distribution ψ has the value π are seen, in the same way, to lie on the lines $I_1\infty$, $I_3\infty$, ...

Let us now find the points which in the new distribution correspond to the axis of y in the old, i.e., points at which $\psi = \frac{\pi}{2}$. The points P' lie on the axis of y, i.e.,

$$y = \frac{b\pi}{2},\ \frac{3b\pi}{2},\ \dots\ \frac{(2n+1)b\pi}{2}, \tag{7}$$

no restriction being placed on the value or sign of x.

The results of this transformation are applied by Clerk Maxwell to two cases in Electrostatics (*Electricity and Magnetism*, Vol. I, pp. 276, &c., 2nd ed.). Of the first of these we propose to give a sketch here.

Take ψ as an electrostatic potential function, and ϕ as the force (or induction) function.

Consider now the limited line $O\infty$, along which $\psi = 0$, and the two unlimited lines above and below it, along which $\psi = \frac{\pi}{2}$. Let these lines become indefinitely thin metallic planes electrified to these potentials. Take any point, A, on the middle plane, $O\infty$, at a distance x from O. Then, since if σ is the surface density at any point on this plane,

$$\sigma = -\frac{1}{4\pi}\frac{d\psi}{dy} = \frac{1}{4\pi}\frac{d\phi}{dx},$$

the quantity on an area whose length is OA and breadth 1 (perpendicular to the plane of the paper) is $\int_0^x \sigma\,dx$, or

$$\frac{1}{4\pi}\phi_A,$$

since $\phi = 0$ at O. Now the transformation which is effected by equation (4) is expressed by the equations

$$2e^{\frac{x}{b}}\cos\frac{y}{b} = (e^{\phi} + e^{-\phi})\cos\psi, \tag{8}$$

$$2e^{\frac{x}{b}}\sin\frac{y}{b} = (e^{\phi} - e^{-\phi})\sin\psi, \tag{9}$$

so that, since for the point A we have $y = 0$, $\psi = 0$, (8) gives

$$\phi = \log\left(e^{\frac{x}{b}} + \sqrt{e^{\frac{2x}{b}} - 1}\right) \tag{10}$$

(the + sign being given to the square root, since when x becomes indefinitely great, ϕ must not be zero).

Assume now that b is very small, i.e., that the parallel metallic planes are at a very small distance apart. Then for points, A, at a considerable distance from O, x will be a very large multiple of b, so that $e^{\frac{2x}{b}} - 1$ is very nearly the same as $e^{\frac{2x}{b}}$; and for such points, we have

$$\phi = \log 2 e^{\frac{x}{b}} = \frac{x}{b} + \log 2. \tag{11}$$

Hence at all such points on the middle (limited) plane, σ is constant and equal to $\frac{1}{4\pi b}$; but this is not the value of σ at points near O.

The charge on the area whose length is OA, with unit breadth, is

$$\frac{1}{4\pi}\left(\frac{x}{b} + \log 2\right);$$

while if σ had at the points near O (the edge of the plate) the value, $\frac{1}{4\pi b}$, which it has at points not near the edge, the charge would be $\frac{x}{4\pi b}$. The actual charge over OA is therefore greater than this, and equal to what it would be over a length $x + b \log 2$, if it were charged with the density which exists at points remote from the edge. If we produce AO (fig. 78) towards the left of the figure to O', so that $OO' = b \log 2$, we have a new intermediate plate whose length exceeds that of $O\infty$ by $b \log 2$, and on this we can imagine a constant density $\frac{1}{4\pi b}$. The actual density at O is easily seen to be infinite; nevertheless the *quantity* on any area measured from O is finite.

Let L be the breadth of the intermediate plate (perpendicular to the plane of the paper) and S its area; then by adding the strip OO' the area of the equivalent plate with constant density is $S + b L \log 2$; and the total charge on the plate is $\dfrac{S + b L \log 2}{4\pi b}$.

If we use B for the distance, πb, between the two planes above and below the middle plate, the total charge on this latter is

$$\frac{1}{4}\left\{\frac{S}{B} + \frac{L}{\pi}\log 2\right\}.$$

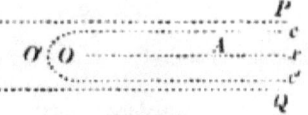

Fig. 78.

The middle plate has, however, some thickness, which is not negligible in comparison with B, the distance between them, so that we cannot assume the potential on it to be that on the plane, Ox, midway between them. To prevent this error, trace one of the equipotential curves, $cO'c'$, (fig. 78) between the plates, for which ψ has some value between the zero value on Ox, and the value, $\dfrac{\pi}{2}$, which it has on the plates P and Q. Assume that the surface of the intermediate plate coincides with this curve; let k be the value of ψ on it; and let O' be the point in which the curve cuts the axis of x. The point O' is found by putting $y = 0$ and $\phi = 0$ in (8); hence $OO' = b \log \sec k$. Again, at points, A, very distant from the edge the ϕ curves are nearly right lines perpendicular to the plates, so that ϕ may be assumed as given by (11); and if β is the thickness of the plate we have $k = \dfrac{\beta}{2b} = \dfrac{\pi\beta}{2B}$ (by considering the value of ψ at points for which x is very great; this gives $\dfrac{y}{b} = \psi$, nearly). The total charge on the plate is easily found to be $\dfrac{1}{4}\left\{\dfrac{S}{B} + \dfrac{L}{\pi}\log_e 2\cos\dfrac{\pi\beta}{2B}\right\}$, since $L \times (Ox + OO') = S$.

3. If ϕ is any function of the co-ordinates (x, y) of a point, and if a double series of curves $\xi = C$, $\eta = C'$, be drawn (by varying the constants C, C') where ξ and η are any two conjugate functions of (x, y), prove that $\displaystyle\int \phi \dfrac{d\phi}{dn} ds$ taken over any one of the curves $\xi = C$ is equal to $\displaystyle\int \phi \dfrac{d\phi}{d\xi} d\eta$ taken over the same curve.

[It can be shown that $\dfrac{d\phi}{dn} = K \dfrac{d\phi}{d\xi}$, and $ds = \dfrac{1}{K}d\eta$, where K is the same as in Art. 140.]

4. Show that the conjugate ellipses and hyperbolas can be deduced from the equation
$$x + y\sqrt{-1} = 2c \sin^2 \tfrac{1}{2}(\psi + \phi\sqrt{-1}).$$
(Mr. Greenhill.)

[Here $\quad x - 2c + y\sqrt{-1} = -2c\cos^2\tfrac{1}{2}(\psi + \phi\sqrt{-1})$;
$x - 2c - y\sqrt{-1} = -2c\cos^2\tfrac{1}{2}(\psi - \phi\sqrt{-1})$;
$\therefore \overline{x - 2c}^2 + y^2 = 4c^2 \cos^2\tfrac{1}{2}(\psi + \phi\sqrt{-1})\cos^2\tfrac{1}{2}(\psi - \phi\sqrt{-1})$.

Similarly $\quad x^2 + y^2 = 4c^2\sin^2\tfrac{1}{2}(\psi + \phi\sqrt{-1})\sin^2\tfrac{1}{2}(\psi - \phi\sqrt{-1})$.

Hence, P being the point (x, y), if $PF = r$, $PF' = r'$, F being the origin and $FF' = 2c$, we have
$$r' + r = 2c\cos\phi\sqrt{-1} = 2c\cosh\phi, \text{ and } r' - r = 2c\cos\psi,$$
so that $\phi = \cosh^{-1}\dfrac{r' + r}{2c}$; $\psi = \cos^{-1}\dfrac{r' - r}{2c}$. See example 2.]

5. Show that the conjugate functions ϕ, ψ determined by the equation
$$x + y\sqrt{-1} = c\operatorname{sn}^2\tfrac{1}{2}(\psi + \phi\sqrt{-1}),$$
may be exhibited in terms of the distances of the point (x, y) from three collinear points A, B, C. (Mr. Greenhill.)

Examples. 253

[If k is the modulus of the elliptic function, take A as origin, take $AB = c$, and $AC = \dfrac{c}{k}$; then denoting PA, PB, PC by r, r', r'', we have

$$r = c \operatorname{sn} \tfrac{1}{2}(\psi + \phi\sqrt{-1}) \operatorname{sn} \tfrac{1}{2}(\psi - \phi\sqrt{-1}),$$

$$r' = c \operatorname{cn} \tfrac{1}{2}(\psi + \phi\sqrt{-1}) \operatorname{cn} \tfrac{1}{2}(\psi - \phi\sqrt{-1}),$$

$$r'' = \dfrac{c}{k} \operatorname{dn} \tfrac{1}{2}(\psi + \phi\sqrt{-1}) \operatorname{dn} \tfrac{1}{2}(\psi - \phi\sqrt{-1}).$$

From these we obtain $r' - r \operatorname{dn}\psi - \operatorname{cn}\psi = 0$, and $r' \operatorname{cn}\phi + r \operatorname{dn}\phi - 1 = 0$, which for constant values of ψ and ϕ denote confocal Cartesians; and as a curve along which ψ is constant cuts one along which ϕ is constant orthogonally, we see that confocal Cartesians cut at right angles. In this way Mr. Greenhill proves Crofton's theorem. See *Proceedings of the Cambridge Philosophical Society*, vol. iv. part ii.]

6. Discuss the electrification which is obtained by taking $n = 2$ in example 1.

[We shall have one electrified surface consisting of a system of two rectangular planes, opposed to another electrified surface consisting of a rectangular hyperbolic cylinder approaching the planes asymptotically. It will be a good exercise for the student to compare an absolute attracted disk electrometer constructed with these surfaces (the moveable disk being in one of the two planes at a great distance from their line of intersection) with the ordinary absolute electrometer which is constructed with two plane surfaces.]

7. If the equipotential curves of an electrostatic distribution (or of a strain) are given by the equation $r^n \sin n\theta = a^n$, where a varies, prove that the displacement at any point $\infty\, r^{n-1}$.

8. Verify that the equation in example 3, p. 165, is equivalent to

$$\dfrac{d^2\phi}{d\rho^2} + \dfrac{d^2\phi}{d\theta^2} = 0,$$

where $\rho = \log\dfrac{r}{a}$.

NOTES.

Note A.

Centrifugal Force.

WHEN a particle of mass m moves in a curved path, it has at each point an acceleration along the normal, towards the *concave*, or inward, side of the path, equal to $\dfrac{v^2}{\rho}$ (p. 56). To produce this acceleration there is required a component, N, of force along the inward normal equal to $\dfrac{mv^2}{\rho}$. This force will be in dynes if m is in grammes, v in centimètres per second, and ρ in centimètres. The particle itself exerts a force which is the exact equal and opposite of this normal force, N, on the agent which produces N, the action of the agent on the particle and of the particle on the agent being propagated by a medium intervening between them. Indeed, simplicity might be gained by saying that the force N is produced on the particle by the medium immediately in contact with it, and the particle reacts on this medium with a force exactly equal and opposite to N. Thus the medium experiences a force N, or $\dfrac{mv^2}{\rho}$, along the normal drawn *outwards*, or towards the convex side of the path. Similarly, of course, for the component, $m\dfrac{dv}{dt}$, along the tangent in the sense in which v increases, the medium experiencing the same force reversed.

Thus, then, the *external* force acting on a moving particle, i.e., the force produced on it immediately by the surrounding medium, though mediately, perhaps, by something else—as when a vortex or a source at any one point is a cause of motion throughout the field—and the reaction of the particle against acceleration, or its *force of inertia* (p. 107), are merely the two aspects which every stress necessarily has (*Statics*, p. 486).

The object of the present note is to guard the student against the erroneous notion which is contained in the expression '*centrifugal* force,' which is applied to the force $\dfrac{mv^2}{\rho}$,—viz., that a particle moving in a circular path is acted upon by a force *outwards* from the centre, (centre-flying). The very reverse is the case; the force acting *on* the particle is wholly towards the *inward* side of the curve; the force exerted *by* the particle on the external agent is exerted in the opposite sense; and similarly for any other curved path.

Take the familiar instance of a particle tied to one end of a string, the other end of which is held in the hand, this latter being kept fixed on a smooth horizontal table, along which the particle is projected. Here the only force acting on the particle is the tension of the string, which is directed *inwards* towards the centre; the hand experiences an equal force outwards; and the numerical value of each is $\dfrac{mv^2}{\rho}$ in absolute measure, or $\dfrac{wv^2}{g\rho}$ in gravitation measure, if w = weight of particle.

Hence will be seen the fallacy of the expression—a very common one—'portions of the revolving solar atmosphere *thrown off by centrifugal force*.' Given a normal component, N, of force, of definite magnitude, and a mass m revolving in a circle of radius r; then the velocity of the body must be equal to $\sqrt{\dfrac{Nr}{m}}$; and if from any cause the velocity is increased above this limit, while the normal force remains the same, the result is that the body must widen its orbit to suit the circumstances. But there is no such thing as 'flinging off by centrifugal force.'

The term 'centrifugal force' is a very unfortunate one, and it ought to be abandoned by physicists.

Note B.

Strain Invariants.

It may be well to add a few remarks on the general, or three-dimensional, strain of a body. With the notation of *Statics* (p. 461), let a, b, c be the elongations along any three rectangular axes (x, y, z) at a point P in the body; $2s_1, 2s_2, 2s_3$ the shears with reference to these axes; and $\omega_1, \omega_2, \omega_3$ the components of the rotation. At P draw another set of rectangular axes whose direction-cosines with respect to the first set are given by the usual scheme here indicated. Then if a', ... s_1', ... ω_1', ...

	x	y	z
x'	l	m	n
y'	l'	m'	n'
z'	l''	m''	n''

denote the components of the displacement produced by the strain, with reference to the new axes, we have the following types of transformation:—

$$\omega_1' = l\omega_1 + m\omega_2 + n\omega_3, \tag{α}$$
$$a' = l^2 a + m^2 b + n^2 c + 2mns_1 + 2nls_2 + 2lms_3, \tag{β}$$
$$s_1' = l\, l''a + m'm''b + n'n''c + (m'n'' + m''n')s_1 + (n'l'' + l'n'')s_2 + (l'm'' + l''m')s_3, \tag{γ}$$

all these resulting from the relations

$$x' = lx + my + nz, \text{ with similar values of } y', z', \tag{α'}$$
$$u' = lu + mv + nw, \text{ with similar values of } v', w'. \tag{α''}$$

Now it can be proved from elementary considerations with regard to the elongation quadric and the line of resultant rotation that the following quantities are invariants:—

$$\theta \equiv a + b + c,$$
$$\Sigma \equiv ab + bc + ca - s_1^2 - s_2^2 - s_3^2,$$
$$\Delta \equiv abc + 2s_1 s_2 s_3 - as_1^2 - bs_2^2 - cs_3^2,$$
$$\Omega^2 \equiv \omega_1^2 + \omega_2^2 + \omega_3^2,$$
$$\Phi \equiv a\omega_1^2 + b\omega_2^2 + c\omega_3^2 + 2s_1\omega_2\omega_3 + 2s_2\omega_3\omega_1 + 2s_3\omega_1\omega_2.$$

Also the vector whose components along the axes are $\dfrac{d\Phi}{d\omega_1}, \dfrac{d\Phi}{d\omega_2}, \dfrac{d\Phi}{d\omega_3}$ is an invariant; for these quantities transform according to the type (α). Again, $\dfrac{d\Delta}{da}$, &c. transform according to the type (β).

In uniplanar strain these invariants are $(a+b)$, $(ab-s^2)$, and ω.

In the general case the elongation along any line is inversely proportional to the square of the radius vector of the elongation quadric in the direction of the line; and it is easy to see that the shear of any two rectangular lines is $\dfrac{2k^2}{RR'}\cos a$, where R and R' are the lengths intercepted from the lines by the quadric the squares of whose axes are $\dfrac{k^2}{\sqrt{e_1}}, \dfrac{k^2}{\sqrt{e_2}}, \dfrac{k^2}{\sqrt{e_3}},$ e_1, e_2, e_3 being the principal elongations, k any constant, and a the angle between the tangent planes to this quadric at the extremities of R and R'.

NOTE C.

Conjugate Functions.

The whole of this work was in type when I became acquainted with a very important and elegant paper on Conjugate Functions by Mr. Routh in the *Proceedings of the London Mathematical Society* (Nos. 170, 171). Mr. Routh proposes to employ the method of Conjugate Functions to solve hydrokinematical problems which are commonly solved by the method of Images.

The main features of this paper, so far as it deals with fluid motion, are as follows:—

(1) Imagine two different motions of a fluid over a plane, the one derived from the other, with the dependence of corresponding points, P, P', explained in (7), p. 229, and let $ABCD$ and $A'B'C'D'$ (fig. 73, p. 240) represent generally any two corresponding boundaries. Then (9), p. 235, shows that velocity of $P = K \times$ velocity of P', by taking for $f(\xi, \eta)$ a velocity potential or a current function.

(2) If a source or a vortex exist at P', there will be a source or a vortex of equal strength at P. If there is a vortex at P', there will also be one at P, but the vortices do not necessarily continue to move so as to occupy corresponding points. However, given the motion of P', the motion of P can be found thus. Let the components of velocity of P' parallel to the axes be expressed as $\dfrac{d\chi'(\xi, \eta)}{d\eta}$ and $-\dfrac{d\chi'(\xi, \eta)}{d\xi}$, so that $\chi'(\xi, \eta)$ is a current function for the vortex P'—but, of course, not the current function whose differential coefficients give the components of velocity at every point of the fluid in which P' is moving. Then the function, $\chi(x, y)$, of x, y whose differential coefficients will give the velocity components of the vortex P is $\chi'(\xi, \eta) - \dfrac{m}{2\pi} \log K$, where m is the strength of the vortex P' and K is the modulus of transformation (p. 234).

[Of course $\chi(x, y)$ is not the current, or flow, function which belongs to every point of the fluid in $ABCD$, but merely that which defines the motion of the vortex P.]

(3) Suppose now that we require the motion of a fluid, in which there is a vortex, P_1, over an infinite area, part of whose (unclosed) boundary is ABC (fig. 73). Imagine the vortex removed, and find, if possible, a steady acyclic irrotational motion of the fluid. Suppose the velocity and stream functions of this motion to be ϕ and ψ; then the boundary being a stream line, ψ has a constant value, k, along it. Now use ϕ and ψ as the ξ and η of transformation, i.e., the co-ordinates of P' are (ϕ, ψ). Hence every stream line of the motion within ABC transforms into an infinite right line parallel to the axis of x; and, in particular, the boundary ABC corresponds to the right line $y = k$.

Now replace the vortex at the point P, inside ABC, and there will be a corresponding vortex at the corresponding point, P_1'; and the motion P_1' is already known (ex. 2, p. 225); P_1' moves parallel to the axis of x with a velocity $\dfrac{m}{2\pi\eta}$, where η is the ordinate of P_1'. Hence

$$\chi'(\xi, \eta) \equiv \frac{m}{2\pi} \log \eta \equiv \frac{m}{2\pi} \log \psi ; \quad \text{so that}$$

$$\chi(x, y) \equiv \frac{m}{2\pi} \log \psi - \frac{m}{2\pi} \log K,$$

and thus the function determining the motion of P_1 is found.

The same method applies if there are more vortices or sources than one within the area ABC, and the solution of the problem will depend on a solution for the case of a rectilinear boundary—i.e., on the simplest case of the method of Images.

Apply this to a particular case. Let there be two infinite right lines, OA, OB, the first being taken as initial line, and let a be the angle between them. Then $\phi \equiv r^n \cos n\theta$, $\psi \equiv r^n \sin n\theta$, where $n = \dfrac{\pi}{a}$ will express a possible steady irrotational motion in a fluid contained between the lines (see p. 246). Hence when a vortex is moving in this fluid we have

$$\chi(x, y) \equiv \frac{m}{2\pi} \log (r^n \sin n\theta) - \frac{m}{2\pi} \log K,$$

where

$$K^2 = \left(\frac{d\psi}{dx}\right)^2 + \left(\frac{d\psi}{dy}\right)^2 = \left(\frac{d\psi}{dr}\right)^2 + \left(\frac{d\psi}{rd\theta}\right)^2 = n^2 r^{2n-2}, \quad \therefore K = nr^{n-1},$$

$$\therefore \chi(x, y) \equiv \frac{m}{2\pi} \log (r \sin n\theta),$$

neglecting the constant $\dfrac{m}{2\pi} \log n$. The path of the vortex is found, of course, by putting χ equal to a constant, so that it is the Cotes Spiral $r \sin n\theta = a$ constant. In this very simple manner Mr. Routh solves Mr. Greenhill's problem (see ex. 3, p. 226).

NOTE D.

Vectors and their Derivatives.

The theorems of Arts. 105 and 106 are particular cases of general results in the theory of three-dimensional displacement. On account of the importance of the general results in the theory of Electromagnetism, it is thought desirable to present them here for the consideration of the more advanced student.

Let (i, j, k) be Hamilton's system of rectangular unit vectors, and let ϖ denote any vector identified with a point, P, in space. Let the components of ϖ in the directions of i, j, k be u, v, w, respectively, so that

$$\varpi = ui + vj + wk. \qquad (1)$$

Let ∇ denote the operator $i\dfrac{d}{dx} + j\dfrac{d}{dy} + k\dfrac{d}{dz}$, where (x, y, z) are the co-ordinates of P referred to fixed axes in the directions of i, j, k.

Then, from the fundamental relations between i, j, k, we have

$$\nabla \varpi = -\left(\frac{du}{dx} + \frac{dv}{dy} + \frac{dw}{dz}\right) + \left(\frac{dw}{dy} - \frac{dv}{dz}\right)i + \left(\frac{du}{dz} - \frac{dw}{dx}\right)j$$
$$+ \left(\frac{dv}{dx} - \frac{du}{dy}\right)k. \qquad (2)$$

Thus, when a vector is operated upon by ∇, the result is a quaternion whose scalar portion is called by Clerk Maxwell (*Electricity and Magnetism*, vol. i. p. 29) the *convergence* of the given vector, and whose vector portion is called the *rotation* of the given vector. Let θ be put for

$$\frac{du}{dx} + \frac{dv}{dy} + \frac{dw}{dz},$$

and 2ω for the vector portion of the right-hand side of (2); then

$$\nabla \varpi = -\theta + 2\omega. \qquad (3)$$

The scalar, θ, and the vector, 2ω, thus derived from a given vector by the operation ∇ are the same whatever be the system of axes of reference; so that if we take any new axes (x', y', z') and calculate the components, $\frac{dw'}{dy'} - \frac{dv'}{dz'}$, &c., of the rotation with respect to them, from the transformation types (a'), (a'') of Note B, we shall find these to be simply equivalent to $\omega \cos \lambda$, &c., where ω is the resultant rotation, $\sqrt{\omega_1^2 + \omega_2^2 + \omega_3^2}$, and λ is the angle between the assumed axis of x' and the line round which ω takes place.

It will be observed that if u, v, w are the velocity-components at P of a moving fluid, θ is what we have called the *expansion*, while ω is the *spin-vector* at P.

Thus, then, in the case of any moving fluid, the condensation at any point is the convergence of the velocity; and the spin is the rotation of the velocity at the point.

Operating on ϖ a second time with ∇, i.e., operating with ∇ on both sides of (2), we have

$$\nabla^2 \varpi = i \nabla^2 u + j \nabla^2 v + k \nabla^2 w, \qquad (4)$$

where on the right-hand side $-\nabla^2$ stands for $\frac{d^2}{dx^2} + \frac{d^2}{dy^2} + \frac{d^2}{dz^2}$. But even at the right-hand side ∇^2 may be understood to mean

$$\left(i\frac{d}{dx} + j\frac{d}{dy} + k\frac{d}{dz} \right)^2,$$

as it does at the left-hand side; for, it is at once found that

$$\left(i\frac{d}{dx} + j\frac{d}{dy} + k\frac{d}{dz} \right)^2 u = -\left(\frac{d^2}{dx^2} + \frac{d^2}{dy^2} + \frac{d^2}{dz^2} \right) u. \qquad (5)$$

Hence (4) shows that *the result of the operation* ∇^2 *on any vector is purely a vector*, and (5) that *the result of the operation* ∇^2 *on any scalar is purely a scalar*; or, in other words, the operator ∇^2 is essentially scalar.

Now imagine any curved surface bounded by an edge of any form. We shall prove that if there be a vector magnitude drawn at every point on the surface and on the edge, its value being any function of the position of the point; and if at each point of the edge the vector be resolved along the tangent to the edge, the line-integral of this tangential component taken all over the edge is equal to the surface-integral of the normal component of the rotation of the vector taken all over the given surface; in other words—

the line-integral of the tangential **component of a** *vector along* **any closed** *curve is equal to the surface-integral of the normal component of its* **rotation** *over any surface whatever having the curve* **for an edge.**   (α)

In fig. 58, p. 155, let A be the given edge—no longer, of course, a plane curve—and break up the surface, or cap, bounded by it into an indefinitely great number of small areas, such as $abb'a'$, which we may assume to be plane areas. Then, obviously, if we prove the theorem to hold for any such small area, we can, exactly as in pp. 155, &c., extend it to any surface. At the mean point, P, of the area draw the normal (axis of ζ), and any two rectangular lines (axes of ξ and η) in the plane of the area; let the components of the given vector, ϖ, at P along these axes be p, q, r, so that $\varpi = pi + qj + rk$, if (i, j, k) denote unit vectors in the directions of the axes: and let Q be any point on the contour of the little area, its co-ordinates with reference to P being (ξ, η). Then the values of p and q at Q are $p + \xi \frac{dp}{dx} + \eta \frac{dp}{dy}$ and $q + \xi \frac{dq}{dx} + \eta \frac{dq}{dy}$; and the sum of the resolved parts of these along the tangent to the curve at Q, multiplied by the element of arc, is

$$-\left(p + \xi \frac{dp}{dx} + \eta \frac{dp}{dy}\right) d\xi + \left(q + \xi \frac{dq}{dx} + \eta \frac{dq}{dy}\right) d\eta,$$

the integral of which, exactly as in p. 155, is $\left(\frac{dq}{dx} - \frac{dp}{dy}\right) dS$, where $dS = \int \xi d\eta =$ the area of the small curve. Now from (2) we see that $\frac{dq}{dx} - \frac{dp}{dy}$ is the normal component of the rotation of ϖ; therefore the proposition is proved.

In particular, if the vector ϖ is the velocity of a moving fluid, the line integral over any closed curve is what we have called the *circulation* round it; and the result is that *the circulation round any curve is equal to twice the surface-integral of normal spin taken over any cap having the curve for an edge.* This is the general case of Art. 105.

Again, imagine any closed surface. We shall prove that if at every point on and inside the surface there be drawn a vector whose value is some function of the position of the point—

the surface-integral of the normal component of the vector, taken all over the closed surface, is equal to the volume-integral of its convergence, taken all through the enclosed volume.   (β)

Break up the enclosed volume into an indefinitely great number of small closed surfaces, or cells, of any shapes. Then, observing that the normal component of the vector at every point on the surface is supposed to be measured constantly *inwards* along the normal, if we prove the proposition to hold for any one small closed surface, it will hold for any closed surface, however large, since by addition, exactly as in pp. 155, &c., the parts of the surface-integrals belonging to the portion of surface common to two adjacent cells will cancel.

At the mean point, P, of any small volume draw rectangular axes of ξ, η, ζ; let the vector ϖ at P be $pi + qj + rk$; let (ξ, η, ζ) be the co-ordinates, referred to P, of any point, Q, on the closed surface; then if dS is the surface element at Q, and (l, m, n) the direction-cosines of the outward normal at Q, the element of the surface-integral contributed by Q is

$$-\left(p + \xi\frac{dp}{dx} + \eta\frac{dp}{dy} + \zeta\frac{dp}{dz}\right) l\,dS - \left(q + \xi\frac{dq}{dx} + \eta\frac{dq}{dy} + \zeta\frac{dq}{dz}\right) m\,dS$$
$$-\left(r + \xi\frac{dr}{dx} + \eta\frac{dr}{dy} + \zeta\frac{dr}{dz}\right) n\,dS.$$

Now observe that $l\,dS = d\eta\,d\zeta$, while $\iint \xi\,d\eta\,d\zeta = dV =$ volume enclosed by the surface $= \iint \eta\,d\zeta\,d\xi = \iint \zeta\,d\xi\,d\eta$, and that all the other integrals, such as $\iint \eta\,d\eta\,d\zeta$, vanish. Then the surface-integral comes to

$$-\left(\frac{dp}{dx} + \frac{dq}{dy} + \frac{dr}{dz}\right) dV,$$

which is the convergence of the vector ϖ at P multiplied by the elementary volume.

Hence, as observed above, the theorem enunciated is true for any closed surface and its enclosed volume.

When the vector, ϖ, is one deduced from any scalar function, ϕ, by the operation ∇, the theorem is at once deducible from Green's equation (*Statics*, p. 443); for p, q, r are then $\frac{d\phi}{dx}$, $\frac{d\phi}{dy}$, $\frac{d\phi}{dz}$, and the convergence becomes $-\nabla^2\phi$. Making, then, the second function in Green's equation constant, we deduce the result.

In particular, if ϖ is the velocity of a moving fluid, *the surface-integral of outward normal velocity over any closed surface is equal to the volume-integral of the expansion taken throughout its volume*; or if v denote the outward normal velocity at any point on the surface where the element of area is dS, and θ the expansion at an internal point where the element of volume is dV,

$$\int v\,dS = \int \theta\,dV.$$

The theorem of Art. 106 is the particular case of this equation for uniplanar motion; for the closed surface is in that case a cylinder of unit (or any) height described on the curve (fig. 59) in the plane of motion, and the element, dS, of surface becomes the element, ds, of arc of the bounding curve multiplied by the height of the cylinder; while the element, dV, of volume becomes the element of area (represented by dS in Art. 106) multiplied by the height of the cylinder.

By assigning different forms to the vector in theorems (a) and (β), we arrive at properties of the motion of a fluid and of the strain of a solid.

Thus, for irrotational fluid motion, let $\varpi \equiv \rho\phi(u\mathbf{i}+v\mathbf{j}+w\mathbf{k})$, where ρ is the density at any point, ϕ the velocity potential, and u, v, w the components of velocity. Then the component of ϖ normal to any surface is $\rho\phi\frac{d\phi}{dn}$, and if T is the energy of the enclosed fluid, (β) gives

$$T = \tfrac{1}{2}\int \rho\phi\frac{d\phi}{dn}dS + \tfrac{1}{2}\int \phi\frac{d\rho}{dt}dV.$$

Again, by putting $\varpi = X\mathbf{i}+Y\mathbf{j}+Z\mathbf{k}$, where X, Y, Z are the components of force, per unit mass, at any point, (β) gives

$$\int N\,dS = \int \left(\frac{dX}{dx}+\frac{dY}{dy}+\frac{dZ}{dz}\right)dV,$$

N being the component of force along the normal to the surface at any point. The fundamental result (*Statics*, Art. 252) follows when X, Y, Z are the differential coefficients of a potential function for the law of inverse square. If we take for ϖ the acceleration vector at any point of a moving fluid, i.e., if $\varpi \equiv a\mathbf{i}+\beta\mathbf{j}+\gamma\mathbf{k}$, where $a = \frac{du}{dt}+u\frac{du}{dx}+v\frac{du}{dy}+w\frac{du}{dz}$, with similar values of β and γ, we find the scalar of $\nabla\varpi = -D_t\theta-\theta^2+2\Sigma-2\omega^2$, where Σ is the invariant given in Note B, and θ and ω, as usual, the expansion and vortical spin; so that theorem (β) gives

$$\int a_n\,dS = \int(D_t\theta+\theta^2+2\omega^2-2\Sigma)dV,$$

where a denotes the normal component of acceleration at any point of any closed surface.

Note E.

Flow of Electricity in one plane.

The theory of the flow of electricity in a plane has been treated at length by Prof. W. Robertson Smith (*Proceedings of Edin. Roy. Soc.*, 1869-70, pp. 79-99), and also by Prof. W. G. Adams (*Proceedings of Roy. Soc.*, 1875, XXIV).

Note F.

Current Power.

The term 'power,' which has been employed in Art. 137, signifies *time-rate of doing work*, and it is already in practical use in the expression 'horse power,' which stands for 33,000 foot-pounds per minute. In the C. G. S. system the unit power is one erg per second. If C and E are the current-strength and electromotive force of a battery in C.G.S. units, the power of the battery is CE ergs per second; and if the battery is not doing any external work—such as decomposing an electrolyte which is placed in the circuit, or working a magnetoelectric machine—the power of the battery is spent in the production of heat, so that in this case the equation $CE = (B+R)C^2$, given in p. 222, holds. If, however, the current is doing any external work at the rate W units of work per unit of time, this equation no longer holds, and it must be replaced by the equation

$$CE = \rho C^2 + W, \qquad (a)$$

where ρ is the whole resistance in circuit. We may write this equation in the form $C = \dfrac{E - \dfrac{W}{C}}{\rho}$, and comparing this latter form with the expression of Ohm's Law, we see that when external work is being done, the resulting current-strength is the same as it would be with a diminished electromotive force equal to $E - \dfrac{W}{C}$ in a circuit of the same total resistance, ρ, in which no external work is being done. If the sole object aimed at is the production of external work, the portion ρC^2 of the power is merely waste power.

The equation (a) may be put into another useful form. Suppose, for definiteness, that the battery is used to decompose an electrolyte. Let a current of unit strength decompose ζ grammes of zinc, or other substance, in the working battery, and let θ units of heat be evolved in the battery by the decomposition of 1 gramme; then $\theta\zeta$ units of heat will be evolved by unit current, and $\theta\zeta C$ units of heat by the passage of a current of strength C. If, then, J (Joule's equivalent) is the number of ergs equivalent to one heat unit, the power of the battery in ergs per second is $J\theta\zeta C$. Similarly if the decomposition of 1 gramme of the substance set free in the electrolyte requires the absorption of θ' heat units, and if a weight ζ' grammes is decomposed per unit of current, the power absorbed in the electrolytic cell is $J\theta'\zeta' C$ when a current of strength C passes. Hence (a) becomes

$$J\theta\zeta C = \rho C^2 + J\theta'\zeta' C,$$

therefore
$$C = \frac{J(\theta\zeta - \theta'\zeta')}{\rho}. \qquad (\beta)$$

Thus the electromotive force of a battery can be expressed in the form $J\theta\zeta$, and if its current is used to effect decomposition in a series of electrolytic cells, the resultant electromotive force of the whole arrangement is

$$J(\theta\zeta - \theta'\zeta' - \theta''\zeta'' - \dots).$$

INDEX.

Acceleration along normal, etc., 55.
— along changing direction, 58.
— of any order, general theorem of, 70.
— resultant, when tangential to a curve, 76.
— in fluid motion, 192.
Amsler's planimeter, 92.
— integrometer, 94.

Barriers, method of, 196.
Boundary condition, 162.

Central acceleration, 59.
Centrifugal force (note A).
Centrodes, body and space, 39.
— — equations of, 43.
— acceleration, 68.
— rolling of, 80.
Change of velocity, 58.
Circulation, 154.
— theorem on, 155.
Clerk Maxwell's graphic transformation of fluid motions, etc., 237.
Clifford, on harmonic motion, 20.
Compressibility, measure of, 138.
Conjugate functions, 226.
'Continuity,' equation of, 142.
— — in polar co-ordinates, 164.
Crank and connecting rod, 47.
Current, electrical, strength of, 219.
— power of, 222, (note F).
Cusps, circle of, 99.

Diagram of space described, 2.
— accumulated velocity, 54.
Dilatation, 126.
Direct distance, orbit in case of, 62.
Displacement, electrical, 234.
Distributions, electrostatic, derivation of, 229, etc.

Electromotive force, 220, (note F).
Elliott, E. B., on Holditch's theorem, 91.
Ellipses and hyperbolas, 168, 247.

Ellipses of Cassini, 171.
Elliptic compass, 80.
Elongation conic, 125.
Energy of a particle, 108.
— general theorem on, 116.
— of moving liquid, 181, 183, 185.
— minimum, theorem of, 199.
— minimum, of derived currents, 223.
Envelope of a right line, 44.
Epicycloidal motion, 79.
Equipotential curves, closure of, 159.
— graphic superposition of, 178.
Eulerian and Lagrangian methods, 139.
Expansion, 146.
— volume-integral of, 157, (note D).
— graphic representation of, 189.

Flow of electricity in a sheet of tinfoil, 211, etc.
— linear, 218.
— lines of, 150.
— in bounded plane, 208.
Fluid, perfect, definition of, 138.
Flux across a curve, 141.
Foster, Prof. G. C., on electrical flow, 209.
Fourier's theorem, 13.

Generalised co-ordinates, 188.
Glissette, 104.
Green's equation, 182.
— modified by Thomson, 198.
Greenhill, problems on vortices, etc., 226, 253.
— problem of, solved by Routh (note C).

Hamilton, Sir W., on curvature of Hodograph, 78.
Harmonic motion, 7.
Helmholtz, on fluid velocity components, 152.
Hodograph, 62.
Holditch, theorem of, 86, 88.
'Hurry,' 54.

Images, general method of, 208.
Inertia, force of, 107.
— forces of, for rigid body, 112.
Inflexions, circle of, 99.
Instantaneous axis, centre, 39.
— centre, acceleration of, 64.
— acceleration centre, 66.
Integration, mechanical, 92–96.
Inverse square of distance, orbit, 64.
Inversion, electric, 240.

Joule's equivalent (note F).

Kempe, theorem of, 86.
Kirchhoff's corollaries to Ohm's Law, 221.

Lagrangian method in fluid motion, 185.
Line-integral of a vector (note D).
— Roulette, 96.
Lodge, Prof., on electrical flow, 209.

McCay, W. S., on Holditch's theorem, 87.
— Kempe's theorem, 89.
— theorem of, 99.
— on lengths of roulettes, 101.
Molecular velocity, mean square of, 119.
Momentum, 106.
— -system of rigid body, 109.
Multiply-connected spaces, 193.

Newton, on force of inertia, 107.

Ohm's Law, 220.
Oscillating cylinder, 45.

Plates, two parallel electrified, 246.
— any number of parallel, 249.
Potential of strain, 134.
— velocity, 148.
— in multiply-connected space, 195.

Relative motion, 26.
Resistance of a flow channel, 184.
— sheet of tin-foil, 212, etc.
Rolling, acceleration in, 71.
Rotation, pure, 37.
— produced by strain, 127.
Roulette, area of, 82.
— generalised, 90.

Roulette, curvature of, 99.
— length of, 101.
Routh, E. J., on conjugate functions (note C).

Screw motion in a liquid, 244.
Shear, graphic representation of, 189.
Source, strength of, 167.
Spin, vortical, 147.
— — graphic representation of, 189.
Spirals, equiangular, 173.
Squirt, 167.
Steady motion, 140.
Steiner on pedals, 86.
Strain, resolution of, 124.
— invariants of, 124, 136, (note P).
— ellipse, 129.
— shearing, 129, 133.
— pure, or irrotational, 130.
Stream lines, 150.
— — closure of, 159.
— — graphic superposition of, 178.
— function, 151.
Surface-integral, 140, (note D).

Thomson, Sir W., tidal clock, 13.
— on Green's equation, 198.
Townsend, Prof., 120, 136.
Tubes of flow, 158.

Velocity, fluid, in terms of potential and stream-functions, 154.
— system determined from expansion and spin, 160.
— infinite, in liquid motion, 207.
Vibrations, composition of, 10.
— elliptic, 19.
— resolution of, 21.
Vortex motion, 147.
— — invariability of, 190.
— motion in liquid due to, 201.
— electrical equivalent of, 203.
Vortical centre, 225.
Vortices, plane, motion of, 224, (note C).

Wave disturbance, 16.
Wheatstone's bridge, principle of, 224.
Whirl, 167.
Wolstenholme, Prof., on motion of lamina, 73.

THE END.

Works in Mathematics and Physical Science, &c., recently published by the Clarendon Press.

A Treatise on Statics. By G. M. Minchin, M.A., Professor of Applied Mathematics in the Indian Engineering College, Cooper's Hill. *Second Edition, Revised and Enlarged.* 1879. 8vo. *cloth*, 14s.

A Treatise on the Kinetic Theory of Gases. By Henry William Watson, M.A., formerly Fellow of Trinity College, Cambridge. 1876. 8vo. *cloth*, 3s. 6d.

A Treatise on the Application of Generalised Coordinates to the Kinetics of a Material System. By H. W. Watson, M.A., and S. H. Burbury, M.A. 1879. 8vo. *cloth*, 6s.

A Treatise on Heat, with numerous Woodcuts and Diagrams. By Balfour Stewart, LL.D., F.R.S., Professor of Natural Philosophy in Owens College, Manchester. *Fourth Edition.* 1881. Extra fcap. 8vo. *cloth*, 7s. 6d.

Lessons on Thermodynamics. By R. E. Baynes, M.A., Senior Student of Christ Church, Oxford, and Lee's Reader in Physics. 1878. Crown 8vo. *cloth*, 7s. 6d.

An Elementary Treatise on Electricity. By J. Clerk Maxwell, M.A., F.R.S., Professor of Experimental Physics in the University of Cambridge. Edited by William Garnett, M.A. Demy 8vo. *cloth*, 7s. 6d.

A Treatise on Electricity and Magnetism. By the same Author. *Second Edition.* 2 vols. Demy 8vo. *cloth*, 1l. 11s. 6d.

Geodesy. By Colonel Alexander Ross Clarke, C.B., R.E. 1880. 8vo. *cloth*, 12s. 6d.

A Cycle of Celestial Objects. Observed, Reduced, and Discussed by Admiral W. H. Smyth, R.N. Revised, condensed, and greatly enlarged by G. F. Chambers, F.R.A.S. 1881. 8vo. *cloth*, 21s.

A Handbook of Descriptive Astronomy. By the same author. *Third Edition.* 1877. Demy 8vo. *cloth*, 28s.

Clarendon Press, Oxford.

Exercises in Practical Chemistry. Vol. I. Elementary Exercises. By A. G. Vernon Harcourt, M.A.; and H. G. Madan, M.A. *Third Edition.* Revised by H. G. Madan, M.A. Crown 8vo. *cloth*, 9s.

Tables of Qualitative Analysis. Arranged by H. G. Madan, M.A. Large 4to. *paper covers*, 4s. 6d.

Chemistry for Students. By A. W. Williamson, Phil. Doc., F.R.S., Professor of Chemistry, University College, London. *A new Edition, with Solutions.* 1873. Extra fcap. 8vo. *cloth*, 8s. 6d.

A Treatise on Rivers and Canals, relating to the Control and Improvement of Rivers, and the Design, Construction, and Development of Canals. By Leveson Francis Vernon-Harcourt, M.A., Balliol College, Oxford, Member of the Institution of Civil Engineers. 2 vols. (Vol. I, Text. Vol. II, Plates). 8vo. *cloth*, 21s.

Müller (J.). On certain Variations in the Vocal Organs of the Passeres that have hitherto escaped notice. Translated by F. J. Bell, B.A., and edited, with an Appendix, by A. H. Garrod, M.A., F.R.S. With Plates. 1878. 4to. *paper covers*, 7s. 6d.

Vesuvius. By John Phillips, M.A., F.R.S., Professor of Geology, Oxford. 1869. Crown 8vo. *cloth*, 10s. 6d.

Geology of Oxford and the Valley of the Thames. By the same Author. 1871. 8vo. *cloth*, 21s.

Thesaurus Entomologicus Hopeianus, or a Description of the rarest Insects in the Collection given to the University by the Rev. William Hope. By J. O. Westwood, M.A., F.L.S. With 40 Plates. 1874. Small folio, *half morocco*, 7l. 10s.

Text-Book of Botany, Morphological and Physiological. By Dr. Julius von Sachs, Professor of Botany in the University of Würzburg. Edited, with an Appendix, by S. H. Vines, M.A., D. Sc., F.L.S. Royal 8vo. *Second Edition.* 1882. *Half morocco*, 31s. 6d.

HENRY FROWDE

OXFORD UNIVERSITY PRESS WAREHOUSE

7 PATERNOSTER ROW, LONDON.

March, 1888.

Clarendon Press, Oxford.

A SELECTION OF

BOOKS

PUBLISHED FOR THE UNIVERSITY BY

HENRY FROWDE,

AT THE OXFORD UNIVERSITY PRESS WAREHOUSE,

AMEN CORNER, LONDON.

ALSO TO BE HAD AT THE

CLARENDON PRESS DEPOSITORY, OXFORD.

[*Every book is bound in cloth, unless otherwise described.*]

LEXICONS, GRAMMARS, ORIENTAL WORKS, &c.

ANGLO-SAXON.—*An Anglo-Saxon Dictionary*, based on the MS. Collections of the late Joseph Bosworth, D.D., Professor of Anglo-Saxon, Oxford. Edited and enlarged by Prof. T. N. Toller, M.A. (To be completed in four parts.) Parts I–III. A—SAR. 4to. 15*s*. each.

ARABIC.—*A Practical Arabic Grammar.* Part I. Compiled by A. O. Green, Brigade Major, Royal Engineers, Author of 'Modern Arabic Stories.' Second Edition, Enlarged and Revised. Crown 8vo. 7*s*. 6*d*.

CHINESE.—*A Handbook of the Chinese Language.* By James Summers. 1863. 8vo. half bound, 1*l*. 8*s*.

—— *A Record of Buddhistic Kingdoms*, by the Chinese Monk FÂ-HIEN. Translated and annotated by James Legge, M.A., LL.D. Crown 4to. cloth back, 10*s*. 6*d*.

ENGLISH.—*A New English Dictionary, on Historical Principles:* founded mainly on the materials collected by the Philological Society. Edited by James A. H. Murray, LL.D., with the assistance of many Scholars and men of Science. Part I. A—ANT. Part II. ANT—BATTEN. Part III. BATTER—BOZ. Imperial 4to. 12*s*. 6*d*. each.

B

ENGLISH.—*An Etymological Dictionary of the English Language.* By W. W. Skeat, Litt.D. *Second Edition.* 1884. 4to. 2*l.* 4*s.*

—— Supplement to the First Edition of the above. 4to. 2*s.* 6*d.*

—— *A Concise Etymological Dictionary of the English Language.* By W. W. Skeat, Litt.D. *Third Edition.* 1887. Crown 8vo. 5*s.* 6*d.*

GREEK.—*A Greek-English Lexicon,* by Henry George Liddell, D.D., and Robert Scott, D.D. Seventh Edition, Revised and Augmented throughout. 1883. 4to. 1*l.* 16*s.*

—— *A Greek-English Lexicon,* abridged from Liddell and Scott's 4to. edition, chiefly for the use of Schools. Twenty-first Edition. 1884. Square 12mo. 7*s.* 6*d.*

—— *A copious Greek-English Vocabulary,* compiled from the best authorities. 1850. 24mo. 3*s.*

—— *A Practical Introduction to Greek Accentuation,* by H. W. Chandler, M.A. Second Edition. 1881. 8vo. 10*s.* 6*d.*

HEBREW.—*The Book of Hebrew Roots,* by Abu 'l-Walîd Marwân ibn Janâh, otherwise called Rabbî Yônâh. Now first edited, with an Appendix, by Ad. Neubauer. 1875. 4to. 2*l.* 7*s.* 6*d.*

—— *A Treatise on the use of the Tenses in Hebrew.* By S. R. Driver, D.D. Second Edition. 1881. Extra fcap. 8vo. 7*s.* 6*d.*

—— *Hebrew Accentuation of Psalms, Proverbs, and Job.* By William Wickes, D.D. 1881. Demy 8vo. 5*s.*

—— *A Treatise on the Accentuation of the twenty-one so-called Prose Books of the Old Testament.* By William Wickes, D.D. 1887. Demy 8vo. 10*s.* 6*d.*

ICELANDIC.—*An Icelandic-English Dictionary,* based on the MS. collections of the late Richard Cleasby. Enlarged and completed by G. Vigfússon, M.A. With an Introduction, and Life of Richard Cleasby, by G. Webbe Dasent, D.C.L. 1874. 4to. 3*l.* 7*s.*

—— *A List of English Words the Etymology of which is illustrated by comparison with Icelandic.* Prepared in the form of an APPENDIX to the above. By W. W. Skeat, Litt.D. 1876. stitched, 2*s.*

—— *An Icelandic Primer,* with Grammar, Notes, and Glossary. By Henry Sweet, M.A. Extra fcap. 8vo. 3*s.* 6*d.*

—— *An Icelandic Prose Reader,* with Notes, Grammar and Glossary, by Dr. Gudbrand Vigfússon and F. York Powell, M.A. 1879. Extra fcap. 8vo. 10*s.* 6*d.*

LATIN.—*A Latin Dictionary,* founded on Andrews' edition of Freund's Latin Dictionary, revised, enlarged, and in great part rewritten by Charlton T. Lewis, Ph.D., and Charles Short, LL.D. 1879. 4to. 1*l.* 5*s.*

CLARENDON PRESS, OXFORD.

MELANESIAN.—*The Melanesian Languages.* By R. H. Codrington, D.D., of the Melanesian Mission. 8vo. 18s.

SANSKRIT.—*A Practical Grammar of the Sanskrit Language*, arranged with reference to the Classical Languages of Europe, for the use of English Students, by Sir M. Monier-Williams, M.A. Fourth Edition. 8vo. 15s.

—— *A Sanskrit-English Dictionary*, Etymologically and Philologically arranged, with special reference to Greek, Latin, German, Anglo-Saxon, English, and other cognate Indo-European Languages. By Sir M. Monier-Williams, M.A. 1888. 4to. 4l. 14s. 6d.

—— *Nalopákhyánam.* Story of Nala, an Episode of the Mahá-Bhárata: the Sanskrit text, with a copious Vocabulary, and an improved version of Dean Milman's Translation, by Sir M. Monier-Williams, M.A. Second Edition, Revised and Improved. 1879. 8vo. 15s.

—— *Sakuntalā.* A Sanskrit Drama, in Seven Acts. Edited by Sir M. Monier-Williams, M.A. Second Edition, 1876. 8vo. 21s.

SYRIAC.—*Thesaurus Syriacus:* collegerunt Quatremère, Bernstein, Lorsbach, Arnoldi, Agrell, Field, Roediger: edidit R. Payne Smith, S.T.P. Fasc. I-VI. 1868-83. sm. fol. each, 1l. 1s. Fasc. VII. 1l. 11s. 6d.
Vol. I, containing Fasc. I-V, sm. fol. 5l. 5s.

—— *The Book of Kalīlah and Dimnah.* Translated from Arabic into Syriac. Edited by W. Wright, LL.D. 1884. 8vo. 21s.

GREEK CLASSICS, &c.

Aristophanes: A Complete Concordance to the Comedies and Fragments. By Henry Dunbar, M.D. 4to. 1l. 1s.

Aristotle: The Politics, with Introductions, Notes, etc., by W. L. Newman, M.A., Fellow of Balliol College, Oxford. Vols. I. and II. Medium 8vo. 28s. *Just Published.*

Aristotle: The Politics, translated into English, with Introduction, Marginal Analysis, Notes, and Indices, by B. Jowett, M.A. Medium 8vo. 2 vols. 21s.

Catalogus Codicum Graecorum Sinaiticorum. Scripsit V. Gardthausen Lipsiensis. With six pages of Facsimiles. 8vo. *linen*, 25s.

Heracliti Ephesii Reliquiae. Recensuit I. Bywater, M.A. Appendicis loco additae sunt Diogenis Laertii Vita Heracliti, Particulae Hippocratei De Diaeta Libri Primi, Epistolae Heracliteae. 1877. 8vo. 6s.

Herculanensium Voluminum Partes II. 1824. 8vo. 10s.

Fragmenta Herculanensia. A Descriptive Catalogue of the Oxford copies of the Herculanean Rolls, together with the texts of several papyri, accompanied by facsimiles. Edited by Walter Scott, M.A., Fellow of Merton College, Oxford. Royal 8vo. *cloth,* 21*s.*

Homer: A Complete Concordance to the Odyssey and Hymns of Homer; to which is added a Concordance to the Parallel Passages in the Iliad, Odyssey, and Hymns. By Henry Dunbar, M.D. 1880. 4to. 1*l.* 1*s.*

—— *Scholia Graeca in Iliadem.* Edited by Professor W. Dindorf, after a new collation of the Venetian MSS. by D. B. Monro, M.A., Provost of Oriel College. 4 vols. 8vo. 2*l.* 10*s.* Vols. V and VI. *In the Press.*

—— *Scholia Graeca in Odysseam.* Edidit Guil. Dindorfius. Tomi II. 1855. 8vo. 15*s.* 6*d.*

Plato: Apology, with a revised Text and English Notes, and a Digest of Platonic Idioms, by James Riddell, M.A. 1878. 8vo. 8*s.* 6*d.*

—— *Philebus,* with a revised Text and English Notes, by Edward Poste, M.A. 1860. 8vo. 7*s.* 6*d.*

—— *Sophistes and Politicus,* with a revised Text and English Notes, by L. Campbell, M.A. 1867. 8vo. 18*s.*

—— *Theaetetus,* with a revised Text and English Notes, by L. Campbell, M.A. Second Edition. 8vo. 10*s.* 6*d.*

—— *The Dialogues,* translated into English, with Analyses and Introductions, by B. Jowett, M.A. A new Edition in 5 volumes, medium 8vo. 1875. 3*l.* 10*s.*

—— *The Republic,* translated into English, with an Analysis and Introduction, by B. Jowett, M.A. Medium 8vo. 12*s.* 6*d.*

Thucydides: Translated into English, with Introduction, Marginal Analysis, Notes, and Indices. By B. Jowett, M.A. 2 vols. 1881. Medium 8vo. 1*l.* 12*s.*

THE HOLY SCRIPTURES, &c.

STUDIA BIBLICA.—Essays in Biblical Archæology and Criticism, and kindred subjects. By Members of the University of Oxford. 8vo. 10*s.* 6*d.*

ENGLISH.—*The Holy Bible in the earliest English Versions,* made from the Latin Vulgate by John Wycliffe and his followers · edited by the Rev. J. Forshall and Sir F. Madden. 4 vols. 1850. Royal 4to. 3*l.* 3*s.*

[Also reprinted from the above, with Introduction and Glossary by W. W. Skeat, Litt. D.

—— *The Books of Job, Psalms, Proverbs, Ecclesiastes, and the Song of Solomon:* according to the Wycliffite Version made by Nicholas de Hereford, about A.D. 1381, and Revised by John Purvey, about A.D. 1388. Extra fcap. 8vo. 3s. 6d.

—— *The New Testament in English,* according to the Version by John Wycliffe, about A.D. 1380, and Revised by John Purvey, about A.D. 1388. Extra fcap. 8vo. 6s.]

ENGLISH.—*The Holy Bible:* an exact reprint, page for page, of the Authorised Version published in the year 1611. Demy 4to. half bound, 1l. 1s.

—— *The Psalter, or Psalms of David, and certain Canticles,* with a Translation and Exposition in English, by Richard Rolle of Hampole. Edited by H. R. Bramley, M.A., Fellow of S. M. Magdalen College, Oxford. With an Introduction and Glossary. Demy 8vo. 1l. 1s.

—— *Lectures on the Book of Job.* Delivered in Westminster Abbey by the Very Rev. George Granville Bradley, D.D., Dean of Westminster. Crown 8vo. 7s. 6d.

—— *Lectures on Ecclesiastes.* By the same Author. Crown 8vo. 4s. 6d.

GOTHIC.—*The Gospel of St. Mark in Gothic,* according to the translation made by Wulfila in the Fourth Century. Edited with a Grammatical Introduction and Glossarial Index by W. W. Skeat, Litt. D. Extra fcap. 8vo. 4s.

GREEK.—*Vetus Testamentum* ex Versione Septuaginta Interpretum secundum exemplar Vaticanum Romae editum. Accedit potior varietas Codicis Alexandrini. Tomi III. Editio Altera. 18mo. 18s. The volumes may be had separately, price 6s. each.

—— *Origenis Hexaplorum* quae supersunt; sive, Veterum Interpretum Graecorum in totum Vetus Testamentum Fragmenta. Edidit Fridericus Field, A.M. 2 vols. 1875. 4to. 5l. 5s.

—— *The Book of Wisdom:* the Greek Text, the Latin Vulgate, and the Authorised English Version; with an Introduction, Critical Apparatus, and a Commentary. By William J. Deane, M.A. Small 4to. 12s. 6d.

—— *Novum Testamentum Graece.* Antiquissimorum Codicum Textus in ordine parallelo dispositi. Accedit collatio Codicis Sinaitici. Edidit E. H. Hansell, S.T.B. Tomi III. 1864. 8vo. 24s.

—— *Novum Testamentum Graece.* Accedunt parallela S. Scripturae loca, etc. Edidit Carolus Lloyd, S.T.P.R. 18mo. 3s.

On writing paper, with wide margin, 10s.

GREEK.—*Novum Testamentum Graece* juxta Exemplar Millianum. 18mo. 2*s*. 6*d*. On writing paper, with wide margin, 9*s*.

—— *Evangelia Sacra Graece*. Fcap. 8vo. limp, 1*s*. 6*d*.

—— *The Greek Testament*, with the Readings adopted by the Revisers of the Authorised Version:—

(1) Pica type, with Marginal References. Demy 8vo. 10*s*. 6*d*.
(2) Long Primer type. Fcap. 8vo. 4*s*. 6*d*.
(3) The same, on writing paper, with wide margin, 15*s*.

—— *The Parallel New Testament*, Greek and English; being the Authorised Version, 1611; the Revised Version, 1881; and the Greek Text followed in the Revised Version. 8vo. 12*s*. 6*d*.

The Revised Version is the joint property of the Universities of Oxford and Cambridge.

—— *Canon Muratorianus:* the earliest Catalogue of the Books of the New Testament. Edited with Notes and a Facsimile of the MS. in the Ambrosian Library at Milan, by S. P. Tregelles, LL.D. 1867. 4to. 10*s*. 6*d*.

—— *Outlines of Textual Criticism* applied to the *New Testament*. By C. E. Hammond, M.A. Fourth Edition. Extra fcap. 8vo. 3*s*. 6*d*.

HEBREW, etc.—*Notes on the Hebrew Text of the Book of Genesis*. With Two Appendices. By G. J. Spurrell, M.A. Crown 8vo. 10*s*. 6*d*.

—— *The Psalms in Hebrew without points.* 1879. Crown 8vo. Price reduced to 2*s*., in stiff cover.

—— *A Commentary on the Book of Proverbs.* Attributed to Abraham Ibn Ezra. Edited from a MS. in the Bodleian Library by S. R. Driver, M.A. Crown 8vo. paper covers, 3*s*. 6*d*.

—— *The Book of Tobit.* A Chaldee Text, from a unique MS. in the Bodleian Library; with other Rabbinical Texts, English Translations, and the Itala. Edited by Ad. Neubauer, M.A. 1878. Crown 8vo. 6*s*.

—— *Horae Hebraicae et Talmudicae*, a J. Lightfoot. A new Edition, by R. Gandell, M.A. 4 vols. 1859. 8vo. 1*l*. 1*s*.

LATIN.—*Libri Psalmorum* Versio antiqua Latina, cum Paraphrasi Anglo-Saxonica. Edidit B. Thorpe, F.A.S. 1835. 8vo. 10*s*. 6*d*.

—— *Old-Latin Biblical Texts: No. I.* The Gospel according to St. Matthew from the St. Germain MS. (g₁). Edited with Introduction and Appendices by John Wordsworth, D.D. Small 4to., stiff covers, 6*s*.

—— *Old-Latin Biblical Texts: No. II.* Portions of the Gospels according to St. Mark and St. Matthew, from the Bobbio MS. (k), &c. Edited by John Wordsworth, D.D., W. Sanday, M.A., D.D., and H. J. White, M.A. Small 4to., stiff covers, 21*s*.

CLARENDON PRESS, OXFORD.

LATIN.—*Old-Latin Biblical Texts: No. III.* The Four Gospels, from the Munich MS. (q), now numbered Lat. 6224 in the Royal Library at Munich. With a Fragment from St. John in the Hof-Bibliothek at Vienna (Cod. Lat. 502). Edited, with the aid of Tischendorf's transcript (under the direction of the Bishop of Salisbury), by H. J. White, M.A. Small 4to. stiff covers, 12s. 6d.

OLD-FRENCH.—*Libri Psalmorum* Versio antiqua Gallica e Cod. MS. in Bibl. Bodleiana adservato, una cum Versione Metrica aliisque Monumentis pervetustis. Nunc primum descripsit et edidit Franciscus Michel, Phil. Doc. 1860. 8vo. 10s. 6d.

FATHERS OF THE CHURCH, &c.

St. Athanasius: Historical Writings, according to the Benedictine Text. With an Introduction by William Bright, D.D. 1881. Crown 8vo. 10s. 6d.

—— *Orations against the Arians.* With an Account of his Life by William Bright, D.D. 1873. Crown 8vo. 9s.

St. Augustine: Select Anti-Pelagian Treatises, and the Acts of the Second Council of Orange. With an Introduction by William Bright, D.D. Crown 8vo. 9s.

Canons of the First Four General Councils of Nicaea, Constantinople, Ephesus, and Chalcedon. 1877. Crown 8vo. 2s. 6d.

—— *Notes on the Canons of the First Four General Councils.* By William Bright, D.D. 1882. Crown 8vo. 5s. 6d.

Cyrilli Archiepiscopi Alexandrini in XII Prophetas. Edidit P. E. Pusey, A.M. Tomi II. 1868. 8vo. cloth, 2l. 2s.

—— *in D. Joannis Evangelium.* Accedunt Fragmenta varia necnon Tractatus ad Tiberium Diaconum duo. Edidit post Aubertum P. E. Pusey, A.M. Tomi III. 1872. 8vo. 2l. 5s.

—— *Commentarii in Lucae Evangelium* quae supersunt Syriace. E MSS. apud Mus. Britan. edidit R. Payne Smith, A.M. 1858. 4to. 1l. 2s.

—— Translated by R. Payne Smith, M.A. 2 vols. 1859. 8vo. 14s.

Ephraemi Syri, Rabulae Episcopi Edesseni, Balaei, aliorumque Opera Selecta. E Codd. Syriacis MSS. in Museo Britannico et Bibliotheca Bodleiana asservatis primus edidit J. J. Overbeck. 1865. 8vo. 1l. 1s.

Eusebius' Ecclesiastical History, according to the text of Burton, with an Introduction by William Bright, D.D. 1881. Crown 8vo. 8s. 6d.

Irenaeus: The Third Book of St. Irenaeus, Bishop of Lyons, against Heresies. With short Notes and a Glossary by H. Deane, B.D. 1874. Crown 8vo. 5s. 6d.

Patrum Apostolicorum, S. Clementis Romani, S. Ignatii, S. Polycarpi, quae supersunt. Edidit Guil. Jacobson, S.T.P.R. Tomi II. Fourth Edition, 1863. 8vo. 1l. 1s.

Socrates' Ecclesiastical History, according to the Text of Hussey, with an Introduction by William Bright, D.D. 1878. Crown 8vo. 7s. 6d.

ECCLESIASTICAL HISTORY, BIOGRAPHY, &c.

Ancient Liturgy of the Church of England, according to the uses of Sarum, York, Hereford, and Bangor, and the Roman Liturgy arranged in parallel columns, with preface and notes. By William Maskell, M.A. Third Edition. 1882. 8vo. 15s.

Baedae Historia Ecclesiastica. Edited, with English Notes, by G. H. Moberly, M.A. 1881. Crown 8vo. 10s. 6d.

Bright (W.). Chapters of Early English Church History. 1878. 8vo. 12s.

Burnet's History of the Reformation of the Church of England. A new Edition. Carefully revised, and the Records collated with the originals, by N. Pocock, M.A. 7 vols. 1865. 8vo. *Price reduced to* 1l. 10s.

Councils and Ecclesiastical Documents relating to Great Britain and Ireland. Edited, after Spelman and Wilkins, by A. W. Haddan, B.D., and W. Stubbs, M.A. Vols. I. and III. 1869-71. Medium 8vo. each 1l. 1s.

    Vol. II. Part I. 1873. Medium 8vo. 10s. 6d.

    Vol. II. Part II. 1878. Church of Ireland; Memorials of St. Patrick. Stiff covers, 3s. 6d.

Hamilton (John, Archbishop of St. Andrews). The Catechism of. Edited, with Introduction and Glossary, by Thomas Graves Law. With a Preface by the Right Hon. W. E. Gladstone. 8vo. 12s. 6d.

Hammond (C. E.). Liturgies, Eastern and Western. Edited, with Introduction, Notes, and Liturgical Glossary. 1878. Crown 8vo. 10s. 6d.

    An Appendix to the above. 1879. Crown 8vo. paper covers, 1s. 6d.

John, Bishop of Ephesus. *The Third Part of his Ecclesiastical History*. [In Syriac.] Now first edited by William Cureton, M.A. 1853. 4to. 1*l*. 12*s*.

—— Translated by R. Payne Smith, M.A. 1860. 8vo. 10*s*.

Leofric Missal, The, as used in the Cathedral of Exeter during the Episcopate of its first Bishop, A.D. 1050-1072; together with some Account of the Red Book of Derby, the Missal of Robert of Jumièges, and a few other early MS. Service Books of the English Church. Edited, with Introduction and Notes, by F. E. Warren, B.D. 4to. half morocco, 35*s*.

Monumenta Ritualia Ecclesiae Anglicanae. The occasional Offices of the Church of England according to the old use of Salisbury, the Prymer in English, and other prayers and forms, with dissertations and notes. By William Maskell, M.A. Second Edition. 1882. 3 vols. 8vo. 2*l*. 10*s*.

Records of the Reformation. The Divorce, 1527-1533. Mostly now for the first time printed from MSS. in the British Museum and other libraries. Collected and arranged by N. Pocock, M.A. 1870. 2 vols. 8vo. 1*l*. 16*s*.

Shirley (W. W.). Some Account of the Church in the Apostolic Age. Second Edition, 1874. Fcap. 8vo. 3*s*. 6*d*.

Stubbs (W.). Registrum Sacrum Anglicanum. An attempt to exhibit the course of Episcopal Succession in England. 1858. Small 4to. 8*s*. 6*d*.

Warren (F. E.). Liturgy and Ritual of the Celtic Church. 1881. 8vo. 14*s*.

ENGLISH THEOLOGY.

Bampton Lectures, 1886. *The Christian Platonists of Alexandria*. By Charles Bigg, D.D. 8vo. 10*s*. 6*d*.

Butler's Works, with an Index to the Analogy. 2 vols. 1874. 8vo. 11*s*.

Also separately,

Sermons, 5*s*. 6*d*.   *Analogy of Religion*, 5*s*. 6*d*.

Greswell's Harmonia Evangelica. Fifth Edition. 8vo. 9*s*. 6*d*.

Heurtley's Harmonia Symbolica: Creeds of the Western Church. 1858. 8vo. 6*s*. 6*d*.

Homilies appointed to be read in Churches. Edited by J. Griffiths, M.A. 1859. 8vo. 7*s*. 6*d*.

Hooker's Works, with his life by Walton, arranged by John Keble, M.A. Seventh Edition. *Revised by R. W. Church, M.A., D.C.L., Dean of St. Paul's, and F. Paget, D.D.* 3 vols. medium 8vo. 36*s*.

Hooker's Works, the text as arranged by John Keble, M.A.
2 vols. 1875. 8vo. 11s.

Jewel's Works. Edited by R. W. Jelf, D.D. 8 vols. 1848.
8vo. 1l. 10s.

Pearson's Exposition of the Creed. Revised and corrected by
E. Burton, D.D. Sixth Edition, 1877. 8vo. 10s. 6d.

Waterland's Review of the Doctrine of the Eucharist, with
a Preface by the late Bishop of London. Crown 8vo. 6s. 6d.

—— *Works,* with Life, by Bp. Van Mildert. A new Edition,
with copious Indexes. 6 vols. 1856. 8vo. 2l. 11s.

Wheatly's Illustration of the Book of Common Prayer. A new
Edition, 1846. 8vo. 5s.

Wyclif. A Catalogue of the Original Works of John Wyclif,
by W. W. Shirley, D.D. 1865. 8vo. 3s. 6d.

—— *Select English Works.* By T. Arnold, M.A. 3 vols.
1869-1871. 8vo. 1l. 1s.

—— *Trialogus.* With the Supplement now first edited.
By Gotthard Lechler. 1869. 8vo. 7s.

HISTORICAL AND DOCUMENTARY WORKS.

British Barrows, a Record of the Examination of Sepulchral
Mounds in various parts of England. By William Greenwell, M.A., F.S.A.
Together with Description of Figures of Skulls, General Remarks on Pre-
historic Crania, and an Appendix by George Rolleston, M.D., F.R.S. 1877.
Medium 8vo. 25s.

Clarendon's History of the Rebellion and Civil Wars in
England. 7 vols. 1839. 18mo. 1l. 1s.

Clarendon's History of the Rebellion and Civil Wars in
England. Also his Life, written by himself, in which is included a Con-
tinuation of his History of the Grand Rebellion. With copious Indexes.
In one volume, royal 8vo. 1842. 1l. 2s.

Clinton's Epitome of the Fasti Hellenici. 1851. 8vo. 6s. 6d.

—— *Epitome of the Fasti Romani.* 1854. 8vo. 7s.

Corpvs Poeticvm Boreale. The Poetry of the Old Northern
Tongue, from the Earliest Times to the Thirteenth Century. Edited, clas-
sified, and translated, with Introduction, Excursus, and Notes, by Gudbrand
Vigfússon, M.A., and F. York Powell, M.A. 2 vols. 1883. 8vo. 42s.

Freeman (E. A.). History of the Norman Conquest of England; its Causes and Results. In Six Volumes. 8vo. 5*l.* 9*s.* 6*d.*

—— *The Reign of William Rufus and the Accession of* Henry the First. 2 vols. 8vo. 1*l.* 16*s.*

Gascoigne's Theological Dictionary ("Liber Veritatum"): Selected Passages, illustrating the condition of Church and State, 1403–1458. With an Introduction by James E. Thorold Rogers, M.A. Small 4to. 10*s.* 6*d.*

Johnson (Samuel, LL.D.), Boswell's Life of; including Boswell's Journal of a Tour to the Hebrides, and Johnson's Diary of a Journey into North Wales. Edited by G. Birkbeck Hill, D.C.L. In six volumes, medium 8vo. With Portraits and Facsimiles of Handwriting. Half bound, 3*l.* 3*s.* (See p. 21.)

Magna Carta, a careful Reprint. Edited by W. Stubbs, D.D. 1879. 4to. stitched, 1*s.*

Passio et Miracula Beati Olaui. Edited from a Twelfth-Century MS. in the Library of Corpus Christi College, Oxford, with an Introduction and Notes, by Frederick Metcalfe, M.A. Small 4to. stiff covers, 6*s.*

Protests of the Lords, including those which have been expunged, from 1624 to 1874; with Historical Introductions. Edited by James E. Thorold Rogers, M.A. 1875. 3 vols. 8vo. 2*l.* 2*s.*

Rogers (J. E. T.). History of Agriculture and Prices in England, A.D. 1259–1793.

  Vols. I and II (1259–1400). 1866. 8vo. 2*l.* 2*s.*
  Vols. III and IV (1401–1582). 1882. 8vo. 2*l.* 10*s.*
  Vols. V and VI (1583–1702). 8vo. 2*l.* 10*s. Just Published.*

—— *The First Nine Years of the Bank of England.* 8vo. 8*s.* 6*d.*

Saxon Chronicles (Two of the) parallel, with Supplementary Extracts from the Others. Edited, with Introduction, Notes, and a Glossarial Index, by J. Earle, M.A. 1865. 8vo. 16*s.*

Stubbs (W., D.D.). Seventeen Lectures on the Study of Medieval and Modern History, &c., delivered at Oxford 1867–1884. Crown 8vo. 8*s.* 6*d.*

Sturlunga Saga, including the Islendinga Saga of Lawman Sturla Thordsson and other works. Edited by Dr. Gudbrand Vigfússon. In 2 vols. 1878. 8vo. 2*l.* 2*s.*

York Plays. The Plays performed by the Crafts or Mysteries of York on the day of Corpus Christi in the 14th, 15th, and 16th centuries. Now first printed from the unique MS. in the Library of Lord Ashburnham. Edited with Introduction and Glossary by Lucy Toulmin Smith. 8vo. 21*s.*

Manuscript Materials relating to the History of Oxford.
Arranged by F. Madan, M.A. 8vo. 7s. 6d.

Statutes made for the University of Oxford, and for the Colleges
and Halls therein, by the University of Oxford Commissioners. 1882. 8vo. 12s. 6d.

Statuta Universitatis Oxoniensis. 1887. 8vo. 5s.

The Oxford University Calendar for the year 1888. Crown 8vo. 4s. 6d.

The present Edition includes all Class Lists and other University distinctions for the eight years ending with 1887.

Also, supplementary to the above, price 5s. (pp. 606),

The Honours Register of the University of Oxford. A complete Record of University Honours, Officers, Distinctions, and Class Lists; of the Heads of Colleges, &c., &c., from the Thirteenth Century to 1883.

The Examination Statutes for the Degrees of B.A., B. Mus.,
B.C.L., and B.M. Revised to the end of Michaelmas Term, 1887. 8vo. sewed, 1s.

The Student's Handbook to the University and Colleges of Oxford. Ninth Edition. Crown 8vo. 2s. 6d.

MATHEMATICS, PHYSICAL SCIENCE, &c.

Acland (H. W., M.D., F.R.S.). Synopsis of the Pathological
Series in the Oxford Museum. 1867. 8vo. 2s. 6d.

Burdon-Sanderson (J., M.D., F.R.SS. L. and E.). Transla-
tions of Foreign Biological Memoirs. I. Memoirs on the Physiology of Nerve, of Muscle, and of the Electrical Organ. Medium 8vo. 21s.

De Bary (Dr. A.). Comparative Anatomy of the Vegetative
Organs of the Phanerogams and Ferns. Translated and Annotated by F. O. Bower, M.A., F.L.S., and D. H. Scott, M.A., Ph.D., F.L.S. With 241 woodcuts and an Index. Royal 8vo., half morocco, 1l. 2s. 6d.

Goebel (Dr. K.). Outlines of Classification and Special Mor-
phology of Plants. A New Edition of Sachs' Text-Book of Botany, Book II. English Translation by H. E. F. Garnsey, M.A. Revised by I. Bayley Balfour, M.A., M.D., F.R.S. With 407 Woodcuts. Royal 8vo. half morocco, 21s.

Sachs (Julius von). Lectures on the Physiology of Plants.
Translated by H. Marshall Ward, M.A. With 445 Woodcuts. Royal 8vo. half morocco, 1l. 11s. 6d.

De Bary (Dr. A). Comparative Morphology and Biology of the Fungi, Mycetozoa and Bacteria. Authorised English Translation by Henry E. F. Garnsey, M.A. Revised by Isaac Bayley Balfour, M.A., M.D., F.R.S. With 198 Woodcuts. Royal 8vo., half morocco, 1*l*. 2*s*. 6*d*.

—— *Lectures on Bacteria.* Second improved edition. Authorised translation by H. E. F. Garnsey, M.A. Revised by Isaac Bayley Balfour, M.A., M.D., F.R.S. With 20 Woodcuts. Crown 8vo. 6*s*.

Annals of Botany. Edited by Isaac Bayley Balfour, M.A., M.D., F.R.S., Sydney H. Vines, D.Sc., F.R.S., and William Gilson Farlow, M.D., Professor of Cryptogamic Botany in Harvard University, Cambridge, Mass., U.S.A., and other Botanists. Royal 8vo.
    Vol. I. No. 1. Price 8*s*. 6*d*. Vol. I. No. 2. Price 7*s*. 6*d*.

Müller (J.). On certain Variations in the Vocal Organs of the Passeres that have hitherto escaped notice. Translated by F. J. Bell, B.A., and edited, with an Appendix, by A. H. Garrod, M.A., F.R.S. With Plates. 1878. 4to. paper covers, 7*s*. 6*d*.

Price (Bartholomew, M.A., F.R.S.). Treatise on Infinitesimal Calculus.
    Vol. I. Differential Calculus. Second Edition. 8vo. 14*s*. 6*d*.
    Vol. II. Integral Calculus, Calculus of Variations, and Differential Equations. Second Edition, 1865. 8vo. 18*s*.
    Vol. III. Statics, including Attractions; Dynamics of a Material Particle. Second Edition, 1868. 8vo. 16*s*.
    Vol. IV. Dynamics of Material Systems; together with a chapter on Theoretical Dynamics, by W. F. Donkin, M.A., F.R.S. 1862. 8vo. 16*s*.

Pritchard (C., D.D., F.R.S.). Uranometria Nova Oxoniensis. A Photometric determination of the magnitudes of all Stars visible to the naked eye, from the Pole to ten degrees south of the Equator. 1885. Royal 8vo. 8*s*. 6*d*.

—— *Astronomical Observations* made at the University Observatory, Oxford, under the direction of C. Pritchard, D.D. No. 1. 1878. Royal 8vo. paper covers, 3*s*. 6*d*.

Rigaud's Correspondence of Scientific Men of the 17th Century, with Table of Contents by A. de Morgan, and Index by the Rev. J. Rigaud, M.A. 2 vols. 1841–1862. 8vo. 18*s*. 6*d*.

Rolleston (George, M.D., F.R.S.). Forms of Animal Life. A Manual of Comparative Anatomy, with descriptions of selected types. Second Edition. Revised and enlarged by W. Hatchett Jackson, M.A. Medium, 8vo. cloth extra, 1*l*. 16*s*.

—— *Scientific Papers and Addresses.* Arranged and Edited by William Turner, M.B., F.R.S. With a Biographical Sketch by Edward Tylor, F.R.S. With Portrait, Plates, and Woodcuts. 2 vols. 8vo. 1*l*. 4*s*.

Westwood (J. O., M.A., F.R.S.). Thesaurus Entomologicus Hopeianus, or a Description of the rarest Insects in the Collection given to the University by the Rev. William Hope. With 40 Plates. 1874. Small folio, half morocco, 7*l*. 10*s*.

The Sacred Books of the East.

TRANSLATED BY VARIOUS ORIENTAL SCHOLARS, AND EDITED BY
F. MAX MÜLLER.

[Demy 8vo. cloth.]

Vol. I. The Upanishads. Translated by F. Max Müller.
Part I. The *Kh*ândogya-upanishad, The Talavakâra-upanishad, The Aitareya-âra*n*yaka, The Kaushîtaki-brâhma*n*a-upanishad, and The Vâgasaneyi-sa*m*hitâ-upanishad. 10*s*. 6*d*.

Vol. II. The Sacred Laws of the Âryas, as taught in the Schools of Âpastamba, Gautama, Vâsish*th*a, and Baudhâyana. Translated by Prof. Georg Bühler. Part I. Âpastamba and Gautama. 10*s*. 6*d*.

Vol. III. The Sacred Books of China. The Texts of Con- fucianism. Translated by James Legge. Part I. The Shû King, The Religious portions of the Shih King, and The Hsiâo King. 12*s*. 6*d*.

Vol. IV. The Zend-Avesta. Translated by James Darme- steter. Part I. The Vendîdâd. 10*s*. 6*d*.

Vol. V. The Pahlavi Texts. Translated by E. W. West.
Part I. The Bundahi*s*, Bahman Ya*s*t, and Shâyast lâ-shâyast. 12*s*. 6*d*.

Vols. VI and IX. The Qur'ân. Parts I and II. Translated by E. H. Palmer. 21*s*.

Vol. VII. The Institutes of Vish*n*u. Translated by Julius Jolly. 10*s*. 6*d*.

Vol. VIII. The Bhagavadgîtâ, with The Sanatsugâtîya, and The Anugîtâ. Translated by Kâshinâth Trimbak Telang. 10*s*. 6*d*.

Vol. X. The Dhammapada, translated from Pâli by F. Max Müller; and The Sutta-Nipâta, translated from Pâli by V. Fausböll; being Canonical Books of the Buddhists. 10*s*. 6*d*.

Vol. XI. Buddhist Suttas. Translated from Pâli by T. W.
Rhys Davids. 1. The Mahâparinibbâna Suttanta; 2. The Dhamma-*k*akka-ppavattana Sutta; 3. The Tevigga Suttanta; 4. The Akankheyya Sutta; 5. The *K*etokhila Sutta; 6. The Mahâ-sudassana Suttanta; 7. The Sabbâsava Sutta. 10s. 6d.

Vol. XII. The Satapatha-Brâhma*n*a, according to the Text of the Mâdhyandina School. Translated by Julius Eggeling. Part I. Books I and II. 12s. 6d.

Vol. XIII. Vinaya Texts. Translated from the Pâli by T. W. Rhys Davids and Hermann Oldenberg. Part I. The Pâtimokkha. The Mahâvagga, I-IV. 10s. 6d.

Vol. XIV. The Sacred Laws of the Âryas, as taught in the Schools of Âpastamba, Gautama, Vâsish*th*a and Baudhâyana. Translated by Georg Bühler. Part II. Vâsish*th*a and Baudhâyana. 10s. 6d.

Vol. XV. The Upanishads. Translated by F. Max Müller. Part II. The Ka*th*a-upanishad, The Mu*nd*aka-upanishad, The Taittirîyaka-upanishad, The B*ri*hadâra*ny*aka-upanishad, The *S*veta*s*vatara-upanishad, The Pra*sn*a-upanishad, and The Maitrâya*n*a-Brâhma*n*a-upanishad. 10s. 6d.

Vol. XVI. The Sacred Books of China. The Texts of Confucianism. Translated by James Legge. Part II. The Yî King. 10s. 6d.

Vol. XVII. Vinaya Texts. Translated from the Pâli by T. W. Rhys Davids and Hermann Oldenberg. Part II. The Mahâvagga, V-X. The *K*ullavagga, I-III. 10s. 6d.

Vol. XVIII. Pahlavi Texts. Translated by E. W. West. Part II. The Dâ*d*istân-î Dînîk and The Epistles of Mânû*sk*îhar. 12s. 6d.

Vol. XIX. The Fo-sho-hing-tsan-king. A Life of Buddha by A*s*vaghosha Bodhisattva, translated from Sanskrit into Chinese by Dharmaraksha, A.D. 420, and from Chinese into English by Samuel Beal. 10s. 6d.

Vol. XX. Vinaya Texts. Translated from the Pâli by T. W. Rhys Davids and Hermann Oldenberg. Part III. The *K*ullavagga, IV-XII. 10s. 6d.

Vol. XXI. The Saddharma-pu*nd*arîka; or, the Lotus of the True Law. Translated by H. Kern. 12s. 6d.

Vol. XXII. *G*aina-Sûtras. Translated from Prâkrit by Hermann Jacobi. Part I. The Â*k*ârânga-Sûtra. The Kalpa-Sûtra. 10s. 6d.

Vol. XXIII. The Zend-Avesta. Translated by James Darmesteter. Part II. The Sîrôzahs, Ya*s*ts, and Nyâyi*s*. 10*s*. 6*d*.

Vol. XXIV. Pahlavi Texts. Translated by E. W. West. Part III. Dînâ-î Maînôg-î Khirad, *S*ikand-gûmânîk, and Sad-Dar. 10*s*. 6*d*.

Second Series.

Vol. XXV. Manu. Translated by Georg Bühler. 21*s*.

Vol. XXVI. The *S*atapatha-Brâhma*n*a. Translated by Julius Eggeling. Part II. 12*s*. 6*d*.

Vols. XXVII and XXVIII. The Sacred Books of China. The Texts of Confucianism. Translated by James Legge. Parts III and IV. The Lî *K*î, or Collection of Treatises on the Rules of Propriety, or Ceremonial Usages. 25*s*.

Vols. XXIX and XXX. The G*ri*hya-Sûtras, Rules of Vedic Domestic Ceremonies. Translated by Hermann Oldenberg.

    Part I (Vol. XXIX), 12*s*. 6*d*. *Just Published*.
    Part II (Vol. XXX). *In the Press*.

Vol. XXXI. The Zend-Avesta. Part III. The Yasna, Visparad, Âfrînagân, and Gâhs. Translated by L. H. Mills. 12*s*. 6*d*.

The following Volumes are in the Press:—

Vol. XXXII. Vedic Hymns. Translated by F. Max Müller. Part I.

Vol. XXXIII. Nârada, and some Minor Law-books. Translated by Julius Jolly. [*Preparing*.]

Vol. XXXIV. The Vedânta-Sûtras, with *S*ankara's Commentary. Translated by G. Thibaut. [*Preparing*.]

    ⁂ The Second Series will consist of Twenty-Four Volumes.

Clarendon Press Series.

I. ENGLISH, &c.

A First Reading Book. By Marie Eichens of Berlin; and edited by Anne J. Clough. Extra fcap. 8vo. stiff covers, 4*d*.

Oxford Reading Book, Part I. For Little Children. Extra fcap. 8vo. stiff covers, 6*d*.

Oxford Reading Book, Part II. For Junior Classes. Extra fcap. 8vo. stiff covers, 6*d*.

An Elementary English Grammar and Exercise Book. By O. W. Tancock, M.A. Second Edition. Extra fcap. 8vo. 1*s*. 6*d*.

An English Grammar and Reading Book, for Lower Forms in Classical Schools. By O. W. Tancock, M.A. Fourth Edition. Extra fcap. 8vo. 3*s*. 6*d*.

Typical Selections from the best English Writers, with Introductory Notices. Second Edition. In 2 vols. Extra fcap. 8vo. 3*s*. 6*d*. each.
Vol. I. Latimer to Berkeley.   Vol. II. Pope to Macaulay.

Shairp (J. C., LL.D.). Aspects of Poetry; being Lectures delivered at Oxford. Crown 8vo. 10*s*. 6*d*.

A Book for the Beginner in Anglo-Saxon. By John Earle, M.A. Third Edition. Extra fcap. 8vo. 2*s*. 6*d*.

An Anglo-Saxon Reader. In Prose and Verse. With Grammatical Introduction, Notes, and Glossary. By Henry Sweet, M.A. Fourth Edition, Revised and Enlarged. Extra fcap. 8vo. 8*s*. 6*d*.

A Second Anglo-Saxon Reader. By the same Author. Extra fcap. 8vo. 4*s*. 6*d*.

An Anglo-Saxon Primer, with Grammar, Notes, and Glossary. By the same Author. Second Edition. Extra fcap. 8vo. 2*s*. 6*d*.

Old English Reading Primers; edited by Henry Sweet, M.A.
I. Selected Homilies of Ælfric. Extra fcap. 8vo., stiff covers, 1*s*. 6*d*.
II. Extracts from Alfred's Orosius. Extra fcap. 8vo., stiff covers, 1*s*. 6*d*.

First Middle English Primer, with Grammar and Glossary. By the same Author. Extra fcap. 8vo. 2*s*.

Second Middle English Primer. Extracts from Chaucer, with Grammar and Glossary. By the same Author. Extra fcap. 8vo. 2*s*.

c

Principles of English Etymology. First Series. *The Native Element.* By W. W. Skeat, Litt.D. Crown 8vo. 9s.

The Philology of the English Tongue. By J. Earle, M.A. Fourth Edition. Extra fcap. 8vo. 7s. 6d.

An Icelandic Primer, with Grammar, Notes, and Glossary. By Henry Sweet, M.A. Extra fcap. 8vo. 3s. 6d.

An Icelandic Prose Reader, with Notes, Grammar, and Glossary. By G. Vigfússon, M.A., and F. York Powell, M.A. Ext. fcap. 8vo. 10s. 6d.

A Handbook of Phonetics, including a Popular Exposition of the Principles of Spelling Reform. By H. Sweet, M.A. Ext. fcap. 8vo. 4s. 6d.

Elementarbuch des Gesprochenen Englisch. Grammatik, Texte und Glossar. Von Henry Sweet. *Second Edition.* Extra fcap. 8vo., stiff covers, 2s. 6d.

The Ormulum; with the Notes and Glossary of Dr. R. M. White. Edited by R. Holt, M.A. 1878. 2 vols. Extra fcap. 8vo. 21s.

Specimens of Early English. A New and Revised Edition. With Introduction, Notes, and Glossarial Index. By R. Morris, LL.D., and W. W. Skeat, Litt.D.

> Part I. From Old English Homilies to King Horn (A.D. 1150 to A.D. 1300). Second Edition. Extra fcap. 8vo. 9s.
>
> Part II. From Robert of Gloucester to Gower (A.D. 1298 to A.D. 1393). Third Edition. Extra fcap. 8vo. 7s. 6d.

Specimens of English Literature, from the 'Ploughmans Crede' to the 'Shepheardes Calender' (A.D. 1394 to A.D. 1579). With Introduction, Notes, and Glossarial Index. By W. W. Skeat, Litt.D. Fourth Edition. Extra fcap. 8vo. 7s. 6d.

The Vision of William concerning Piers the Plowman, in three Parallel Texts; together with *Richard the Redeless.* By William Langland (about 1362–1399 A.D.). Edited from numerous Manuscripts, with Preface, Notes, and a Glossary, by W. W. Skeat, Litt.D. 2 vols. 8vo. 31s. 6d.

The Vision of William concerning Piers the Plowman, by William Langland. Edited, with Notes, by W. W. Skeat, Litt.D. Fourth Edition. Extra fcap. 8vo. 4s. 6d.

Chaucer. I. *The Prologue to the Canterbury Tales;* the Knightes Tale; The Nonne Prestes Tale. Edited by R. Morris, LL.D. Sixty-sixth thousand. Extra fcap. 8vo. 2s. 6d.

—— II. *The Prioresses Tale; Sir Thopas; The Monkes Tale; The Clerkes Tale; The Squieres Tale,* &c. Edited by W. W. Skeat, Litt.D. Third Edition. Extra fcap. 8vo. 4s. 6d.

Chaucer. III. *The Tale of the Man of Lawe;* The Pardoneres
Tale; The Second Nonnes Tale; The Chanouns Yemannes Tale. By the
same Editor. *New Edition, Revised.* Extra fcap. 8vo. 4*s.* 6*d.*

Gamelyn, The Tale of. Edited with Notes, Glossary, &c., by
W. W. Skeat, Litt.D. Extra fcap. 8vo. Stiff covers, 1*s.* 6*d.*

Minot (Laurence). Poems. Edited, with Introduction and
Notes, by Joseph Hall, M.A., Head Master of the Hulme Grammar School,
Manchester. Extra fcap. 8vo. 4*s.* 6*d.*

Spenser's Faery Queene. Books I and II. Designed chiefly
for the use of Schools. With Introduction and Notes by G. W. Kitchin, D.D.,
and Glossary by A. L. Mayhew, M.A. Extra fcap. 8vo. 2*s.* 6*d.* each.

Hooker. Ecclesiastical Polity, Book I. Edited by R. W.
Church, M.A. Second Edition. Extra fcap. 8vo. 2*s.*

OLD ENGLISH DRAMA.

The Pilgrimage to Parnassus with *The Two Parts of the
Return from Parnassus.* Three Comedies performed in St. John's College,
Cambridge, A.D. MDXCVII–MDCI. Edited from MSS. by the Rev. W. D.
Macray, M.A., F.S.A. Medium 8vo. Bevelled Boards, Gilt top, 8*s.* 6*d.*

*Marlowe and Greene. Marlowe's Tragical History of Dr.
Faustus,* and *Greene's Honourable History of Friar Bacon and Friar Bungay.*
Edited by A. W. Ward, M.A. *New and Enlarged Edition.* Extra fcap.
8vo. 6*s.* 6*d.*

Marlowe. Edward II. With Introduction, Notes, &c. By
O. W. Tancock, M.A. Extra fcap. 8vo. Paper covers, 2*s.* Cloth 3*s.*

SHAKESPEARE.

Shakespeare. Select Plays. Edited by W. G. Clark, M.A.,
and W. Aldis Wright, M.A. Extra fcap. 8vo. stiff covers.

The Merchant of Venice. 1*s.* Macbeth. 1*s.* 6*d.*
Richard the Second. 1*s.* 6*d.* Hamlet. 2*s.*

Edited by W. Aldis Wright, M.A.

The Tempest. 1*s.* 6*d.* Midsummer Night's Dream. 1*s.* 6*d.*
As You Like It. 1*s.* 6*d.* Coriolanus. 2*s.* 6*d.*
Julius Cæsar. 2*s.* Henry the Fifth. 2*s.*
Richard the Third. 2*s.* 6*d.* Twelfth Night. 1*s.* 6*d.*
King Lear. 1*s.* 6*d.* King John. 1*s.* 6*d.*

Shakespeare as a Dramatic Artist; a popular Illustration of
the Principles of Scientific Criticism. By R. G. Moulton, M.A. Crown 8vo. 5*s.*

Bacon. I. *Advancement of Learning.* Edited by W. Aldis Wright, M.A. Third Edition. Extra fcap. 8vo. 4s. 6d.

—— II. *The Essays.* With Introduction and Notes. By S. H. Reynolds, M.A., late Fellow of Brasenose College. *In Preparation.*

Milton. I. *Areopagitica.* With Introduction and Notes. By John W. Hales, M.A. Third Edition. Extra fcap. 8vo. 3s.

—— II. *Poems.* Edited by R. C. Browne, M.A. 2 vols. Fifth Edition. Extra fcap. 8vo. 6s. 6d. Sold separately, Vol. I. 4s.; Vol. II. 3s.

In paper covers:—

Lycidas, 3d.   L'Allegro, 3d.   Il Penseroso, 4d.   Comus, 6d.

—— III. *Paradise Lost.* Book I. Edited by H. C. Beeching. Extra fcap. 8vo. stiff cover, 1s. 6d.; in white Parchment, 3s. 6d.

—— IV. *Samson Agonistes.* Edited with Introduction and Notes by John Churton Collins. Extra fcap. 8vo. stiff covers, 1s.

Bunyan. I. *The Pilgrim's Progress, Grace Abounding, Relation of the Imprisonment of Mr. John Bunyan.* Edited, with Biographical Introduction and Notes, by E. Venables, M.A. 1879. Extra fcap. 8vo. 5s. In ornamental Parchment, 6s.

—— II. *Holy War, &c.* Edited by E. Venables, M.A. In the Press.

Clarendon. *History of the Rebellion.* Book VI. Edited by T. Arnold, M.A. Extra fcap. 8vo. 4s. 6d.

Dryden. *Select Poems.* Stanzas on the Death of Oliver Cromwell; Astræa Redux; Annus Mirabilis; Absalom and Achitophel; Religio Laici; The Hind and the Panther. Edited by W. D. Christie, M.A. Second Edition. Extra fcap. 8vo. 3s. 6d.

Locke's *Conduct of the Understanding.* Edited, with Introduction, Notes, &c., by T. Fowler, D.D. Second Edition. Extra fcap. 8vo. 2s.

Addison. *Selections from Papers in the Spectator.* With Notes. By T. Arnold, M.A. Extra fcap. 8vo. 4s. 6d. In ornamental Parchment, 6s.

Steele. *Selections from the Tatler, Spectator, and Guardian.* Edited by Austin Dobson. Extra fcap. 8vo. 4s. 6d. In white Parchment, 7s. 6d.

Pope. With Introduction and Notes. By Mark Pattison, B.D.

—— I. *Essay on Man.* Extra fcap. 8vo. 1s. 6d.

—— II. *Satires and Epistles.* Extra fcap. 8vo. 2s.

Parnell. The Hermit. Paper covers, 2*d.*

Gray. Selected Poems. Edited by Edmund Gosse. Extra fcap. 8vo. Stiff covers, 1*s.* 6*d.* In white Parchment, 3*s.*

—— *Elegy and Ode on Eton College.* Paper covers, 2*d.*

Goldsmith. Selected Poems. Edited, with Introduction and Notes, by Austin Dobson. Extra fcap. 8vo. 3*s.* 6*d.* In white Parchment, 4*s.* 6*d.*

—— *The Deserted Village.* Paper covers, 2*d.*

Johnson. I. *Rasselas; Lives of Dryden and Pope.* Edited by Alfred Milnes, M.A. (London). Extra fcap. 8vo. 4*s.* 6*d.*, or *Lives of Dryden and Pope* only, stiff covers, 2*s.* 6*d.*

—— II. *Rasselas.* Edited, with Introduction and Notes, by G. Birkbeck Hill, D.C.L. Extra fcap. 8vo. Bevelled boards, 3*s.* 6*d.* In white Parchment, 4*s.* 6*d.*

—— III. *Vanity of Human Wishes.* With Notes, by E. J. Payne, M.A. Paper covers, 4*d.*

—— IV. *Life of Milton.* By C. H. Firth, M.A. Preparing.

—— V. *Wit and Wisdom of Samuel Johnson.* Edited by G. Birkbeck Hill, D.C.L. Crown 8vo. 7*s.* 6*d.*

—— VI. *Boswell's Life of Johnson. With the Journal of a Tour to the Hebrides.* Edited, with copious Notes, Appendices, and Index, by G. Birkbeck Hill, D.C.L., Pembroke College. With Portraits and Facsimiles. 6 vols. Medium 8vo. *Half bound,* 3*l.* 3*s.*

Cowper. Edited, with Life, Introductions, and Notes, by H. T. Griffith, B.A.

—— I. *The Didactic Poems of* 1782, with Selections from the Minor Pieces, A.D. 1779-1783. Extra fcap. 8vo. 3*s.*

—— II. *The Task, with Tirocinium,* and Selections from the Minor Poems, A.D. 1784-1799. Second Edition. Extra fcap. 8vo. 3*s.*

Burke. Select Works. Edited, with Introduction and Notes, by E. J. Payne, M.A.

—— I. *Thoughts on the Present Discontents; the two Speeches on America.* Second Edition. Extra fcap. 8vo. 4*s.* 6*d.*

—— II. *Reflections on the French Revolution.* Second Edition. Extra fcap. 8vo. 5*s.*

—— III. *Four Letters on the Proposals for Peace with the* Regicide Directory of France. Second Edition. Extra fcap. 8vo. 5*s.*

Keats. Hyperion, Book I. With Notes by W. T. Arnold, B.A. Paper covers, 4*d*.

Byron. Childe Harold. Edited, with Introduction and Notes, by H. F. Tozer, M.A. Extra fcap. 8vo. 3*s*. 6*d*. In white Parchment, 5*s*.

Scott. Lay of the Last Minstrel. Edited with Preface and Notes by W. Minto, M.A. With Map. Extra fcap. 8vo. Stiff covers, 2*s*. Ornamental Parchment, 3*s*. 6*d*.

——— *Lay of the Last Minstrel.* Introduction and Canto I, with Preface and Notes, by the same Editor. 6*d*.

II. LATIN.

Rudimenta Latina. Comprising Accidence, and Exercises of a very Elementary Character, for the use of Beginners. By John Barrow Allen, M.A. Extra fcap. 8vo. 2*s*.

An Elementary Latin Grammar. By the same Author. Fifty-Seventh Thousand. Extra fcap. 8vo. 2*s*. 6*d*.

A First Latin Exercise Book. By the same Author. Fourth Edition. Extra fcap. 8vo. 2*s*. 6*d*.

A Second Latin Exercise Book. By the same Author. Extra fcap. 8vo. 3*s*. 6*d*.

Reddenda Minora, or Easy Passages, Latin and Greek, for Unseen Translation. For the use of Lower Forms. Composed and selected by C. S. Jerram, M.A. Extra fcap. 8vo. 1*s*. 6*d*.

Anglice Reddenda, or Extracts, Latin and Greek, for Unseen Translation. By C. S. Jerram, M.A. Third Edition, Revised and Enlarged. Extra fcap. 8vo. 2*s*. 6*d*.

Anglice Reddenda. Second Series. By the same Author. Extra fcap. 8vo. 3*s*.

Passages for Translation into Latin. For the use of Passmen and others. Selected by J. Y. Sargent, M.A. Seventh Edition. Extra fcap. 8vo. 2*s*. 6*d*.

Exercises in Latin Prose Composition; with Introduction, Notes, and Passages of Graduated Difficulty for Translation into Latin. By G. G. Ramsay, M.A., LL.D. Second Edition. Extra fcap. 8vo. 4*s*. 6*d*.

Hints and Helps for Latin Elegiacs. By H. Lee-Warner, M.A. Extra fcap. 8vo. 3*s*. 6*d*.

First Latin Reader. By T. J. Nunns, M.A. Third Edition. Extra fcap. 8vo. 2*s*.

Caesar. The Commentaries (for Schools). With Notes and Maps. By Charles E. Moberly, M.A.
    Part I. *The Gallic War.* Second Edition. Extra fcap. 8vo. 4*s.* 6*d.*
    Part II. *The Civil War.* Extra fcap. 8vo. 3*s.* 6*d.*
    The Civil War. Book I. Second Edition. Extra fcap. 8vo. 2*s.*

Cicero. Speeches against Catilina. By E. A. Upcott, M.A., Assistant Master in Wellington College. In one or two Parts. Extra fcap. 8vo. 2*s.* 6*d.*

Cicero. Selection of interesting and descriptive passages. With Notes. By Henry Walford, M.A. In three Parts. Extra fcap. 8vo. 4*s.* 6*d.*
    Each Part separately, limp, 1*s.* 6*d.*
    Part I. Anecdotes from Grecian and Roman History. Third Edition.
    Part II. Omens and Dreams: Beauties of Nature. Third Edition.
    Part III. Rome's Rule of her Provinces. Third Edition.

Cicero. De Senectute. Edited, with Introduction and Notes, by L. Huxley, M.A. In one or two Parts. Extra fcap. 8vo. 2*s.*

Cicero. Selected Letters (for Schools). With Notes. By the late C. E. Prichard, M.A., and E. R. Bernard, M.A. Second Edition. Extra fcap. 8vo. 3*s.*

Cicero. Select Orations (for Schools). In Verrem I. De Imperio Gn. Pompeii. Pro Archia. Philippica IX. With Introduction and Notes by J. R. King, M.A. Second Edition. Extra fcap. 8vo. 2*s.* 6*d.*

Cicero. In Q. Caecilium Divinatio, and *In C. Verrem Actio Prima.* With Introduction and Notes, by J. R. King, M.A. Extra fcap. 8vo. limp, 1*s.* 6*d.*

Cicero. Speeches against Catilina. With Introduction and Notes, by E. A. Upcott, M.A. In one or two Parts. Extra fcap. 8vo. 2*s.* 6*d.*

Cornelius Nepos. With Notes. By Oscar Browning, M.A. Second Edition. Extra fcap. 8vo. 2*s.* 6*d.*

Horace. Selected Odes. With Notes for the use of a Fifth Form. By E. C. Wickham, M.A. In one or two Parts. Extra fcap. 8vo. cloth, 2*s.*

Livy. Selections (for Schools). With Notes and Maps. By H. Lee-Warner, M.A. Extra fcap. 8vo. In Parts, limp, each 1*s.* 6*d.*
    Part I. The Caudine Disaster. Part II. Hannibal's Campaign in Italy. Part III. The Macedonian War.

Livy. Books V–VII. With Introduction and Notes. By A. R. Cluer, B.A. Second Edition. Revised by P. E. Matheson, M.A. (In one or two Parts.) Extra fcap. 8vo. 5*s.*

Livy. Books XXI, XXII, and XXIII. With Introduction and Notes. By M. T. Tatham, M.A. Extra fcap. 8vo. 4*s.* 6*d.*

Ovid. Selections for the use of Schools. With Introductions and Notes, and an Appendix on the Roman Calendar. By W. Ramsay, M.A. Edited by G. G. Ramsay, M.A. Third Edition. Extra fcap. 8vo. 5s. 6d.

Ovid. Tristia. Book I. The Text revised, with an Introduction and Notes. By S. G. Owen, B.A. Extra fcap. 8vo. 3s. 6d.

Plautus. Captivi. Edited by W. M. Lindsay, M.A. Extra fcap. 8vo. (In one or two Parts.) 2s. 6d.

Plautus. The Trinummus. With Notes and Introductions. (Intended for the Higher Forms of Public Schools.) By C. E. Freeman, M.A., and A. Sloman, M.A. Extra fcap. 8vo. 3s.

Pliny. Selected Letters (for Schools). With Notes. By the late C. E. Prichard, M.A., and E. R. Bernard, M.A. Extra fcap. 8vo. 3s.

Sallust. With Introduction and Notes. By W. W. Capes, M.A. Extra fcap. 8vo. 4s. 6d.

Tacitus. The Annals. Books I–IV. Edited, with Introduction and Notes (for the use of Schools and Junior Students), by H. Furneaux, M.A. Extra fcap. 8vo. 5s.

Tacitus. The Annals. Book I. With Introduction and Notes, by the same Editor. Extra fcap. 8vo. limp, 2s.

Terence. Andria. With Notes and Introductions. By C. E. Freeman, M.A., and A. Sloman, M.A. Extra fcap. 8vo. 3s.

—— *Adelphi.* With Notes and Introductions. (Intended for the Higher Forms of Public Schools.) By A. Sloman, M.A. Extra fcap. 8vo. 3s.

—— *Phormio.* With Notes and Introductions. By A. Sloman, M.A. Extra fcap. 8vo. 3s.

Tibullus and Propertius. Selections. Edited by G. G. Ramsay, M.A. Extra fcap. 8vo. (In one or two vols.) 6s.

Virgil. With Introduction and Notes. By T. L. Papillon, M.A. Two vols. Crown 8vo. 10s. 6d. The Text separately, 4s. 6d.

Virgil. Bucolics. Edited by C. S. Jerram, M.A. In one or two Parts. Extra fcap. 8vo. 2s. 6d.

Virgil. Aeneid I. With Introduction and Notes, by C. S. Jerram, M.A. Extra fcap. 8vo. limp, 1s. 6d.

Virgil. Aeneid IX. Edited, with Introduction and Notes, by A. E. Haigh, M.A., late Fellow of Hertford College, Oxford. Extra fcap. 8vo. limp, 1s. 6d. In two Parts, 2s.

Avianus, The Fables of. Edited, with Prolegomena, Critical Apparatus, Commentary, etc. By Robinson Ellis, M.A., LL.D. Demy 8vo. 8s. 6d.

Catulli Veronensis Liber. Iterum recognovit, apparatum criticum prolegomena appendices addidit, Robinson Ellis. A.M. 1878. Demy 8vo. 16s.

—— *A Commentary on Catullus.* By Robinson Ellis, M.A. 1876. Demy 8vo. 16s.

Catulli Veronensis Carmina Selecta, secundum recognitionem Robinson Ellis, A.M. Extra fcap. 8vo. 3s. 6d.

Cicero de Oratore. With Introduction and Notes. By A. S. Wilkins, M.A.
   Book I. 1879. 8vo. 6s. Book II. 1881. 8vo. 5s.

—— *Philippic Orations.* With Notes. By J. R. King, M.A. Second Edition. 1879. 8vo. 10s. 6d.

Cicero. Select Letters. With English Introductions, Notes, and Appendices. By Albert Watson, M.A. Third Edition. Demy 8vo. 18s.

—— *Select Letters.* Text. By the same Editor. Second Edition. Extra fcap. 8vo. 4s.

—— *pro Cluentio.* With Introduction and Notes. By W. Ramsay, M.A. Edited by G. G. Ramsay, M.A. 2nd Ed. Ext. fcap. 8vo. 3s. 6d.

Horace. With a Commentary. Volume I. The Odes, Carmen Seculare, and Epodes. By Edward C. Wickham, M.A. Second Edition. 1877. Demy 8vo. 12s.

—— A reprint of the above, in a size suitable for the use of Schools. In one or two Parts. Extra fcap. 8vo. 6s.

Livy, Book I. With Introduction, Historical Examination, and Notes. By J. R. Seeley, M.A. Second Edition. 1881. 8vo. 6s.

Ovid. P. Ovidii Nasonis Ibis. Ex Novis Codicibus edidit, Scholia Vetera Commentarium cum Prolegomenis Appendice Indice addidit, R. Ellis, A.M. 8vo. 10s. 6d.

Persius. The Satires. With a Translation and Commentary. By John Conington, M.A. Edited by Henry Nettleship, M.A. Second Edition. 1874. 8vo. 7s. 6d.

Juvenal. XIII Satires. Edited, with Introduction and Notes, by C. H. Pearson, M.A., and Herbert A. Strong, M.A., LL.D., Professor of Latin in Liverpool University College, Victoria University. In two Parts. Crown 8vo. Complete, 6s.
 Also separately, Part I. Introduction, Text, etc., 3s. Part II. Notes, 3s. 6d.

Tacitus. The Annals. Books I-VI. Edited, with Introduction and Notes, by H. Furneaux, M.A. 8vo. 18s.

Nettleship (H., M.A.). Lectures and Essays on Subjects connected with Latin Scholarship and Literature. Crown 8vo. 7s. 6d.

—— *The Roman Satura.* 8vo. sewed, 1s.

—— *Ancient Lives of Vergil.* 8vo. sewed, 2s.

Papillon (T. L., M.A.). A Manual of Comparative Philology. Third Edition, Revised and Corrected. 1882. Crown 8vo. 6s.

Pinder (North, M.A.). Selections from the less known Latin Poets. 1869. 8vo. 15s.

Sellar (W. Y., M.A.). Roman Poets of the Augustan Age. VIRGIL. New Edition. 1883. Crown 8vo. 9s.

—— *Roman Poets of the Republic.* New Edition, Revised and Enlarged. 1881. 8vo. 14s.

Wordsworth (J., M.A.). Fragments and Specimens of Early Latin. With Introductions and Notes. 1874. 8vo. 18s.

III. GREEK.

A Greek Primer, for the use of beginners in that Language. By Charles Wordsworth, D.C.L. Seventh Edition. Extra fcap. 8vo. 1s. 6d.

A Greek Testament Primer. An Easy Grammar and Reading Book for the use of Students beginning Greek. By the Rev. E. Miller, M.A. Extra fcap. 8vo. 3s. 6d.

Easy Greek Reader. By Evelyn Abbott, M.A. In one or two Parts. Extra fcap. 8vo. 3s.

Graecae Grammaticae Rudimenta in usum Scholarum. Auctore Carolo Wordsworth, D.C.L. Nineteenth Edition, 1882. 12mo. 4s.

A Greek-English Lexicon, abridged from Liddell and Scott's 4to. edition, chiefly for the use of Schools. Twenty-first Edition. 1886. Square 12mo. 7s. 6d.

Greek Verbs, Irregular and Defective. By W. Veitch. Fourth Edition. Crown 8vo. 10s. 6d.

The Elements of Greek Accentuation (for Schools): abridged from his larger work by H. W. Chandler, M.A. Extra fcap. 8vo. 2s. 6d.

A SERIES OF GRADUATED GREEK READERS:—

First Greek Reader. By W. G. Rushbrooke, M.L. Second Edition. Extra fcap. 8vo. 2s. 6d.

Second Greek Reader. By A. M. Bell, M.A. Extra fcap. 8vo. 3s. 6d.

Fourth Greek Reader; being Specimens of Greek Dialects. With Introductions, etc. By W. W. Merry, D.D. Extra fcap. 8vo. 4s. 6d.

Fifth Greek Reader. Selections from Greek Epic and Dramatic Poetry, with Introductions and Notes. By Evelyn Abbott, M.A. Extra fcap. 8vo. 4s. 6d.

The Golden Treasury of Ancient Greek Poetry: being a Collection of the finest passages in the Greek Classic Poets, with Introductory Notices and Notes. By R. S. Wright, M.A. Extra fcap. 8vo. 8s. 6d.

A Golden Treasury of Greek Prose, being a Collection of the finest passages in the principal Greek Prose Writers, with Introductory Notices and Notes. By R. S. Wright, M.A., and J. E. L. Shadwell, M.A. Extra fcap. 8vo. 4s. 6d.

Aeschylus. Prometheus Bound (for Schools). With Introduction and Notes, by A. O. Prickard, M.A. Second Edition. Extra fcap. 8vo. 2s.

—— *Agamemnon.* With Introduction and Notes, by Arthur Sidgwick, M.A. Third Edition. In one or two parts. Extra fcap. 8vo. 3s.

—— *Choephoroi.* With Introduction and Notes by the same Editor. Extra fcap. 8vo. 3s.

—— *Eumenides.* With Introduction and Notes, by the same Editor. In one or two Parts. Extra fcap. 8vo. 3s.

Aristophanes. In Single Plays. Edited, with English Notes, Introductions, &c., by W. W. Merry, D.D. Extra fcap. 8vo.
    I. The Clouds, Second Edition, 2s.
    II. The Acharnians, Third Edition. In one or two parts, 3s.
    III. The Frogs, Second Edition. In one or two parts, 3s.
    IV. The Knights. In one or two parts, 3s.

Cebes. Tabula. With Introduction and Notes. By C. S. Jerram, M.A. Extra fcap. 8vo. 2s. 6d.

Demosthenes. Orations against Philip. With Introduction and Notes, by Evelyn Abbott, M.A., and P. E. Matheson, M.A. Vol. I. Philippic I. Olynthiacs I–III. In one or two Parts. Extra fcap. 8vo. 3s.

Euripides. Alcestis (for Schools). By C. S. Jerram, M.A. Extra fcap. 8vo. 2s. 6d.

—— *Helena.* Edited, with Introduction, Notes, etc., for Upper and Middle Forms. By C. S. Jerram, M.A. Extra fcap. 8vo. 3s.

—— *Iphigenia in Tauris.* Edited, with Introduction, Notes, etc., for Upper and Middle Forms. By C. S. Jerram, M.A. Extra fcap. 8vo. cloth, 3s.

—— *Medea.* By C. B. Heberden, M.A. In one or two Parts. Extra fcap. 8vo. 2s.

Herodotus, Book IX. Edited, with Notes, by Evelyn Abbott, M.A. In one or two Parts. Extra fcap. 8vo. 3*s*.

Herodotus, *Selections from*. Edited, with Introduction, Notes, and a Map, by W. W. Merry, D.D. Extra fcap. 8vo. 2*s*. 6*d*.

Homer. *Odyssey*, Books I–XII (for Schools). By W. W. Merry, D.D. Fortieth Thousand. (In one or two Parts.) Extra fcap. 8vo. 5*s*.

    Books I, and II, *separately*, each 1*s*. 6*d*.

—— *Odyssey*, Books XIII–XXIV (for Schools). By the same Editor. Second Edition. Extra fcap. 8vo. 5*s*.

—— *Iliad*, Book I (for Schools). By D. B. Monro, M.A. Second Edition. Extra fcap. 8vo. 2*s*.

—— *Iliad*, Books I–XII (for Schools). With an Introduction, a brief Homeric Grammar, and Notes. By D. B. Monro, M.A. Second Edition. Extra fcap. 8vo. 6*s*.

—— *Iliad*, Books VI and XXI. With Introduction and Notes. By Herbert Hailstone, M.A. Extra fcap. 8vo. 1*s*. 6*d*. each.

Lucian. *Vera Historia* (for Schools). By C. S. Jerram, M.A. Second Edition. Extra fcap. 8vo. 1*s*. 6*d*.

Lysias. *Epitaphios*. Edited, with Introduction and Notes, by F. J. Snell, B.A. (In one or two Parts.) Extra fcap. 8vo. 2*s*.

Plato. *Meno*. With Introduction and Notes. By St. George Stock, M.A., Pembroke College. (In one or two Parts.) Extra fcap. 8vo. 2*s*. 6*d*.

Plato. *The Apology*. With Introduction and Notes. By St. George Stock, M.A. (In one or two Parts.) Extra fcap. 8vo. 2*s*. 6*d*.

Sophocles. For the use of Schools. Edited with Introductions and English Notes. By Lewis Campbell, M.A., and Evelyn Abbott, M.A. *New and Revised Edition*. 2 Vols. Extra fcap. 8vo. 10*s*. 6*d*.

    Sold separately, Vol. I, Text, 4*s*. 6*d*.; Vol. II, Explanatory Notes, 6*s*.

Sophocles. In Single Plays, with English Notes, &c. By Lewis Campbell, M.A., and Evelyn Abbott, M.A. Extra fcap. 8vo. limp.

    Oedipus Tyrannus, Philoctetes. New and Revised Edition, 2*s*. each.
    Oedipus Coloneus, Antigone, 1*s*. 9*d*. each.
    Ajax, Electra, Trachiniae, 2*s*. each.

—— *Oedipus Rex:* Dindorf's Text, with Notes by the present Bishop of St. David's. Extra fcap. 8vo. limp, 1*s*. 6*d*.

Theocritus (for Schools). With Notes. By H. Kynaston, D.D. (late Snow). Third Edition. Extra fcap. 8vo. 4*s*. 6*d*.

Xenophon. Easy Selections (for Junior Classes). With a Vocabulary, Notes, and Map. By J. S. Phillpotts, B.C.L., and C. S. Jerram, M.A. Third Edition. Extra fcap. 8vo. 3s. 6d.

Xenophon. Selections (for Schools). With Notes and Maps. By J. S. Phillpotts, B.C.L. Fourth Edition. Extra fcap. 8vo. 3s. 6d.

—— *Anabasis*, Book I. Edited for the use of Junior Classes and Private Students. With Introduction, Notes, etc. By J. Marshall, M.A., Rector of the Royal High School, Edinburgh. Extra fcap. 8vo. 2s. 6d.

—— *Anabasis*, Book II. With Notes and Map. By C. S. Jerram, M.A. Extra fcap. 8vo. 2s.

—— *Cyropaedia*, Books IV and V. With Introduction and Notes by C. Bigg, D.D. Extra fcap. 8vo. 2s. 6d.

Aristotle's Politics. With an Introduction, Essays, and Notes. By W. L. Newman, M.A., Fellow of Balliol College. Vols. I and II. Medium 8vo. 28s. *Just Published.*

Aristotelian Studies. I. On the Structure of the Seventh Book of the Nicomachean Ethics. By J. C. Wilson, M.A. 8vo. stiff, 5s.

Aristotelis Ethica Nicomachea, ex recensione Immanuelis Bekkeri. Crown 8vo. 5s.

Demosthenes and Aeschines. The Orations of Demosthenes and Æschines on the Crown. With Introductory Essays and Notes. By G. A. Simcox, M.A., and W. H. Simcox, M.A. 1872. 8vo. 12s.

Head (Barclay V.). Historia Numorum: A Manual of Greek Numismatics. Royal 8vo. half-bound. 2l. 2s.

Hicks (E. L., M.A.). A Manual of Greek Historical Inscriptions. Demy 8vo. 10s. 6d.

Homer. Odyssey, Books I–XII. Edited with English Notes, Appendices, etc. By W. W. Merry, D.D., and the late James Riddell, M.A. 1886. Second Edition. Demy 8vo. 16s.

Homer. A Grammar of the Homeric Dialect. By D. B. Monro, M.A. Demy 8vo. 10s. 6d.

Sophocles. The Plays and Fragments. With English Notes and Introductions, by Lewis Campbell, M.A. 2 vols.
    Vol. I. Oedipus Tyrannus. Oedipus Coloneus. Antigone. 8vo. 16s.
    Vol. II. Ajax. Electra. Trachiniae. Philoctetes. Fragments. 8vo. 16s.

IV. FRENCH AND ITALIAN.

Brachet's Etymological Dictionary of the French Language. Translated by G. W. Kitchin, D.D. Third Edition. Crown 8vo. 7s. 6d.

—— *Historical Grammar of the French Language.* Translated by G. W. Kitchin, D.D. Fourth Edition. Extra fcap. 8vo. 3s. 6d.

Works by GEORGE SAINTSBURY, M.A.

Primer of French Literature. Extra fcap. 8vo. 2s.

Short History of French Literature. Crown 8vo. 10s. 6d.

Specimens of French Literature, from Villon to Hugo. Crown 8vo. 9s.

MASTERPIECES OF THE FRENCH DRAMA.

Corneille's Horace. Edited, with Introduction and Notes, by George Saintsbury, M.A. Extra fcap. 8vo. 2s. 6d.

Molière's Les Précieuses Ridicules. Edited, with Introduction and Notes, by Andrew Lang, M.A. Extra fcap. 8vo. 1s. 6d.

Racine's Esther. Edited, with Introduction and Notes, by George Saintsbury, M.A. Extra fcap. 8vo. 2s.

Beaumarchais' Le Barbier de Séville. Edited, with Introduction and Notes, by Austin Dobson. Extra fcap. 8vo. 2s. 6d.

Voltaire's Mérope. Edited, with Introduction and Notes, by George Saintsbury. Extra fcap. 8vo. cloth, 2s.

Musset's On ne badine pas avec l'Amour, and *Fantasio.* Edited, with Prolegomena, Notes, etc., by Walter Herries Pollock. Extra fcap. 8vo. 2s.

    The above six Plays may be had in ornamental case, and bound in Imitation Parchment, price 12s. 6d.

Perrault's Popular Tales. Edited from the Original Editions, with Introduction, etc., by Andrew Lang, M.A. Small 4to. Hand-made paper, vellum back, gilt top, 15s.

Sainte-Beuve. Selections from the Causeries du Lundi. Edited by George Saintsbury, M.A. Extra fcap. 8vo. 2s.

Quinet's Lettres à sa Mère. Selected and edited by George Saintsbury, M.A. Extra fcap. 8vo. 2s.

Gautier, Théophile. Scenes of Travel. Selected and Edited by George Saintsbury, M.A. Extra fcap. 8vo. 2s.

L'Éloquence de la Chaire et de la Tribune Françaises. Edited by Paul Blouët, B.A. Vol. I. Sacred Oratory. Extra fcap. 8vo. 2s. 6d.

Edited by GUSTAVE MASSON, B.A.

Corneille's Cinna. With Notes, Glossary, etc. Extra fcap. 8vo.
cloth, 2s. Stiff covers, 1s. 6d.

Louis XIV and his Contemporaries; as described in Extracts from the best Memoirs of the Seventeenth Century. With English Notes, Genealogical Tables, &c. Extra fcap. 8vo. 2s. 6d.

Maistre, Xavier de. Voyage autour de ma Chambre. Ourika, by *Madame de Duras;* Le Vieux Tailleur, by *MM. Erckmann-Chatrian;* La Veillée de Vincennes, by *Alfred de Vigny;* Les Jumeaux de l'Hôtel Corneille, by *Edmond About;* Mésaventures d'un Écolier, by *Rodolphe Töpffer.* Third Edition, Revised and Corrected. Extra fcap. 8vo. 2s. 6d.

—— *Voyage autour de ma Chambre.* Limp, 1s. 6d.

Molière's Les Fourberies de Scapin, and *Racine's Athalie.* With Voltaire's Life of Molière. Extra fcap. 8vo. 2s. 6d.

Molière's Les Fourberies de Scapin. With Voltaire's Life of Molière. Extra fcap. 8vo. stiff covers, 1s. 6d.

Molière's Les Femmes Savantes. With Notes, Glossary, etc. Extra fcap. 8vo. *cloth,* 2s. Stiff covers, 1s. 6d.

Racine's Andromaque, and *Corneille's Le Menteur.* With Louis Racine's Life of his Father. Extra fcap. 8vo. 2s. 6d.

Regnard's Le Joueur, and *Brueys and Palaprat's Le Grondeur.* Extra fcap. 8vo. 2s. 6d.

Sévigné, Madame de, and her chief Contemporaries, Selections from the Correspondence of. Intended more especially for Girls' Schools. Extra fcap. 8vo. 3s.

Dante. Selections from the Inferno. With Introduction and Notes. By H. B. Cotterill, B.A. Extra fcap. 8vo. 4s. 6d.

Tasso. La Gerusalemme Liberata. Cantos i, ii. With Introduction and Notes. By the same Editor. Extra fcap. 8vo. 2s. 6d.

V. GERMAN.

Scherer (W.). A History of German Literature. Translated from the Third German Edition by Mrs. F. Conybeare. Edited by F. Max Müller. 2 vols. 8vo. 21s.

Max Müller. The German Classics, from the Fourth to the Nineteenth Century. With Biographical Notices, Translations into Modern German, and Notes. By F. Max Müller, M.A. A New Edition, Revised, Enlarged, and Adapted to Wilhelm Scherer's 'History of German Literature,' by F. Lichtenstein. 2 vols. crown 8vo. 21s.

GERMAN COURSE. By HERMANN LANGE.

The Germans at Home; a Practical Introduction to German Conversation, with an Appendix containing the Essentials of German Grammar. Third Edition. 8vo. 2s. 6d.

The German Manual; a German Grammar, Reading Book, and a Handbook of German Conversation. 8vo. 7s. 6d.

Grammar of the German Language. 8vo. 3s. 6d.

German Composition; A Theoretical and Practical Guide to the Art of Translating English Prose into German. Ed. 2. 8vo. 4s. 6d.

German Spelling; A Synopsis of the Changes which it has undergone through the Government Regulations of 1880. Paper covers, 6d.

Lessing's Laokoon. With Introduction, English Notes, etc. By A. Hamann, Phil. Doc., M.A. Extra fcap. 8vo. 4s. 6d.

Schiller's Wilhelm Tell. Translated into English Verse by E. Massie, M.A. Extra fcap. 8vo. 5s.

Also, Edited by C. A. BUCHHEIM, Phil. Doc.

Becker's Friedrich der Grosse. With an Historical Sketch of the Rise of Prussia and of the Times of Frederick the Great. With Map. Extra fcap. 8vo. 3s. 6d.

Goethe's Egmont. With a Life of Goethe, &c. Third Edition. Extra fcap. 8vo. 3s.

—— *Iphigenie auf Tauris.* A Drama. With a Critical Introduction and Notes. Second Edition. Extra fcap. 8vo. 3s.

Heine's Prosa, being Selections from his Prose Works. With English Notes, etc. Second Edition. Extra fcap. 8vo. 4s. 6d.

Heine's Harzreise. With Life of Heine, Descriptive Sketch of the Harz, and Index. Extra fcap. 8vo. paper covers, 1s. 6d.; cloth, 2s. 6d.

Lessing's Minna von Barnhelm. A Comedy. With a Life of Lessing, Critical Analysis, etc. Extra fcap. 8vo. 3s. 6d.

—— *Nathan der Weise.* With Introduction, Notes, etc. Second Edition. Extra fcap. 8vo. 4s. 6d.

Schiller's Historische Skizzen; Egmont's Leben und Tod, and *Belagerung von Antwerpen.* With a Map. Extra fcap. 8vo. 2s. 6d.

—— *Wilhelm Tell.* With a Life of Schiller; an historical and critical Introduction, Arguments, and a complete Commentary and Map. Sixth Edition. Extra fcap. 8vo. 3s. 6d.

—— *Wilhelm Tell.* School Edition. With Map. 2s.

Modern German Reader. A Graduated Collection of Extracts in Prose and Poetry from Modern German writers :—
Part I. With English Notes, a Grammatical Appendix, and a complete Vocabulary Fourth Edition. Extra fcap. 8vo. 2s. 6d.
Part II. With English Notes and an Index. Extra fcap. 8vo. 2s. 6d.

Niebuhr's Griechische Heroen-Geschichten. Tales of Greek Heroes. Edited with English Notes and a Vocabulary, by Emma S. Buchheim. School Edition. Extra fcap. 8vo., *cloth*, 2s.

A Middle High German Primer. With Grammar, Notes, and Glossary. By Joseph Wright, Ph.D. Extra fcap. 8vo. 3s. 6d.

VI. MATHEMATICS, PHYSICAL SCIENCE, &c.

By LEWIS HENSLEY, M.A.

Figures made Easy: a first Arithmetic Book. Crown 8vo. 6d.

Answers to the Examples in Figures made Easy, together with two thousand additional Examples, with Answers. Crown 8vo. 1s.

The Scholar's Arithmetic. Crown 8vo. 2s. 6d.

Answers to the Examples in the Scholar's Arithmetic. 1s. 6d.

The Scholar's Algebra. Crown 8vo. 2s. 6d.

Aldis (W. S., M.A.). A Text-Book of Algebra: with Answers to the Examples. Crown 8vo. 7s. 6d.

Baynes (R. E., M.A.). Lessons on Thermodynamics. 1878. Crown 8vo. 7s. 6d.

Chambers (G. F., F.R.A.S.). A Handbook of Descriptive Astronomy. Third Edition. 1877. Demy 8vo. 28s.

Clarke (Col. A. R., C.B., R.E.). Geodesy. 1880. 8vo. 12s. 6d.

Cremona (Luigi). Elements of Projective Geometry. Translated by C. Leudesdorf, M.A. 8vo. 12s. 6d.

Donkin. Acoustics. Second Edition. Crown 8vo. 7s. 6d.

Euclid Revised. Containing the Essentials of the Elements of Plane Geometry as given by Euclid in his first Six Books. Edited by R. C. J. Nixon, M.A. Crown 8vo. 7s. 6d.

Sold separately as follows,
Book I. 1s.   Books I, II. 1s. 6d.
Books I–IV. 3s. 6d. Books V, VI. 3s.

Euclid.—Geometry in Space. Containing parts of Euclid's Eleventh and Twelfth Books. By the same Editor. Crown 8vo. 3s. 6d.

D

Galton (Douglas, C.B., F.R.S.). The Construction of Healthy Dwellings. Demy 8vo. 10s. 6d.

Hamilton (Sir R. G. C.), and J. Ball. Book-keeping. New and enlarged Edition. Extra fcap. 8vo. limp cloth, 2s.
  Ruled Exercise books adapted to the above may be had, price 2s.

Harcourt (A. G. Vernon, M.A.), and H. G. Madan, M.A. Exercises in Practical Chemistry. Vol. I. Elementary Exercises. Fourth Edition. Crown 8vo. 10s. 6d.

Maclaren (Archibald). A System of Physical Education: Theoretical and Practical. Extra fcap. 8vo. 7s. 6d.

Madan (H. G., M.A.). Tables of Qualitative Analysis. Large 4to. paper, 4s. 6d.

Maxwell (J. Clerk, M.A., F.R.S.). A Treatise on Electricity and Magnetism. Second Edition. 2 vols. Demy 8vo. 1l. 11s. 6d.

—— *An Elementary Treatise on Electricity.* Edited by William Garnett, M.A. Demy 8vo. 7s. 6d.

Minchin (G. M., M.A.). A Treatise on Statics with Applications to Physics. Third Edition, Corrected and Enlarged. Vol. I. *Equilibrium of Coplanar Forces.* 8vo. 9s. Vol. II. *Statics.* 8vo. 16s.

—— *Uniplanar Kinematics of Solids and Fluids.* Crown 8vo. 7s. 6d.

Phillips (John, M.A., F.R.S.). Geology of Oxford and the Valley of the Thames. 1871. 8vo. 21s.

—— *Vesuvius.* 1869. Crown 8vo. 10s. 6d.

Prestwich (Joseph, M.A., F.R.S.). Geology, Chemical, Physical, and Stratigraphical. In two Volumes.
  Vol. I. Chemical and Physical. Royal 8vo. 25s.
  Vol. II. Stratigraphical and Physical. With a new Geographical Map of Europe. Royal 8vo. 36s. *Just published.*

Rolleston (George, M.D., F.R.S.). Forms of Animal Life. A Manual of Comparative Anatomy, with descriptions of selected types. Second Edition. Revised and enlarged by W. Hatchett Jackson, M.A. Medium, 8vo. cloth extra, 1l. 16s.

Smyth. A Cycle of Celestial Objects. Observed, Reduced, and Discussed by Admiral W. H. Smyth, R.N. Revised, condensed, and greatly enlarged by G. F. Chambers, F.R.A.S. 1881. 8vo. 12s.

Stewart (Balfour, LL.D., F.R.S.). A Treatise on Heat, with numerous Woodcuts and Diagrams. Fourth Edition. Extra fcap. 8vo. 7s. 6d.

Vernon-Harcourt (L. F., M.A.). A Treatise on Rivers and Canals, relating to the Control and Improvement of Rivers, and the Design, Construction, and Development of Canals. 2 vols. (Vol. I, Text. Vol. II, Plates.) 8vo. 21s.

—— *Harbours and Docks;* their Physical Features, History, Construction, Equipment, and Maintenance; with Statistics as to their Commercial Development. 2 vols. 8vo. 25s.

Walker (James, M.A.) The Theory of a Physical Balance. 8vo. stiff cover, 3s. 6d.

Watson (H. W., M.A.). A Treatise on the Kinetic Theory of Gases. 1876. 8vo. 3s. 6d.

Watson (H. W., D. Sc., F.R.S.), and S. H. Burbury, M.A.
 I. *A Treatise on the Application of Generalised Coördinates to the Kinetics of a Material System.* 1879. 8vo. 6s.
 II. *The Mathematical Theory of Electricity and Magnetism.* Vol. I. Electrostatics. 8vo. 10s. 6d.

Williamson (A. W., Phil. Doc., F.R.S.). Chemistry for Students. A new Edition, with Solutions. 1873. Extra fcap. 8vo. 8s. 6d.

VII. HISTORY.

Bluntschli (J. K.). The Theory of the State. By J. K. Bluntschli, late Professor of Political Sciences in the University of Heidelberg. Authorised English Translation from the Sixth German Edition. Demy 8vo. half bound, 12s. 6d.

Finlay (George, LL.D.). A History of Greece from its Conquest by the Romans to the present time, B.C. 146 to A.D. 1864. A new Edition, revised throughout, and in part re-written, with considerable additions, by the Author, and edited by H. F. Tozer, M.A. 7 vols. 8vo. 3l. 10s.

Fortescue (Sir John, Kt.). The Governance of England: otherwise called The Difference between an Absolute and a Limited Monarchy. A Revised Text. Edited, with Introduction, Notes, and Appendices, by Charles Plummer, M.A. 8vo. half bound, 12s. 6d.

Freeman (E.A., D.C.L.). A Short History of the Norman Conquest of England. Second Edition. Extra fcap. 8vo. 2s. 6d.

George (H.B., M.A.). Genealogical Tables illustrative of Modern History. Third Edition, Revised and Enlarged. Small 4to. 12s.

Hodgkin (T.). Italy and her Invaders. Illustrated with Plates and Maps. Vols. I–IV, A.D 376–553. 8vo. 3l. 8s.

Hughes (Alfred). Geography for Schools. With Diagrams.
Part I. Practical Geography. Crown 8vo. 2s. 6d. *Just Published.*
Part II. General Geography. *In preparation.*

Kitchin (G. W., D.D.). A History of France. With numerous Maps, Plans, and Tables. In Three Volumes. *Second Edition.* Crown 8vo. each 10s. 6d.

Vol. I. Down to the Year 1453.
Vol. II. From 1453-1624.   Vol. III. From 1624-1793.

Lucas (C. P.). Introduction to a Historical Geography of the British Colonies. With Eight Maps. Crown 8vo. 4s. 6d.

Payne (E. J., M.A.). A History of the United States of America. In the Press.

Ranke (L. von). A History of England. principally in the Seventeenth Century. Translated by Resident Members of the University of Oxford, under the superintendence of G. W. Kitchin, D.D., and C. W. Boase, M.A. 1875. 6 vols. 8vo. 3l. 3s.

Rawlinson (George, M.A.). A Manual of Ancient History. Second Edition. Demy 8vo. 14s.

Ricardo. Letters of David Ricardo to Thomas Robert Malthus (1810-1823). Edited by James Bonar, M.A. Demy 8vo. 10s. 6d.

Rogers (J. E. Thorold, M.A.). The First Nine Years of the Bank of England. 8vo. 8s. 6d.

Select Charters and other Illustrations of English Constitutional History, from the Earliest Times to the Reign of Edward I. Arranged and edited by W. Stubbs, D.D. Fifth Edition. 1883. Crown 8vo. 8s. 6d.

Stubbs (W., D.D.). The Constitutional History of England, in its Origin and Development. Library Edition. 3 vols. demy 8vo. 2l. 8s.
Also in 3 vols. crown 8vo. price 12s. each.

—— *Seventeen Lectures on the Study of Medieval and Modern History,* &c., delivered at Oxford 1867-1884. Crown 8vo. 8s. 6d.

Wellesley. A Selection from the Despatches, Treaties, and other Papers of the Marquess Wellesley, K.G., during his Government of India. Edited by S. J. Owen, M.A. 1877. 8vo. 1l. 4s.

Wellington. A Selection from the Despatches, Treaties, and other Papers relating to India of Field-Marshal the Duke of Wellington, K.G. Edited by S. J. Owen, M.A. 1880. 8vo. 24s.

A History of British India. By S. J. Owen, M.A., Reader in Indian History in the University of Oxford. In preparation.

VIII. LAW.

Alberici Gentilis, I.C.D., I.C., De Iure Belli Libri Tres.
Edidit T. E. Holland, I.C.D. 1877. Small 4to. half morocco, 21*s*.

Anson (Sir William R., Bart., D.C.L.). *Principles of the English Law of Contract, and of Agency in its Relation to Contract.* Fourth Edition. Demy 8vo. 10*s*. 6*d*.

—— *Law and Custom of the Constitution.* Part I. Parliament. Demy 8vo. 10*s*. 6*d*.

Bentham (Jeremy). *An Introduction to the Principles of Morals and Legislation.* Crown 8vo. 6*s*. 6*d*.

Digby (Kenelm E., M.A.). *An Introduction to the History of the Law of Real Property.* Third Edition. Demy 8vo. 10*s*. 6*d*.

Gaii Institutionum Juris Civilis Commentarii Quattuor; or, Elements of Roman Law by Gaius. With a Translation and Commentary by Edward Poste, M.A. Second Edition. 1875. 8vo. 18*s*.

Hall (W. E., M.A.). International Law. Second Ed. 8vo. 21*s*.

Holland (T. E., D.C.L.). The Elements of Jurisprudence. Fourth Edition. Demy 8vo. 10*s*. 6*d*.

—— *The European Concert in the Eastern Question*, a Collection of Treaties and other Public Acts. Edited, with Introductions and Notes, by Thomas Erskine Holland, D.C.L. 8vo. 12*s*. 6*d*.

Imperatoris Iustiniani Institutionum Libri Quattuor; with Introductions, Commentary, Excursus and Translation. By J. B. Moyle, B.C.L., M.A. 2 vols. Demy 8vo. 21*s*.

Justinian, The Institutes of, edited as a recension of the Institutes of Gaius, by Thomas Erskine Holland, D.C.L. Second Edition, 1881. Extra fcap. 8vo. 5*s*.

Justinian, Select Titles from the Digest of. By T. E. Holland, D.C.L., and C. L. Shadwell, B.C.L. 8vo. 14*s*.

Also sold in Parts, in paper covers, as follows:—
Part I. Introductory Titles. 2*s*. 6*d*.   Part II. Family Law. 1*s*.
Part III. Property Law. 2*s*. 6*d*.   Part IV. Law of Obligations (No. 1). 3*s*. 6*d*.
Part IV. Law of Obligations (No. 2). 4*s*. 6*d*.

Lex Aquilia. The Roman Law of Damage to Property: being a Commentary on the Title of the Digest 'Ad Legem Aquiliam' (ix. 2). With an Introduction to the Study of the Corpus Iuris Civilis. By Erwin Grueber, Dr. Jur., M.A. Demy 8vo. 10*s*. 6*d*.

Markby (W., D.C.L.). Elements of Law considered with reference to Principles of General Jurisprudence. Third Edition. Demy 8vo. 12s. 6d.

Stokes (Whitley, D.C.L.). The Anglo-Indian Codes.
    Vol. I. *Substantive Law.* 8vo. 30s.
    Vol. II. *Adjective Law.* In the Press.

Twiss (Sir Travers, D.C.L.). The Law of Nations considered as Independent Political Communities.

Part I. On the Rights and Duties of Nations in time of Peace. A new Edition, Revised and Enlarged. 1884. Demy 8vo. 15s.

Part II. On the Rights and Duties of Nations in Time of War. Second Edition, Revised. 1875. Demy 8vo. 21s.

IX. MENTAL AND MORAL PHILOSOPHY, &c.

Bacon's Novum Organum. Edited, with English Notes, by G. W. Kitchin, D.D. 1855. 8vo. 9s. 6d.

—— Translated by G. W. Kitchin, D.D. 1855. 8vo. 9s. 6d.

Berkeley. The Works of George Berkeley, D.D., formerly Bishop of Cloyne; including many of his writings hitherto unpublished. With Prefaces, Annotations, and an Account of his Life and Philosophy, by Alexander Campbell Fraser, M.A. 4 vols. 1871. 8vo. 2l. 18s.
    The Life, Letters, &c. 1 vol. 16s.

Berkeley. Selections from. With an Introduction and Notes. For the use of Students in the Universities. By Alexander Campbell Fraser, LL.D. Third Edition. Crown 8vo. 7s. 6d.

Fowler (T., D.D.). The Elements of Deductive Logic, designed mainly for the use of Junior Students in the Universities. Ninth Edition, with a Collection of Examples. Extra fcap. 8vo. 3s. 6d.

—— *The Elements of Inductive Logic,* designed mainly for the use of Students in the Universities. Fourth Edition. Extra fcap. 8vo. 6s.

—— and Wilson (J. M., B.D.). *The Principles of Morals* (Introductory Chapters). 8vo. *boards,* 3s. 6d.

—— *The Principles of Morals.* Part II. (Being the Body of the Work.) 8vo. 10s. 6d.

Edited by T. FOWLER, D.D.

Bacon. Novum Organum. With Introduction, Notes, &c. 1878. 8vo. 14s.

Locke's Conduct of the Understanding. Second Edition. Extra fcap. 8vo. 2s.

Danson (J. T.). *The Wealth of Households.* Crown 8vo. 5s.

Green (T. H., M.A.). *Prolegomena to Ethics.* Edited by A. C. Bradley, M.A. Demy 8vo. 12s. 6d.

Hegel. *The Logic of Hegel;* translated from the Encyclopaedia of the Philosophical Sciences. With Prolegomena by William Wallace, M.A. 1874. 8vo. 14s.

Lotze's Logic, in Three Books; of Thought, of Investigation, and of Knowledge. English Translation; Edited by B. Bosanquet, M.A., Fellow of University College, Oxford. 8vo. *cloth,* 12s. 6d.

—— *Metaphysic,* in Three Books; Ontology, Cosmology, and Psychology. English Translation; Edited by B. Bosanquet, M.A. Second Edition. 2 vols. Crown 8vo. 12s.

Martineau (James, D.D.). *Types of Ethical Theory.* Second Edition. 2 vols. Crown 8vo. 15s.

—— *A Study of Religion: its Sources and Contents.* 2 vols. 8vo. 24s.

Rogers (J. E. Thorold, M.A.). *A Manual of Political Economy,* for the use of Schools. Third Edition. Extra fcap. 8vo. 4s. 6d.

Smith's Wealth of Nations. A new Edition, with Notes, by J. E. Thorold Rogers, M.A. 2 vols. 8vo. 1880. 21s.

X. FINE ART.

Butler (A. J., M.A., F.S.A.) *The Ancient Coptic Churches of Egypt.* 2 vols. 8vo. 30s.

Head (Barclay V.). *Historia Numorum. A Manual of Greek* Numismatics. Royal 8vo. *half morocco,* 42s.

Hullah (John). *The Cultivation of the Speaking Voice.* Second Edition. Extra fcap. 8vo. 2s. 6d.

Jackson (T. G., M.A.). *Dalmatia, the Quarnero and Istria;* with Cettigne in Montenegro and the Island of Grado. By T. G. Jackson, M.A., Author of 'Modern Gothic Architecture.' In 3 vols. 8vo. With many Plates and Illustrations. *Half bound,* 42s.

Ouseley (Sir F. A. Gore, Bart.). A Treatise on Harmony. Third Edition. 4to. 10s.

——— *A Treatise on Counterpoint, Canon, and Fugue*, based upon that of Cherubini. Second Edition. 4to. 16s.

——— *A Treatise on Musical Form and General Composition.* Second Edition. 4to. 10s.

Robinson (J. C., F.S.A.). A Critical Account of the Drawings by Michel Angelo and Raffaello in the University Galleries, Oxford. 1870. Crown 8vo. 4s.

Troutbeck (J., M.A.) and R. F. Dale, M.A. A Music Primer (for Schools). Second Edition. Crown 8vo. 1s. 6d.

Tyrwhitt (R. St. J., M.A.). A Handbook of Pictorial Art. With coloured Illustrations, Photographs, and a chapter on Perspective by A. Macdonald. Second Edition. 1875. 8vo. half morocco, 18s.

Upcott (L. E., M.A.). An Introduction to Greek Sculpture. Crown 8vo. 4s. 6d.

Vaux (W. S. W., M.A.). Catalogue of the Castellani Collection of Antiquities in the University Galleries, Oxford. Crown 8vo. 1s.

The Oxford Bible for Teachers, containing Supplementary HELPS TO THE STUDY OF THE BIBLE, including Summaries of the several Books, with copious Explanatory Notes and Tables illustrative of Scripture History and the characteristics of Bible Lands; with a complete Index of Subjects, a Concordance, a Dictionary of Proper Names, and a series of Maps. Prices in various sizes and bindings from 3s. to 2l. 5s.

Helps to the Study of the Bible, taken from the OXFORD BIBLE FOR TEACHERS, comprising Summaries of the several Books, with copious Explanatory Notes and Tables illustrative of Scripture History and the Characteristics of Bible Lands; with a complete Index of Subjects, a Concordance, a Dictionary of Proper Names, and a series of Maps. Crown 8vo. cloth, 3s. 6d.; 16mo. cloth, 1s.

LONDON: HENRY FROWDE,
OXFORD UNIVERSITY PRESS WAREHOUSE, AMEN CORNER,

OXFORD: CLARENDON PRESS DEPOSITORY,
116 HIGH STREET.

☞ *The* DELEGATES OF THE PRESS *invite suggestions and advice from all persons interested in education; and will be thankful for hints, &c. addressed to the* SECRETARY TO THE DELEGATES, *Clarendon Press, Oxford.*

www.ingramcontent.com/pod-product-compliance
Lightning Source LLC
Chambersburg PA
CBHW030806230426
43667CB00008B/1090